W0254699

Mathematische Leitfäden

Herausgegeben von
em. o. Prof. Dr. phil. Dr. h. c. G. Köthe, Universität Frankfurt/M., und
o. Prof. Dr. rer. nat. G. Trautmann, Universität Kaiserslautern

Real Variable and Integration
With Historical Notes
by J. J. BENEDETTO, Prof. at the University of Maryland
280 pages. Paper DM 48,—

Spectral Synthesis
by J. J. BENEDETTO, Prof. at the University of Maryland
280 pages. Paper DM 68,—

Partial Differential Equations
An Introduction
by Dr. rer. nat. G. HELLWIG, o. Prof. at the Technische Hochschule Aachen
2nd Edition. xi, 259 pages with 35 figures. Paper 48,—

Einführung in die mathematische Logik
Klassische Prädikatenlogik
Von Dr. rer. nat. H. HERMES, o. Prof. an der Universität Freiburg i. Br.
4. Auflage. 206 Seiten. Kart. DM 34,—

Funktionalanalysis
Von Dr. rer. nat. H. HEUSER, o. Prof. an der Universität Karlsruhe
416 Seiten mit 6 Bildern, 462 Aufgaben und 50 Beispielen. Kart. DM 58,—

Lineare Integraloperatoren
Von Prof. Dr. rer. nat. K. JÖRGENS
224 Seiten mit 6 Bildern, 222 Aufgaben und zahlreichen Beispielen. Kart. DM 48,—

Moduln und Ringe
Von Dr. rer. nat. F. KASCH, o. Prof. an der Universität München
328 Seiten mit 177 Übungen und zahlreichen Beispielen. Kart. DM 52,—

Gewöhnliche Differentialgleichungen
Von Dr. rer. nat. H. W. KNOBLOCH, o. Prof. an der Universität Würzburg und
Dr. phil. F. KAPPEL, o. Prof. an der Universität Graz
332 Seiten mit 29 Bildern und 98 Aufgaben. Kart. DM 48,—

Garbentheorie
Von Dr. rer. nat. R. KULTZE, Prof. an der Universität Frankfurt/M.
179 Seiten mit 77 Aufgaben und zahlreichen Beispielen. Kart. DM 42,—

Differentialgeometrie
Von Dr. rer. nat. D. LAUGWITZ, Prof. an der Technischen Hochschule Darmstadt
3. Auflage. 183 Seiten mit 44 Bildern. Ln. DM 44,—

Fortsetzung dritte Umschlagseite

Springer Fachmedien Wiesbaden GmbH

Mathematische Leitfäden

Herausgegeben von
em. o. Prof. Dr. phil. Dr. h.c. G. Köthe, Universität Frankfurt/M., und
o. Prof. Dr. rer. nat. G. Trautmann, Universität Kaiserslautern

Partial Differential Equations

An Introduction

By Dr. rer. nat. Günter Hellwig
o. Professor at the Technische Hochschule Aachen

2nd Edition. With 35 figures

Springer Fachmedien Wiesbaden GmbH 1977

Prof. Dr. rer. nat. Günter Hellwig

Born 1926. Studied from 1945 until 1949 mathematics and physics at the Universität Göttingen, 1949 degree in mathematics. Then assistant, senior assistant, and later "docent" at the Technische Universität Berlin, 1951 Dr. rer. nat. and 1952 "habilitation". 1955/56 Research Associate at the Institute of Mathematical Sciences (later the Courant Institute) of New York University. 1958 full professor and director of the Mathematisches Institut at the Technische Universität Berlin and permanent lecturer at the 1. Mathematisches Institut of the Freie Universität Berlin. Since 1966 full professor at the Technische Hochschule Aachen. Working fields: Partial differential equations and spectral theory of differential operators in Hilbert spaces.

CIP-Kurztitelaufnahme der Deutschen Bibliothek

Hellwig, Günter
Partial differential equations : an introd. –
2. ed. – Stuttgart : Teubner, 1977.
  (Mathematische Leitfäden)
  Einheitssacht.: Partielle Differential-
  gleichungen ⟨engl.⟩

  ISBN 978-3-519-12213-5   ISBN 978-3-663-11002-6 (eBook)
  DOI 10.1007/978-3-663-11002-6

Cover design: W. Koch, Sindelfingen

AUTHOR'S PREFACE

This book is intended to give an introduction to the field of partial differential equations. The presentation is intentionally not too brief so that graduate students should be able to read it without serious difficulty. In addition to requiring a thorough knowledge of differential and integral calculus as well as of the theory of ordinary differential equations, it presupposes a few results from complex variables, and, in its last part, a few from functional analysis and real variables.

The goals of the book necessitated a careful selection of material. Of course, in the framework of this "guidebook," problems that today stand in the foreground of scientific development could only be taken into consideration peripherally. However, the author hopes that his efforts to present some of these can be felt.

In Part I, simple examples are treated, namely, the wave, potential, and heat equations. There the Gauss integral theorem in R_n appears as an important tool. Part II deals with the normal forms and characteristic manifolds for partial differential equations of the second order and for systems of partial differential equations of the first order in more than one unknown function. Here normal forms are given that can be obtained by very elementary means. In Part III questions of uniqueness for various initial-value and boundary-value problems are discussed, by means of the maximum-minimum principle and the energy-integral method, respectively. Since such considerations are much simpler than questions of existence, they are treated first; dealing with them first often brings with it the right point of view for the questions of existence which are to be treated in the following two parts. Different means of proof are purposely selected each time in order to provide the reader with at least a modest insight into the variety of methods. In Part IV, the method of successive iteration and the use of the characteristic relations are discussed for hyperbolic equations and systems, while the Laplace transform calculus is used for initial- and boundary-value problems in hyperbolic and parabolic equations. For boundary-value problems in elliptic equations the theory of weak solutions, together with an extended version of Weyl's lemma, is used. The delicate question about the assumption of the boundary values is treated by means of a new method due to E. Wienholtz, which—though not published so far—he has kindly made available for this book. The last part deals with questions of existence for elliptic equations and systems, using simple tools from functional analysis. It outlines Schauder's technique of proof and the treatment of the eigenvalue problem, and concludes with an introduction to the boundary-value problems for elliptic systems of the first order in two unknown functions. It is interesting here that for such problems the Fredholm alternative does not hold. A part on singular problems, which had been planned, has been postponed for the time being.

Partial differential equations of the first order in one unknown function have not been included in the book, since their theory can be reduced to the theory of ordinary differential equations; therefore their treatment perhaps belongs in a textbook on ordinary differential equations.

A few exercises are scattered through the text, among which the more difficult ones are indicated by an asterisk. Their solutions are given at the end of the book. References to the literature have been kept brief intentionally, since a small textbook is not in a position to provide a survey of the enormous wealth of literature in this field. Fortunately there are excellent summarizing reports that should be accessible to the reader after reading this book.

Formulas are numbered by section; for example, by (IV-3.19) we mean Formula 19 of Chapter 3 in Part IV. In references to places in the same part, the number of the part is omitted.

In particular, I must express my gratitude to my revered teacher Professor Haack for important stimulations and suggestions. Our joint investigations and numerous seminars on this subject have had much influence upon this book. My stay at the Institute of Mathematical Sciences, New York University, during the academic year 1954–55 has been another influencing factor. For many new points of view I have to thank Professors L. Asgeirsson, L. Bers, R. Courant, K. O. Friedrichs, F. John, P. Lax, and L. Nirenberg.

I also wish to express my cordial gratitude to my coworker Dr. E. Wienholtz for many hints and valuable suggestions which made many presentations clearer.

Further, I have to thank my secretary Mrs. L. Schröder for her cooperation in the preparation of the manuscript, Mr. H. Drucks for offering help with the proof reading and preparing the index, Mr. K.-H. Diener and Dr. K. Jörgens for many useful remarks, and Mr. H. Zehle for drawing the original figures.

Last, but not least, I must thank the editor of this series of books, Professor G. Köthe, who encouraged me to write this book, and to the publishers for their patient compliance with my wishes.

G. HELLWIG

Berlin—Charlottenburg, July 1959.

TRANSLATOR'S NOTE

In line with the common practice in partial differential equations, coordinates of vectors and points have been given lower indices. This is practiced uniformly except in Section II-2.6, where concepts and notations from differential geometry are used.

With the exception of the above and a few other inconsequential changes in notation, no change has been made in Professor Hellwig's text.

E. GERLACH

PREFACE TO THE SECOND EDITION

After the English edition had been out of print for some time and since the original edition is now also no longer available, publishers and author decided to reprint the English edition.

It contains numerous additions as compared with the German edition, and it conforms in its notation better to the present standard.

For an introductory course into the whole subject particularly Part 1 and Part 3 seem especially suitable. These two parts take into account the various types of partial differential equations and are independent of Part 2.

Precisely the uniqueness questions which are dealt with in Part 3 give the proper perspective concerning the existence problems without being encumbered with the usual difficulties of the latter.

In remembrance of our joint efforts in this field this edition will be dedicated to my honoured teacher Professor Dr. Dr. h. c. W. Haack on the occasion of his 75th birthday in April 1977.

Aachen, April 1977

G. HELLWIG

CONTENTS

Part 1. Examples

Part 2. Classification into Types, Theory of Characteristics, and Normal Form

Part 3. Questions of Uniqueness

Part 4. Questions of Existence

Part 5. Simple Tools from Functional Analysis Applied to Questions of Existence

Contents

xi

Part I.   Examples

1

INTRODUCTION

1.1 Definitions

A relation of the form

$$F(x_1, x_2, \ldots, x_n, u, u_{x_1}, u_{x_2}, \ldots, u_{x_n}, u_{x_1 x_1}, u_{x_1 x_2}, \ldots, u_{x_n x_n}) = 0 \qquad (1.1)$$

where $n > 1$, will be called a *partial differential equation of the second order*. Here (1.1) is considered in a suitable domain $\mathfrak{D}$ of the *n-dimensional* space R_n in the independent variables $x_1, x_2, \ldots, x_n$. We look for functions $u = u(x_1, x_2, \ldots, x_n)$ which satisfy (1.1) identically in $\mathfrak{D}$. Such functions u are called solutions of (1.1).

In (1.1), the partial derivatives of u are denoted in abbreviated form by indices; that is,

$$u_{x_i} \equiv \frac{\partial u}{\partial x_i}, \qquad u_{x_i x_k} \equiv \frac{\partial^2 u}{\partial x_i \, \partial x_k}. \qquad (1.2)$$

The expression (1.1) is said to be of the *second order* because the highest partial derivatives which appear are of the second order.

If $n = 1$, then (1.1) becomes an *ordinary differential equation* of the second order:

$$F(x_1, u, u', u'') = 0 \quad \text{where} \quad u' \equiv \frac{du}{dx_1}, \qquad u'' = \frac{d^2 u}{dx_1^2}.$$

In general, even an equation of this form has infinitely many solutions $u = u(x_1)$.

From this infinity of possible solutions we attempt to single out a unique one by introducing suitable additional conditions. *Initial conditions* often serve the purpose; these arbitrarily prescribe the value of u and its first derivative at a point a:

$$u(a) = u_0, \qquad u'(a) = u_1.$$

Frequently, *boundary conditions* also suffice; these arbitrarily prescribe the value of u at two points a and b:

$$u(a) = u_0, \qquad u(b) = u_1.$$

It is readily apparent that the analogous problem for the expression (1.1) is substantially more difficult. For the time being, discussion will be restricted to the simplest representatives of (1.1).

To find representatives of (1.1) that are not only simple, but also typical and important, we look to mathematical physics; many of the problems in this field reduce to partial differential equations. Before starting in this direction, we provide some mathematical tools.

1.2   The Gauss Integral Theorem

Let *n-dimensional space* consisting of points $P: (x_1, x_2, \ldots, x_n)$ be denoted by R_n.   Points in R_n will in most cases be represented in vectorial form, where

$$x = (x_1, x_2, \ldots, x_n) \tag{1.3}$$

denotes the vector representing the point P.   The inner product of two vectors x, y and the absolute value or modulus $|x|$ of x are given by

$$(x,y) = \sum_{i=1}^{n} x_i y_i, \qquad |x| = (x,x)^{1/2} = \sqrt{\sum_{i=1}^{n} (x_i)^2}, \tag{1.4}$$

respectively.   In particular, $|x - y|$ is then the distance of the points x and y from one another.

In the following we briefly collect some formulas which can be found in any textbook on differential and integral calculus.   Here it suffices to consider merely R_2 or R_3 and sufficiently simple domains $\mathfrak{D}$.

By a domain $\mathfrak{D}$ we understand an *open connected point set* in R_n.   By $\bar{\mathfrak{D}}$ we denote the closure of $\mathfrak{D}$; by $\dot{\mathfrak{D}}$, the set of boundary points of $\mathfrak{D}$.   Then $\bar{\mathfrak{D}} = \mathfrak{D} + \dot{\mathfrak{D}}$.   As the simplest example for $\mathfrak{D}$, we mention the ball S with center $a = (a_1, a_2, \ldots, a_n)$ and radius r.   In this case $S: |x - a| < r$, $\dot{S}: |x - a| = r$, and $\bar{S}: |x - a| \leqslant r$.   When $n = 2$, we call S a disk.

Frequently it is necessary to form *volume integrals* over $\mathfrak{D}$ and surface integrals over $\dot{\mathfrak{D}}$. The *volume element* is denoted by $dx = dx_1 dx_2 \cdots dx_n$ (dx is not a vector); dS denotes the surface element.   For a function $u(x_1, x_2, \ldots, x_n)$ defined on $\mathfrak{D}$ or $\dot{\mathfrak{D}}$, respectively, we briefly write $u(x)$.   We write $u \in C^0$ in $\mathfrak{D}$ if u is continuous in $\mathfrak{D}$; $u \in C^j$ in $\mathfrak{D}$ if $u(x)$ is j times continuously differentiable in all its variables, including all mixed derivatives up to the jth order.   The notations $u(x) \in C^j$ in $\bar{\mathfrak{D}}$ and $u(x) \in C^j$ on $\dot{\mathfrak{D}}$ are analogous; in the latter, we suppose that $\dot{\mathfrak{D}}$ is described by a parametric representation and consider $u(x)$ as a function of the parameters.   The integrals are of the form

$$\int_{\mathfrak{D}} u(x_1, x_2, \ldots, x_n)\, dx_1 dx_2 \cdots dx_n \equiv \int_{\mathfrak{D}} u(x)\, dx \quad \text{or} \quad \int_{\dot{\mathfrak{D}}} u(x)\, dS, \tag{1.5}$$

respectively.

A domain $\mathfrak{D}$ will be called a normal domain if it is bounded and simply connected and if it admits the application of the Gauss integral theorem; that is, if on $\dot{\mathfrak{D}}$ there is a vector field $v(x)$, where

$$v(x) = (v_1(x), v_2(x), \ldots, v_n(x)) \quad \text{and} \quad (v,v) = 1 \tag{1.6}$$

such that

$$\int_{\mathfrak{D}} u_{x_i}(x)\, dx = \int_{\dot{\mathfrak{D}}} u(x) v_i(x)\, dS; \qquad i = 1, 2, \ldots, n, \tag{1.7}$$

for all $u(x) \in C^1$ in $\bar{\mathfrak{D}}$.   If this is the case, the $v_i(x)$ are such that, at the points $x \in \dot{\mathfrak{D}}$ where $\dot{\mathfrak{D}}$ possesses an outer normal, the vector $v(x)$ of (1.6) coincides with this outer normal.   We commonly omit explicit statement of the independent variable in v and v_i.   In the case of R_2, $\dot{\mathfrak{D}}$ is a closed curve and dS is to be interpreted as ds, where s is the arc length on $\dot{\mathfrak{D}}$. Note that under the stated assumptions we can choose either $\mathfrak{D}$ or $\bar{\mathfrak{D}}$ for the domain of integration in the volume integral in (1.7). Whenever the limits of integration are given explicitly, we indicate the boundary of $\mathfrak{D}$.

Nearly every domain $\mathfrak{D}$ occurring in this book is such a normal domain; thus we always neglect pointing out this fact and make special mention of the exceptions only. The explicit assumptions that must be made about $\dot{\mathfrak{D}}$ so that $\mathfrak{D}$ will be a normal domain are discussed in textbooks on differential and integral calculus. Hypotheses particularly suited to our purposes may be found in O. D. Kellogg[1] and Cl. Müller[2].

If $\dot{\mathfrak{D}}$ is the union of several pairwise disjoint closed components

$$\dot{\mathfrak{D}} = \sum_{j=1}^{N} \dot{\mathfrak{D}}_j$$

so that $\mathfrak{D}$ is not simply connected, then the Gauss integral theorem holds too, provided that every $\mathfrak{D}_j$ is a normal domain. Instead of relation (1.7), we then have

$$\int_{\mathfrak{D}} u_{x_i}(x)\, dx = \sum_{j=1}^{N} \int_{\dot{\mathfrak{D}}_j} u(x) v_i^j(x)\, dS^j, \tag{1.8}$$

where v^j is the outer normal of $\mathfrak{D}$ on $\dot{\mathfrak{D}}_j$, and dS^j is the surface element corresponding to $\dot{\mathfrak{D}}_j$.

If $\dot{\mathfrak{D}}$ is the unit sphere in R_n: $|x| = 1$, we denote its dS by $d\omega$. For this surface,

$$\omega_n = \int_{|x|=1} d\omega = \frac{2[\Gamma(\tfrac{1}{2})]^n}{\Gamma(\tfrac{1}{2}n)} = \frac{2\pi^{n/2}}{\Gamma(n/2)}, \tag{1.9}$$

where $\Gamma(z)$ is the gamma function. Of course, we have the values $\omega_3 = 4\pi$ and $\omega_2 = 2\pi$. Finally, the volume of the unit ball $|x| \leqslant 1$ has the value ω_n/n which can be seen as follows: If, for $\mathfrak{D}$ in (1.7), we choose the unit ball $|x| \leqslant 1$ in R_n and set $u(x) = x_i$, we obtain

$$\int_{|x|\leqslant 1} dx = \int_{|x|=1} x_i v_i\, d\omega.$$

Since $v_i = x_i$ here, summation yields

$$n \int_{|x|\leqslant 1} dx = \int_{|x|=1} \sum_{i=1}^{n} x_i v_i\, d\omega = \int_{|x|=1} \sum_{i=1}^{n} (x_i)^2\, d\omega = \int_{|x|=1} d\omega = \omega_n.$$

If, in (1.7), we merely suppose that $u(x) \in C^0$ in $\mathfrak{D}$ and $u(x) \in C^1$ in $\mathfrak{D}$, then, in general, $\int_{\mathfrak{D}} u_{x_i}(x)\, dx$ will be an improper integral, because u_{x_i} may become infinite on $\dot{\mathfrak{D}}$. If, however, the existence of the integral $\int_{\mathfrak{D}} u_{x_i}(x)\, dx$ is required, (1.7) remains correct. See O. D. Kellogg[3].

1.3 Vector Fields

If the components $u^i(x)$ of a vector field $u(x) = (u^1(x), u^2(x), \ldots, u^n(x))$ belong to C^1 in $\mathfrak{D}$, then the *divergence* of u is defined by

$$\operatorname{div} u = \sum_{i=1}^{n} u_{x_i}^i(x), \tag{1.10}$$

so that under use of (1.7) the Gauss integral theorem can also be written in the form

$$\int_{\mathfrak{D}} \operatorname{div} u(x) \, dx = \int_{\dot{\mathfrak{D}}} (u,\nu) \, dS. \tag{1.11}$$

By forming the *gradient* of a function $u(x) \in C^1$, the vector field

$$\operatorname{grad} u(x) = (u_{x_1}, u_{x_2}, \ldots, u_{x_n}) \tag{1.12}$$

is obtained.

Now for two vector fields, u, v in R_3, there are the notions of *vector product* $u \times v$, and of forming the *rotation (curl)* rot u:

$$u \times v = (u^2 v^3 - u^3 v^2, \; u^3 v^1 - u^1 v^3, \; u^1 v^2 - u^2 v^1), \tag{1.13}$$
$$\operatorname{rot} u = (u^3_{x_2} - u^2_{x_3}, \; u^1_{x_3} - u^3_{x_1}, \; u^2_{x_1} - u^1_{x_2}). \tag{1.14}$$

1.4   The Green Formulas

By the *directional derivative* u_ν of $u(x)$ in direction of the outer normal ν we understand

$$u_\nu = \sum_{i=1}^{n} \nu_i u_{x_i}. \tag{1.15}$$

Further, we put

$$\Delta_n u = \sum_{i=1}^{n} u_{x_i x_i}.$$

If we assume $u(x) \in C^1$ in $\mathfrak{D}$, $v(x) \in C^2$ in $\mathfrak{D}$, we obtain the *first Green formula:*

$$\int_{\mathfrak{D}} u \, \Delta_n v \, dx = \int_{\dot{\mathfrak{D}}} u v_\nu \, dS - \int_{\mathfrak{D}} \sum_{i=1}^{n} u_{x_i} v_{x_i} \, dx \equiv \int_{\dot{\mathfrak{D}}} u v_\nu \, dS - \int_{\mathfrak{D}} (\operatorname{grad} u, \operatorname{grad} v) \, dx. \tag{1.16}$$

Indeed, we have

$$\int_{\mathfrak{D}} u \, \Delta_n v \, dx = \int_{\mathfrak{D}} \sum_{i=1}^{n} \{ (u v_{x_i})_{x_i} - u_{x_i} v_{x_i} \} \, dx.$$

If we apply formula (1.7) to the middle term here and consider (1.15), then (1.16) follows immediately.

According to the remark at the end of Section 1.2, it would suffice to assume $u(x)$, $v(x) \in C^1$ in $\mathfrak{D}$, $v(x) \in C^2$ in $\mathfrak{D}$, and the existence of $\int_{\mathfrak{D}} u \, \Delta_n v \, dx$.

By interchanging u and v in (1.16) and then subtracting the new formula, we obtain the *second Green formula:*

$$\int_{\mathfrak{D}} (u \, \Delta_n v - v \, \Delta_n u) \, dx = \int_{\dot{\mathfrak{D}}} (u v_\nu - v u_\nu) \, dS. \tag{1.17}$$

If $\dot{\mathfrak{D}}$ is the union of several components, then using the hypotheses made for (1.8) we find

$$\int_{\mathfrak{D}} (u \, \Delta_n v - v \, \Delta_n u) \, dx = \sum_{j=1}^{N} \int_{\dot{\mathfrak{D}}_j} (u v_{\nu^j} - v u_{\nu^j}) \, dS^j. \tag{1.17a}$$

Here we have to suppose $u(x)$, $v(x) \in C^2$ in $\mathfrak{D}$, or the corresponding weaker conditions. Setting $v \equiv 1$ in (1.17), we obtain the special case

$$\int_{\mathfrak{D}} \Delta_n u \, dx = \int_{\dot{\mathfrak{D}}} u_\nu \, dS. \tag{1.18}$$

Here we have to suppose $u(x) \in C^2$ in $\mathfrak{D}$, or, in the weakened form, $u(x) \in C^1$ in $\bar{\mathfrak{D}}$, $u(x) \in C^2$ in $\mathfrak{D}$, and the existence of $\int_{\mathfrak{D}} \Delta_n u \, dx$.

1.5   The Maxwell Equations

We begin with the study of partial differential equations arising in mathematical physics.

As is well known, part of the theory of electrodynamics can be deduced from the *Maxwell equations*. In R_3, the vector fields E and H, representing the electric and magnetic fields, which also depend on a parameter t (that is, time) are investigated:

$$E(x,t) = (E^1(x,t),\, E^2(x,t),\, E^3(x,t)), \qquad E^i(x,t) = E^i(x_1,\, x_2,\, x_3,\, t);$$
$$H(x,t) = (H^1(x,t),\, H^2(x,t),\, H^3(x,t)), \qquad H^i(x,t) = H^i(x_1,\, x_2,\, x_3,\, t).$$

We assume they are free of sources: $\operatorname{div} E = \operatorname{div} H = 0$. In the volume element dx, the electric and the magnetic energy is given by

$$du_{\text{el}} = \frac{\epsilon}{8\pi}\,(E,E)\,dx, \quad du_{\text{magn}} = \frac{\mu}{8\pi}\,(H,H)\,dx;$$

the energy flow in the time dt, through the surface element of $\dot{\mathfrak{D}}$ to the outside amounts to $(c/4\pi)(E \times H,\, \nu)\,dS\,dt$. Further, in the time dt, the electric energy $\sigma(E,E)\,dx\,dt$ in dx will be transformed into heat. Here ϵ, μ, σ, c are nonnegative physical constants.

The law of the conservation of energy tells us that the decrease in time of the energy in $\mathfrak{D}$ equals the sum of the energy converted into heat in $\mathfrak{D}$ and of the energy flowing out through $\dot{\mathfrak{D}}$:

$$-\frac{1}{8\pi}\frac{d}{dt}\int_{\mathfrak{D}} \{\epsilon(E,E) + \mu(H,H)\}\,dx = \sigma \int_{\mathfrak{D}} (E,E)\,dx + \frac{c}{4\pi}\int_{\dot{\mathfrak{D}}} (E \times H,\nu)\,dS. \tag{1.19}$$

Using (1.11) we have $\int_{\dot{\mathfrak{D}}} (E \times H,\, \nu)\,dS = \int_{\mathfrak{D}} \operatorname{div} E \times H\,dx$. Further, $\operatorname{div} E \times H = (H,\operatorname{rot} E) - (E,\operatorname{rot} H)$ is a well-known vector identity. If we substitute this in (1.19), we see that (1.19) holds for every $\mathfrak{D}$ whenever the equations

$$\operatorname{rot} E = -\frac{\mu}{c}\,H_t,$$
$$\text{where} \quad \operatorname{div} E = \operatorname{div} H = 0, \tag{1.20}$$
$$\operatorname{rot} H = \frac{\epsilon}{c}\,E_t + \frac{4\pi\sigma}{c}\,E,$$

are satisfied. Relation (1.20) is a special case of the Maxwell equations. It is a system of eight partial differential equations of the first order in the six unknown functions E^1, $E^2, \ldots, H^3$.

The mathematical problem consists of finding solutions E, H of (1.20) for all times t which for $t = 0$ agree with given vector fields $E_0(x)$, $H_0(x)$. Therefore the *initial conditions* for (1.20) read

$$E(x,0) = E_0(x), \qquad H(x,0) = H_0(x) \quad \text{where} \quad \operatorname{div} E_0 = \operatorname{div} H_0 = 0. \tag{1.21}$$

At first glance (1.20) might seem to be overdetermined, since the system contains more equations than unknown functions. However, the last two scalar equations are satisfied for all t as soon as they hold for $t = 0$, which is the case according to (1.21). They can therefore be omitted; namely, if we form the divergence of the second equation (1.20), it follows that

$$0 = \text{div rot } H = \frac{\epsilon}{c} f_t + \frac{4\pi\sigma}{c} f,$$

where div $E(x,t) = f(x,t)$. Integration over t gives

$$f(x,t) = C(x) e^{-(4\pi\sigma/\epsilon)t}$$

with arbitrary $C(x)$. Now $f(x,0) = 0$, so that $C(x) = 0$, which proves that div $E = 0$ for all values of t. Analogously it can be shown that div $H = 0$.

If we make use of the vector identity

$$\text{rot rot } E = \text{grad div } E - \Delta_3 E,$$

and of grad div $E = 0$ (since div $E = 0$), then, when the rotation is formed, the two vector equations in (1.20) go over into

$$\Delta_3 E = \frac{\epsilon\mu}{c^2} E_{tt} + \frac{4\pi\sigma\mu}{c^2} E_t, \qquad \Delta_3 H = \frac{\epsilon\mu}{c^2} H_{tt} + \frac{4\pi\sigma\mu}{c^2} H_t. \tag{1.22}$$

Here, of course, $\Delta_3 E$ stands for $(\Delta_3 E^1, \Delta_3 E^2, \Delta_3 E^3)$. Equations (1.22) are six partial differential equations of the second order for six unknown functions. As initial conditions— for the time being—we have to require conditions (1.21); however, $E_t(x,0)$ and $H_t(x,0)$ can be determined by use of (1.20). Therefore, for (1.22) we obtain the initial conditions

$$E(x,0) = E_0(x), \qquad E_t(x,0) = E_1(x), \quad \text{where} \quad E_1(x) = \frac{c}{\epsilon} \text{ rot } H_0 - \frac{4\pi\sigma}{\epsilon} E_0;$$

$$H(x,0) = H_0(x), \qquad H_t(x,0) = H_1(x), \quad \text{where} \quad H_1(x) = -\frac{c}{\mu} \text{ rot } E_0. \tag{1.23}$$

Obviously we can solve the problem (1.22), (1.23) if we can merely solve the initial-value problem for an equation of the second order

$$\Delta_3 u = \frac{\epsilon\mu}{c^2} u_{tt} + \frac{4\pi\sigma\mu}{c^2} u_t \quad \text{where} \quad u(x,0) = u_0(x), \qquad u_t(x,0) = u_1(x). \tag{1.24}$$

For $\sigma = 0$, (1.24) is called the *wave equation* in three space dimensions and one time dimension, and for $\sigma \neq 0$ it is called the *telegraph equation*.

1.6   The Equations of Gas Dynamics

In R_3 we consider a compressible medium (gas) which is in motion, and whose pressure p, density ρ, and velocity vector $v = (v^1, v^2, v^3)$ are functions of x and t. If we neglect viscous friction, heat conduction, and exterior forces, then *Euler's equation of motion* yields

$$\rho \frac{dv}{dt} = -\text{ grad } p, \tag{1.25}$$

and the theorem of the conservation of mass gives the *equation of continuity*

$$\rho_t + \text{div } (\rho v) = 0. \tag{1.26}$$

If we assume that the pressure p is a one-to-one function of the density ρ: $p = p(\rho)$, $\rho = \rho(p)$, then (1.25) and (1.26) represent a system of four partial differential equations of the first order for the four unknown functions v^1, v^2, v^3, ρ.   Written out with $' \equiv d/d\rho$, it reads

$$v_t^1 + v^1 v_{x_1}^1 + v^2 v_{x_2}^1 + v^3 v_{x_3}^1 + \frac{p'}{\rho}\, \rho_{x_1} = 0,$$

$$v_t^2 + v^1 v_{x_1}^2 + v^2 v_{x_2}^2 + v^3 v_{x_3}^2 + \frac{p'}{\rho}\, \rho_{x_2} = 0,$$

$$v_t^3 + v^1 v_{x_1}^3 + v^2 v_{x_2}^3 + v^3 v_{x_3}^3 + \frac{p'}{\rho}\, \rho_{x_3} = 0,$$

$$\rho_t + \rho(v_{x_1}^1 + v_{x_2}^2 + v_{x_3}^3) + v^1 \rho_{x_1} + v^2 \rho_{x_2} + v^3 \rho_{x_3} = 0. \tag{1.27}$$

The first three equations can also be taken together into one vector equation:

$$v_t + \mathrm{grad}\, \frac{|v|^2}{2} + v \times \mathrm{rot}\, v = -\frac{1}{\rho}\, \mathrm{grad}\, p. \tag{1.28}$$

We shall restrict ourselves to stationary processes ($\partial/\partial t \equiv 0$).   Further, let $\mathrm{rot}\, v = 0$ so that we can set $v = \mathrm{grad}\, u$ with $u = u(x)$, the velocity potential.   Equation (1.28) then simplifies to

$$\mathrm{grad}\, \frac{|v|^2}{2} = -\frac{1}{\rho}\, \mathrm{grad}\, p. \tag{1.29}$$

Therefore the surfaces $p = \mathrm{const}$ are identical with the surfaces $|v| = \mathrm{const}$ so that p can be interpreted as a function $p = p(|v|)$.   But then

$$|v| + \frac{1}{\rho}\frac{dp}{d|v|} = 0$$

is equivalent to (1.29).   If in the usual way we define the local sound velocity a by $a = \sqrt{dp/d\rho}$, then for $\rho = \rho(p)$,

$$\rho_{x_i} = \frac{d\rho}{dp}\frac{dp}{d|v|}\, |v|_{x_i} = -\rho|v| \frac{d\rho}{dp}\, |v|_{x_i} = -\frac{\rho}{2a^2}\, (|v|^2)_{x_i} \tag{1.30}$$

results.   Now $|v| = |\mathrm{grad}\, u|$, so that from (1.30)

$$\rho_{x_i} = -\frac{\rho}{2a^2}\, ((u_{x_1})^2 + (u_{x_2})^2 + (u_{x_3})^2)_{x_i} \tag{1.31}$$

follows.   With this, the first three equations in (1.27) have been used up completely in the *stationary*, *vortex-free* case.   If we substitute (1.31) in the continuity equation (1.26): $\mathrm{div}\, (\rho v) = 0$, then for u only the following partial differential equation of the second order appears:

$$(a^2 - (u_{x_1})^2)u_{x_1 x_1} + (a^2 - (u_{x_2})^2)u_{x_2 x_2} + (a^2 - (u_{x_3})^2)u_{x_3 x_3} - 2u_{x_1}u_{x_2}u_{x_1 x_2} - 2u_{x_1}u_{x_3}u_{x_1 x_3}$$
$$- 2u_{x_2}u_{x_3}u_{x_2 x_3} = 0. \tag{1.32}$$

Here $a = a(|v|) = a(|\mathrm{grad}\, u|)$.   In the special case of *plane* flow all the terms that contain derivatives with respect to x_3 disappear.

If finally in (1.32) we introduce cylindrical coordinates x_1, r, ω and consider only flows which are rotationally symmetric so that $u = u(x_1, r)$, then from (1.32) it follows that

$$(a^2 - (u_{x_1})^2)u_{x_1 x_1} + (a^2 - (u_r)^2)u_{rr} - 2u_{x_1}u_r u_{x_1 r} + \frac{a^2}{r}\, u_r = 0. \tag{1.33}$$

This presentation closely follows that of R. Sauer[4].

1.7   The Heat Equation

In R_3 we consider a body $\mathscr{B}$. At the point $x \in \mathscr{B}$ let the temperature be $u(x,t)$. The difference of temperature in $\mathscr{B}$ generates a heat flow which we suppose given by

$$f(x,t) = -\alpha(x) \text{ grad } u(x,t). \tag{1.34}$$

The amount of heat present in dx is $\beta(x)u(x_1t)\,dx$. Here $\alpha(x)$ and $\beta(x)$ are positive functions. If $\mathfrak{D} \subset \mathscr{B}$ is an arbitrary normal domain, then the heat flow through the surface element dS of $\dot{\mathfrak{D}}$, according to (1.34), is $(f,\nu)\,dS$.

Now the decrease in time of the amount of heat in $\mathfrak{D}$ must be equal to the flow through $\dot{\mathfrak{D}}$:

$$-\frac{d}{dt} \int_{\mathfrak{D}} \beta u\,dx = \int_{\dot{\mathfrak{D}}} (f,\nu)\,dS = - \int_{\dot{\mathfrak{D}}} \alpha(\text{grad } u,\nu)\,dS. \tag{1.35}$$

But then from (1.11),

$$\int_{\mathfrak{D}} \beta u_t\,dx = \int_{\mathfrak{D}} \text{div } (\alpha \text{ grad } u)\,dx \tag{1.36}$$

follows for any $\mathfrak{D} \subset \mathscr{B}$. From this, the *heat equation*

$$\text{div } (\alpha \text{ grad } u) = \beta u_t \quad \text{or} \quad \sum_{i=1}^{3} (\alpha(x)u_{x_i}(x,t))_{x_i} = \beta(x)u_t(x,t) \tag{1.37}$$

results. If α and β are positive constants, then (1.37) reduces to

$$\Delta_3 u = \frac{\beta}{\alpha}\, u_t. \tag{1.38}$$

The physical as well as the mathematical posing of the problem now requires that in addition to (1.38) we give initial and boundary conditions. For the beginning of the experiment, say at time $t = 0$, let the temperature distribution be known:

$$u(x,0) = u_0(x) \quad \text{in} \quad \overline{\mathscr{B}}. \tag{1.39}$$

Further let the temperature distribution be given on $\dot{\mathscr{B}}$ for all times $t \geqslant 0$. It may also be the case that $\dot{\mathscr{B}}$ is isolated relative to the heat flow, that is, $(f,\nu) = 0$ on $\dot{\mathscr{B}}$, so that according to (1.34) the boundary condition $u_\nu = 0$ holds on $\dot{\mathscr{B}}$. Finally, more complicated boundary conditions are possible. They arise from the two previously mentioned conditions by linear combination:

$$\gamma(x)u_\nu + \delta(x)u = \eta(x) \quad \text{for} \quad x \in \dot{\mathscr{B}}, \qquad t \geqslant 0. \tag{1.40}$$

If we consider only *stationary* temperature fields ($\partial/\partial t \equiv 0$), the pure boundary-value problem arises:

$$\Delta_3 u = 0 \quad \text{in} \quad \mathscr{B}, \qquad \gamma(x)u_\nu + \delta(x)u = \eta(x) \quad \text{with} \quad x \in \dot{\mathscr{B}}. \tag{1.41}$$

The relation $\Delta_n u = 0$ is called the potential equation.

After these considerations we may expect that a study of the wave, potential, and heat equations will bring forth some essential features of the theory of partial differential equations.

2

THE WAVE EQUATION

2.1 The Wave Equation in R_1

According to (1.24), the wave equation in R_1 is of the form

$$u_{xx} - \frac{1}{\gamma^2} u_{tt} = 0, \quad \text{where} \quad \gamma = \text{const} > 0. \tag{2.1}$$

The symbol x denotes a real variable here. Introduction of a new time scale $\bar{t} = \gamma t$ shows that we may restrict ourselves to the case $\gamma = 1$. Therefore we base (2.1) on the form

$$u_{xx} - u_{tt} = 0 \quad \text{or} \quad u_{\bar{x}\bar{t}} = 0, \tag{2.2}$$

where the last equation results from the transformation $\bar{x} = x + t, \bar{t} = x - t$. By integration we find all solutions of (2.2):

$$u(x,t) = w_1(\bar{x}) + w_2(\bar{t}) = w_1(x + t) + w_2(x - t), \tag{2.3}$$

where the $w_i \in C^2$ $(i = 1, 2)$ are arbitrary functions.

The expression $w_1(x + t)$ $[w_2(x - t)]$ is called an undistorted wave, progressing to the left (right) with velocity 1, for the following reason: If, for $t = 0$, $w_1(x)$ is given by Figure 1, then Figure 2 shows the process at time $t = 1$. For determining suitable conditions in

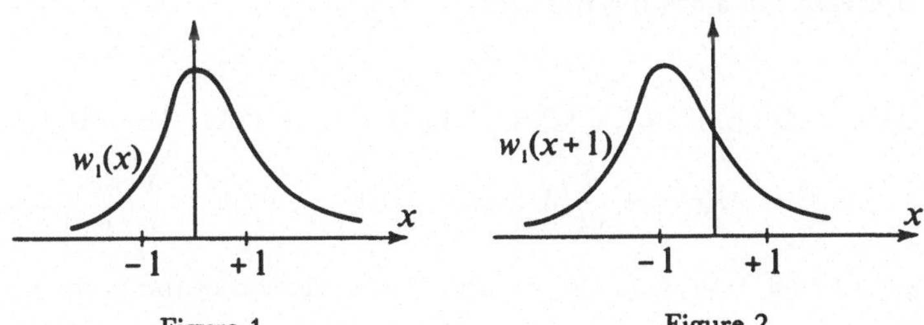

Figure 1   Figure 2

addition to (2.2) it is useful to give a physical interpretation different from the one in Section 1.5. Let a stretched string be of homogeneous mass and suppose that in its equilibrium position it fills out the continuum $0 \leqslant x \leqslant l$ on the x-axis. Let the string be clamped at the points $0, l$. If we laterally displace the string from the equilibrium position and release it at time $t = 0$ and if, in addition, we give it a certain initial velocity, it will oscillate; we wish to determine these oscillations. If $u(x,t)$ denotes the amplitude (displacement from the equilibrium position) of a point x of the string at time t, then u satisfies equation (2.1) [or (2.2)] with the initial conditions

$$u(x,0) = u_0(x), \qquad u_t(x,0) = u_1(x), \qquad 0 \leqslant x \leqslant l, \tag{2.4}$$

where the functions $u_i(x)$ are to be prescribed arbitrarily, and with boundary ("clamp") conditions

$$u(0,t) = 0, \qquad u(l,t) = 0, \qquad 0 \leqslant t < \infty. \tag{2.5}$$

In order that (2.2), (2.4), and (2.5) represent a reasonable *mathematical (physical) problem* it has to satisfy *two (three) requirements*.

1. *Uniqueness Requirement.*   There is at most one $u(x,t)$ which satisfies conditions (2.2), (2.4), and (2.5).

2. *Existence Requirement.*   There is at least one such $u(x,t)$.

3. *Physical Requirement of Continuous Dependence on the Initial Conditions.*   If the $u_i(x)$ are changed by a small amount, then the solution also changes by a small amount.   That is, if $\bar{u}(x,t)$ is the solution of (2.2), (2.4), and (2.5) with the initial conditions $\bar{u}(x,0) = \bar{u}_0(x)$, $\bar{u}_t(x,0) = \bar{u}_1(x)$, then for any $\epsilon > 0$, $|\bar{u}(x,t) - u(x,t)| < \epsilon$ holds whenever $|\bar{u}_i(x) - u_i(x)| < \delta(\epsilon)$.   (In later problems, the last inequality is sometimes to be postulated also for suitable derivatives.)

The physical interpretation here is as follows: The $u_i(x)$ are found by observation, and inaccuracies in observation are always present.   If the mathematical description is to be reasonable, then $u(x,t)$ should change only by a little if the inaccuracies of observation were small.   If x runs through an infinite interval, then corrections of $u_i(x)$ may be carried out only in a finite interval, since the infinite interval is to be considered only as a mathematical idealization.   If we have to deal with boundary-value problems, as we do later in the case of the potential equation, the third requirement is posed in quite an analogous fashion.

In the ideal case of a string of infinite length $(-\infty < x < \infty)$ we shall prove for the initial-value problem the following theorem.

THEOREM.   If $u_0(x)\ \epsilon\ C^2$, $u_1(x)\ \epsilon\ C^1$ in $-\infty < x < \infty$, then the function

$$u(x,t) = M(t)u_1 + \frac{\partial}{\partial t}[M(t)u_0], \quad \text{where} \quad M(t)u_i \equiv \frac{1}{2}\int_{x-t}^{x+t} u_i(\tau)\,d\tau \tag{2.6}$$

belongs to C^2 on $-\infty < x, t < \infty$, and it is a solution of the problem

$$u_{xx} - u_{tt} = 0; \qquad u(x,0) = u_0(x), \qquad u_t(x,0) = u_1(x).$$

This problem satisfies the three requirements.

Proof.   First, from (2.3)

$$u(x,0) = w_1(x) + w_2(x) = u_0(x), \qquad u_t(x,0) = w_1'(x) - w_2'(x) = u_1(x)$$

follows.   Integration of the last equation gives

$$w_1(x) - w_2(x) = \int^x u_1(\tau)\, d\tau,$$

and by adding and subtracting:

$$2w_1(x) = u_0(x) + \int^x u_1(\tau)\, d\tau, \qquad 2w_2(x) = u_0(x) - \int^x u_1(\tau)\, d\tau.$$

Then (2.3) yields

$$u(x,t) = w_1(x + t) + w_2(x - t) = \frac{1}{2}\left\{ u_0(x + t) + u_0(x - t) + \int_{x-t}^{x+t} u_1(\tau)\, d\tau \right\}, \quad (2.7)$$

which coincides with (2.6).   With this, requirements 2 and 1 are satisfied, but the latter only because of the favorable circumstance that in (2.3) we had found all solutions of (2.2) and have singled out a unique one by means of the initial conditions.   If we change u_i in the interval $a \leqslant x \leqslant b$ of length L: $\bar{u}_i(x) - u_i(x) = v_i(x)$ where $|v_i(x)| < \epsilon/(1 + L)$, then from (2.7) it follows immediately that $|\bar{u}(x,t) - u(x,t)| < \epsilon$, so that requirement 3 is satisfied.

2.2   Domain of Dependence; Domain of Determinateness

If (2.6) is considered at the world point (or point in space-time)[1] $x_0,\ t_0$, then (2.7) yields the representation

$$u(x_0,t_0) = \frac{1}{2}\left\{ u_0(x_0 + t_0) - u_0(x_0 - t_0) + \int_{x_0 - t_0}^{x_0 + t_0} u_1(\tau)\, d\tau \right\}.$$

Obviously $u(x_0,t_0)$ does not depend on the total behavior of the initial values $u_i(x)$ but only on those values $u_i(x)$ for which x lies in the interval $[x_0 - t_0,\ x_0 + t_0]$.[2]   This point set is called the *domain of dependence* relative to the point $x_0,\ t_0$ in space-time; it is denoted by $\bar{\mathfrak{a}}(x_0,t_0)$.   Thus we have $\bar{\mathfrak{a}}(x_0,t_0)$: $|x - x_0| \leqslant |t_0|$.   A change of $u_i(x)$ outside of $\bar{\mathfrak{a}}(x_0,t_0)$ has no influence on the value of u at the world point $x_0,\ t_0$.   If on the other hand, we fix an arbitrary interval $\bar{I}$: $|x - a| \leqslant r$ with center a and length $2r$, then the collection of those world points (x,t) at which the solution $u(x,t)$ depends only on the initial values $u_i(x)$ in $\bar{I}$ is called the *domain of determinateness* $\bar{\mathfrak{B}}(\bar{I})$.   According to (2.7), a point in space-time obviously belongs to $\bar{\mathfrak{B}}(\bar{I})$ if and only if $x + t$ and $x - t$ are contained in $\bar{I}$, that is, if $|x + t - a| \leqslant r$ and $|x - t - a| \leqslant r$ are both satisfied.   Written out, this means the validity of the inequalities

$$-r - t \leqslant x - a \leqslant r - t, \qquad -r + t \leqslant x - a \leqslant r + t,$$

which can be combined in $|x - a| \leqslant r - |t|$.   Thus we have

$$\bar{\mathfrak{B}}(\bar{I}): |x - a| \leqslant r - |t|,$$

---

[1] In general a point is called a point in space-time if it is described by $x = (x_1, x_2, \ldots, x_n)$ and the time coordinate t.   It can always be interpreted as a point in R_{n+1} with coordinates $(x_1, x_2, \ldots, x_n, t)$.

[2] In fact from $u_0(x)$ only the values in the end points of the interval enter.

and geometrically $\bar{\mathcal{B}}(\bar{I})$ consists of all points of the closed square in Figure 3. The sides of the square $\dot{\mathcal{B}} = \sum_{i=1}^{4} \dot{\mathcal{B}}_i$ are called characteristic manifolds $\dot{C}$. Hence we have

$$\dot{C}: |x - a| = r - |t|.$$

An easy consequence of the theorem and these concepts is the statement: If $u_i(x) = 0$ in $a - r \leqslant x \leqslant a + r$ (Figure 3), then $u \equiv 0$ in $\bar{\mathcal{B}}$. Along with the initial-value problem, the *characteristic problem* is significant. For this problem, $u \in C^2$ is prescribed arbitrarily on

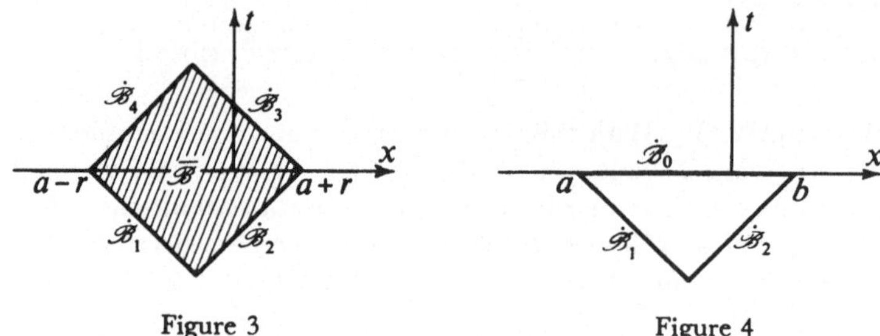

Figure 3   Figure 4

$\dot{\mathcal{B}}_1 + \dot{\mathcal{B}}_2$ (or $\dot{\mathcal{B}}_2 + \dot{\mathcal{B}}_3$, . . .), as illustrated in Figure 3. We ask for solutions in $\bar{\mathcal{B}}$. This problem also satisfies those three requirements.

Problem 1. Verify the foregoing statements.

Problem 2. Solve (2.2) in the triangle bounded by $\dot{\mathcal{B}}_0 + \dot{\mathcal{B}}_1 + \dot{\mathcal{B}}_2$ (Figure 4) if $u \in C^2$ on $\dot{\mathcal{B}}_1$, and $u_t \in C^1$ on $\dot{\mathcal{B}}_0$ are prescribed arbitrarily.

2.3   The Initial Boundary-Value Problem

This consists of (2.2), (2.4), and (2.5). Here the initial values $u_i(x)$ are given in $0 \leqslant x \leqslant l$. If we can continue them so that $u_0(x) \in C^2$, $u_1(x) \in C^1$ in $-\infty < x < \infty$, then (2.6) already satisfies the differential equation and the initial condition properly. If, furthermore, the continuation can be carried out so that the boundary conditions (2.5) are automatically satisfied by (2.6), then (2.6) is the solution of the initial boundary-value problem.

The functions

$$u_0(x) \in C^2, \qquad u_1(x) \in C^1 \quad \text{in} \quad 0 \leqslant x \leqslant l$$

where

$$u_i(0) = u_i'(0) = u_0''(0) = 0, \qquad u_i(l) = u_i'(l) = u_0''(l) = 0, \qquad i = 0, 1$$

are continued as odd periodic functions of period $2l$:

$$u_i(x) = -u_i(-x), \qquad u_i(x + 2l) = u_i(x), \qquad i = 0, 1. \tag{2.8}$$

Then from (2.7) we have

$$u(0,t) = \frac{1}{2} \left\{ u_0(t) + u_0(-t) + \int_0^t u_1(\tau) \, d\tau + \int_{-t}^0 u_1(\tau) \, d\tau \right\}$$

$$= \frac{1}{2} \int_0^t \{ u_1(\tau) + u_1(-\tau) \} \, d\tau = 0,$$

$$u(l,t) = \frac{1}{2}\left\{u_0(l+t) + u_0(l-t) + \int_0^{l+t} u_1(\tau)\,d\tau + \int_{l-t}^0 u_1(\tau)\,d\tau\right\}$$

$$= \frac{1}{2}\left\{u_0(l+t) - u_0(-l+t) + \int_0^{l+t} u_1(\tau)\,d\tau + \int_0^{-l+t} u_1(-\tau)\,d\tau\right\} = 0;$$

this proves everything.

Problem. Solve (2.2), (2.4) with the boundary conditions $u(0,t) = u_x(l,t) = 0$, and state sufficient hypotheses for the $u_i(x)$.

2.4 The Wave Equation in R_3

Because of (1.24), and with a suitable time scale, the wave equation in R_3 may be assumed to have the form

$$\Delta_3 u - u_{tt} = 0, \qquad \Delta_3 u \equiv \sum_{i=1}^3 u_{x_i x_i}. \tag{2.9}$$

We shall treat the initial-value problem. For this the following theorem holds.

THEOREM. If $u_0(x) \in C^3$, $u_1(x) \in C^2$ in $-\infty < x_1, x_2, x_3 < \infty$, then the function[3]

$$u(x,t) = tM(t)u_1 + \frac{\partial}{\partial t}\left[tM(t)u_0\right] \quad \text{where} \quad M(t)u_i \equiv \frac{1}{4\pi}\int_{|\nu|=1} u_i(x+vt)\,d\omega \tag{2.10}$$

belongs to C^2 on $-\infty < x_1, x_2, x_3, t < \infty$, and it is a solution of the problem

$$\Delta_3 u - u_{tt} = 0, \qquad u(x,0) = u_0(x), \qquad u_t(x,0) = u_1(x).$$

This problem satisfies the three requirements.

In (2.10) we have to integrate over the unit sphere. The vector $y = x + vt$ describes the sphere $\dot{S}$: $|y - x| = |t|$ with the center x and radius $|t|$ (Figure 5). The abbreviation $M(t)u_i$

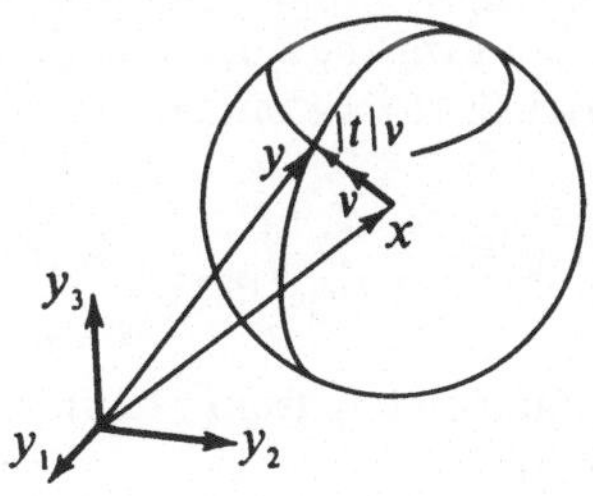

Figure 5

suggests itself, since we are concerned with taking the mean value of u_i over the sphere $|y - x| = |t|$ (spherical mean). If in (2.9) we replace t by $-t$, then (2.9) remains unchanged. Therefore we may suppose $t \geq 0$.

[3] For $(\partial/\partial t)[M(t)u_i]$ we later also write $M_t(t)u_i$.

Proof. Clearly, $u(x,t) \in C^2$. Furthermore,

$$u(x,0) = \lim_{t \to 0} \{M(t)u_0 + t[M(t)u_1 + M_t(t)u_0]\} = M(0)u_0$$

$$= \frac{1}{4\pi} \int_{|\nu|=1} u_0(x)\, d\omega = \frac{u_0(x)}{4\pi} \int_{|\nu|=1} d\omega = u_0(x), \tag{2.11}$$

$$u_t(x,0) = \lim_{t \to 0} \{M(t)u_1 + 2M_t(t)u_0 + t[M_t(t)u_1 + M_{tt}(t)u_0]\}$$

$$= \lim_{t \to 0} \{M(t)u_1 + 2M_t(t)u_0\} = M(0)u_1 = u_1(x), \tag{2.12}$$

for we have

$$M_t(0)u_0 = \frac{1}{4\pi} \sum_{i=1}^{3} \int_{|\nu|=1} u_{0x_i}(x)\nu_i\, d\omega = \frac{1}{4\pi} \sum_{i=1}^{3} u_{0x_i}(x) \int_{|\nu|=1} \nu_i\, d\omega = 0. \tag{2.13}$$

Here $\int_{|\nu|=1} \nu_i\, d\omega = 0$ follows immediately from the Gauss integral theorem. (Namely, put $u(x) \equiv 1$ in (1.7), and choose $\dot{\mathfrak{D}}$: $|\nu| = 1$.)

Furthermore, we show that an expression of the form

$$v(x,t) = tM(t)\varphi \equiv \frac{t}{4\pi} \int_{|\nu|=1} \varphi(x + \nu t)\, d\omega \quad \text{where} \quad \varphi(x) \in C^2 \tag{2.14}$$

satisfies the wave equation (2.9). First it is clear that

$$v_{tt} = 2M_t(t)\varphi + tM_{tt}(t)\varphi. \tag{2.15}$$

To calculate the first term we begin with a change of integrals where the integration has to be carried out over a sphere in y-space, with center x and radius t. Using (1.18) and (1.15) we obtain

$$\frac{1}{4\pi t^2} \int_{|y-x| \leqslant t} \Delta_3\varphi(y)\, dy = \frac{1}{4\pi t^2} \int_{|y-x|=t} \varphi_\nu(y)\, dS = \frac{1}{4\pi t^2} \sum_{i=1}^{3} \int_{|y-x|=t} \nu_i\varphi_{y_i}(y)\, dS. \tag{2.16}$$

Here for ν_i we have explicitly $\nu_i = (y_i - x_i)/t$. Now, from Figure 5, $y = x + \nu t$, and the integration can also be carried out over the surface of the unit sphere $|\nu| = 1$. As is well known, here $dS = t^2\, d\omega$, and from (2.16) we obtain

$$\frac{1}{4\pi t^2} \int_{|y-x| \leqslant t} \Delta_3\varphi(y)\, dy = \frac{1}{4\pi} \sum_{i=1}^{3} \int_{|\nu|=1} \nu_i\varphi_{y_i}(x + \nu t)\, d\omega = M_t(t)\varphi, \tag{2.17}$$

where the last equality follows immediately from (2.10). By a different evaluation of the volume integral in (2.17) we obtain

$$M_t(t)\varphi = \frac{1}{4\pi t^2} \int_{|y-x| \leqslant t} \Delta_3\varphi(y)\, dy = \frac{1}{4\pi t^2} \int_0^t d\rho \int_{|y-x|=\rho} \Delta_3\varphi(y)\, dS. \tag{2.18}$$

This transformation is obtained as follows: If we introduce spherical coordinates

$$y_1 = x_1 + \rho \cos \phi \sin \psi, \qquad y_2 = x_2 + \rho \sin \phi \sin \psi, \qquad y_3 = x_3 + \rho \cos \psi$$

in the sphere $|y - x| \leqslant t$, with $0 \leqslant \rho \leqslant t$, $0 \leqslant \phi < 2\pi$, $0 \leqslant \psi \leqslant \pi$, then using $\Delta_3\varphi(y) \equiv f(y)$ we have

$$\int_{|y-x| \leqslant t} f(y)\, dy$$
$$= \int_0^t \left\{ \int_0^\pi \int_0^{2\pi} f(x_1 + \rho \cos \phi \sin \psi,\, x_2 + \rho \sin \phi \sin \psi,\, x_3 + \rho \cos \psi)\rho^2 \sin \psi\, d\phi\, d\psi \right\} d\rho.$$

The surface element for the spherical surface $|y - x| = \rho$, is well known to be $dS = \rho^2 \sin \psi\, d\phi\, d\psi$. This justifies the transformation in (2.18).

From (2.18) it follows, by differentiation with respect to t, that

$$M_{tt}(t)\varphi = -\frac{1}{2\pi t^3} \int_{|y-x| \leqslant t} \Delta_3\varphi(y)\, dy + \frac{1}{4\pi t^2} \int_{|y-x| = t} \Delta_3\varphi(y)\, dS, \tag{2.19}$$

and because of (2.15), v_{tt} is given by

$$v_{tt} = \frac{1}{4\pi t} \int_{|y-x| = t} \Delta_3\varphi(y)\, dS. \tag{2.20}$$

Now we only have to determine $\Delta_3 v$. From (2.14) we have

$$\Delta_3 v(x,t) = \frac{t}{4\pi} \int_{|\nu| = 1} \Delta_3\varphi(x + \nu t)\, d\omega. \tag{2.21}$$

If we put again $y = x + \nu t$, $\Delta_3\varphi(x + \nu t) = \Delta_3\varphi(y)$ follows immediately. Here the differentiations on the left are to be carried out with respect to x_i, on the right with respect to y_i. Therefore (2.21) can be put into the form

$$\Delta_3 v(x,t) = \frac{1}{4\pi t} \int_{|y-x| = t} \Delta_3\varphi(y)\, dS, \tag{2.22}$$

and comparison of (2.20) with (2.22) shows that (2.14) satisfies the wave equation, and with it so does the first term $tM(t)u_1$ in (2.10). For the second term we reason as follows: First, $z(x,t) = tM(t)u_0$ satisfies the wave equation, and from this

$$0 = (\Delta_3 z - z_{tt})_t = \Delta_3(z_t) - (z_t)_{tt} \tag{2.23}$$

follows. Consequently z_t also satisfies the wave equation.

Thus the existence of a solution of our problem has been proved. Obviously it satisfies the third requirement. Here, in anology to the situation in Section 2.1, we have to change $u_1(x)$ and $u_0(x)$ a little; however, the change of $u_0(x)$ has to be made in such a way that at the same time $u_{0x_i}(x)$ is only changed a little:

$$|\bar{u}_i(x) - u_i(x)| < \delta(\epsilon), \qquad |\bar{u}_{0x_i}(x) - u_{0x_i}(x)| < \delta(\epsilon).$$

By systematically finding the solution we shall show in Section 2.5 that (2.10) is the unique solution of the problem. Therefore for our problem the three requirements are fulfilled.

2.5 Finding the Solution

Let the initial-value problem be written in the coordinates y, t:

$$\Delta_3 u - u_{tt} = 0; \qquad u(y,0) = u_0(y), \qquad u_t(y,0) = u_1(y), \tag{2.24}$$

where $-\infty < y_1, y_2, y_3, t < \infty$. Let $u(y,t) \,\varepsilon\, C^2$ be an arbitrary solution of this problem. Without loss of generality we may again assume $t \geqslant 0$. We integrate (2.24) over the ball $\mathcal{S}: |y - x| \leqslant r$ with center x and radius r. On $\mathcal{S}$ we have $y = x + \nu r$. From (2.24), under use of (1.18) and (1.15),

$$\int_{|y-x|\leqslant r} u_{tt}(y,t)\, dy = \int_{|y-x|\leqslant r} \Delta_3 u(y,t)\, dy = \int_{|y-x|=r} u_\nu(y,t)\, dS$$

$$= \sum_{i=1}^{3} \int_{|y-x|=r} \nu_i u_{y_i}(y,t)\, dS = r^2 \sum_{i=1}^{3} \int_{|\nu|=1} \nu_i u_{y_i}(x + \nu r,\, t)\, d\omega$$

$$= r^2 \int_{|\nu|=1} \frac{\partial}{\partial r} u(x + \nu r,\, t)\, d\omega, \qquad (2.25)$$

results; for indeed, we have

$$\frac{\partial}{\partial r} u(x + \nu r,\, t) = \sum_{i=1}^{3} \nu_i u_{y_i}(x + \nu r,\, t).$$

If for brevity we set

$$\tilde{M}u \equiv \frac{1}{4\pi} \int_{|\nu|=1} u(x + \nu r,\, t)\, d\omega,$$

then (2.25) becomes

$$4\pi r^2 (\tilde{M}u)_r = \int_{|y-x|\leqslant r} u_{tt}(y,t)\, dy = \int_0^r d\rho \int_{|y-x|=\rho} u_{tt}(y,t)\, dS. \qquad (2.26)$$

Differentiation with respect to r—again using $y = x + \nu t$ and $dS = r^2\, d\omega$—gives

$$4\pi (r^2(\tilde{M}u)_r)_r = \int_{|y-x|=r} u_{tt}(y,t)\, dS = \left(r^2 \int_{|\nu|=1} u(x + \nu r,\, t)\, d\omega \right)_{tt} = 4\pi r^2 (\tilde{M}u)_{tt}. \qquad (2.27)$$

Thus, as can be verified immediately, $r\tilde{M}u$ satisfies the wave equation in R_1:

$$(r\tilde{M}u)_{rr} - (r\tilde{M}u)_{tt} = 0. \qquad (2.28)$$

According to (2.3), all solutions of this equation are given by

$$r\tilde{M}u = w_1(r + t) + w_2(r - t). \qquad (2.29)$$

If the limit is taken as $r \to 0$, $w_1(t) = -w_2(-t)$ follows so that (2.29) can be written in the form

$$r\tilde{M}u = w_1(r + t) - w_1(t - r). \qquad (2.30)$$

Differentiation of (2.30) with respect to r and passage to the limit as $r \to 0$ give

$$\lim_{r \to 0} \tilde{M}u = 2w_1'(t). \quad \text{Now} \quad \lim_{r \to 0} \tilde{M}u = \frac{1}{4\pi} \int_{|\nu|=1} u(x,t)\, d\omega = u(x,t). \qquad (2.31)$$

Thus the solution of the initial-value problem has been found systematically if we succeed in determining $w_1'(t)$ uniquely from the initial values.

First, from (2.30) it follows that (the prime here always denotes the derivative with respect to the argument)

$$(r\tilde{M}u)_r = w_1'(r + t) + w_1'(t - r), \qquad (r\tilde{M}u)_t = w_1'(r + t) - w_1'(t - r)$$

or, by addition,

$$(r\tilde{M}u)_r + (r\tilde{M}u)_t = 2w_1'(r + t).$$

Passage to the limit $t \to 0$ gives $2w_1'(r)$ on the right-hand side which, together with (2.31), gives $u(x,r)$.   By writing out $\tilde{M}u$, we find

$$u(x,r) = \left(\frac{r}{4\pi} \int_{|\nu|=1} u(x + \nu r, 0)\, d\omega \right)_r + \frac{r}{4\pi} \int_{|\nu|=1} u_t(x + \nu r, 0)\, d\omega. \qquad (2.32)$$

From (2.24) we have $u(x + \nu t, 0) = u_0(x + \nu r)$, $u_t(x + \nu r, 0) = u_1(x + \nu r)$.   If finally we put $r = t$ in (2.32), it follows that

$$u(x,t) = \frac{t}{4\pi} \int_{|\nu|=1} u_1(x + \nu t)\, d\omega + \left(\frac{t}{4\pi} \int_{|\nu|=1} u_0(x + \nu t)\, d\omega \right)_t, \qquad (2.33)$$

which agrees with (2.10).   Thus the first requirement has been taken care of, since we have shown that an arbitrary solution of the initial-value problem has to coincide with (2.10).

It is amazing how many results in the field of partial differential equations have been obtained by use of such methods of spherical means.   John, F.[5] reports about these in his book.

2.6   The Domains $\bar{\mathfrak{A}}$, $\bar{\mathfrak{B}}$ for the Wave Equation in R_3

We consider (2.10) at an arbitrary point x_0, t_0 in space-time.   We have the representation

$$u(x_0,t_0) = \frac{t_0}{4\pi} \int_{|\nu|=1} u_1(x_0 + \nu t_0)\, d\omega + \lim_{t \to t_0} \left(\frac{t}{4\pi} \int_{|\nu|=1} u_0(x_0 + \nu t)\, d\omega \right)_t. \qquad (2.34)$$

Obviously $u(x_0,t_0)$ does not depend on the total behavior of the $u_i(x)$, but only on the values of $u_1(x)$ on the surface of the sphere with center x_0 and radius $|t_0|$, and on the values of $u_0(x)$ in a neighborhood of this surface.   If as a generalization of Section 2.2—this generalization is necessitated merely by the appearance of a derivative of u_0 in (2.34)—we define the intersection of these neighborhoods as the *domain of dependence* $\bar{\mathfrak{A}}(x_0,t_0)$, then

$$\bar{\mathfrak{A}}(x_0,t_0): |x - x_0| = |t_0|. \qquad (2.35)$$

If, on the other hand, we fix a ball $\bar{S}$: $|x - a| \leqslant r$ in R_3, then the set of points (x,t) in space-time, at which the solution $u(x,t)$ depends merely on the initial values $u_i(x)$ in $\bar{S}$, is called the *domain of determinateness* $\bar{\mathfrak{B}}(\bar{S})$.

From (2.33) it is obvious that a point x, t in space-time belongs to $\bar{\mathfrak{B}}(\bar{S})$ if and only if for any unit vector ν we have $x + \nu|t| \in \bar{S}$, that is,

$$\left| x + \nu|t| - a \right| \leqslant r. \qquad (2.36)$$

If $x \neq a$, from (2.36) it follows that for $\nu = (x - a)/|x - a|$:

$$\left| x + \frac{x - a}{|x - a|}|t| - a \right| = \left| (x - a) \frac{|x - a| + |t|}{|x - a|} \right| = |x - a| + |t| \leqslant r,$$

and therefore

$$|x - a| \leqslant r - |t|. \qquad (2.37)$$

Of course this relation also holds for $x = a$ if only (2.36) is satisfied.   Conversely, from (2.37), for any unit vector ν,

$$r \geqslant |x - a| + |t| = |x - a| + |\nu||t| \geqslant |x - a + \nu|t||$$

follows, which is exactly (2.36). Thus we have shown that relation (2.37) characterizes the domain of determinateness $\mathfrak{B}(\bar{S})$ precisely. This, for arbitrary r, is the double cone (including the lateral surfaces) that is symmetric to the plane $t = 0$ and has its vertices at a, r and $a, -r$. The geometric interpretation is given in Section 2.8.

The lateral surfaces of the cones, according to Section 2.2, are called characteristic manifolds $\dot{C}$. We have

$$\dot{C}: |x - a| = r - |t|. \tag{2.38}$$

2.7 The Wave Equation in R_2

The initial-value problem for the wave equation in R_2 reads

$$\Delta_2 u - u_{tt} = 0, \qquad u(\bar{x},0) = u_0(\bar{x}), \qquad u_t(\bar{x},0) = u_1(\bar{x}) \tag{2.39}$$

where $-\infty < x_1, x_2 < \infty$. Here $\bar{x}$ stands for $\bar{x} = (x_1, x_2)$. The following theorem holds.

THEOREM. If $u_0(\bar{x}) \in C^3$, $u_1(\bar{x}) \in C^2$ in $-\infty < x_1, x_2 < \infty$, then the function

$$u(\bar{x},t) = \bar{M}(t)u_1 + \frac{\partial}{\partial t} \bar{M}(t)u_0$$

$$\text{where} \quad \bar{M}(t)u_i = \frac{1}{2\pi} \int_{|\bar{y}-\bar{x}| \leqslant |t|} \frac{u_i(\bar{y})}{\sqrt{t^2 - |\bar{y} - \bar{x}|^2}} \, d\bar{y} \tag{2.40}$$

belongs to C^2 on $-\infty < x_1, x_2, t < \infty$, and it is a solution of the problem (2.39). This problem satisfies the three requirements.

Here the integration in (2.40) is to be carried out in the $\bar{y}$-plane over the disk with center $\bar{x}$ and radius $|t|$.

Proof. J. Hadamard's method of descent shows that (2.40) is a special case of (2.10). If in (2.10) we assume the initial values $u_i(x)$ as depending on x_1, x_2 only: $u(x,0) = u_0(\bar{x})$, $u_t(x,0) = u_1(\bar{x})$, then from (2.10) $u = u(\bar{x},t)$ also results. But then this u satisfies the wave equation in R_2, for the term $u_{x_3 x_3}$ vanishes identically. Thus it merely remains to rewrite the terms in (2.10). Again we may restrict ourselves to $t \geqslant 0$, and first we have in (2.10):

$$tM(t)u_i = \frac{t}{4\pi} \int_{|\nu|=1} u_i(x + \nu t) \, d\omega = \frac{1}{4\pi t} \int_{|y-x|=t} u_i(y) \, dS. \tag{2.41}$$

We compute dS. Solving $|y - x| = t$ gives

$$y_3 = x_3 \pm \sqrt{t^2 - |\bar{y} - \bar{x}|^2}. \tag{2.42}$$

The sphere $|y - x| = t$ is described by

$$y = (y_1, y_2, x_3 \pm \sqrt{t^2 - |\bar{y} - \bar{x}|^2}). \tag{2.43}$$

First we restrict ourselves to the upper hemisphere (plus sign) and compute

$$dS = \sqrt{EG - F^2} \, dy_1 \, dy_2.$$

Here, in Gauss's notation, $E = (y_{y_1}, y_{y_1})$, $F = (y_{y_1}, y_{y_2})$, $G = (y_{y_2}, y_{y_2})$. We find

$$dS = \frac{t}{\sqrt{t^2 - |\bar{y} - \bar{x}|^2}} \, dy_1 \, dy_2. \tag{2.44}$$

Because of (2.43), y_1 and y_2 in (2.44) run through all values for which $|\bar{y} - \bar{x}| \leqslant t$. Rewriting of (2.41) then gives

$$tM(t)u_i = 2\,\frac{1}{4\pi}\int_{|\bar{y}-\bar{x}|\leqslant t}\frac{u_i(\bar{y})}{\sqrt{t^2 - |\bar{y} - \bar{x}|^2}}\,d\bar{y} \equiv \bar{M}(t)u_i. \tag{2.45}$$

The factor of 2 arises from taking into consideration the lower hemisphere. But (2.45) completes the proof of the theorem.

2.8 The Domains $\bar{\mathfrak{A}}$, $\mathfrak{B}$ for the Wave Equation in R_2

If for $\bar{x}$ we write $x = (x_1,x_2)$ again, then the solution of the initial-value problem (2.39) at the point x_0, t_0 in space-time is given by

$$u(x_0,t_0) = \frac{1}{2\pi}\int_{|x-x_0|\leqslant|t_0|}\frac{u_1(x)}{\sqrt{t_0^2 - |x - x_0|^2}}\,dx$$

$$+ \lim_{t\to t_0}\frac{\partial}{\partial t}\frac{1}{2\pi}\int_{|x-x_0|\leqslant|t|}\frac{u_0(x)}{\sqrt{t^2 - |x - x_0|^2}}\,dx. \tag{2.46}$$

Analogous to that in Section 2.6, the *domain of dependence* is therefore

$$\bar{\mathfrak{A}}(x_0,t_0)\colon |x - x_0| \leqslant |t_0|. \tag{2.47}$$

If we ask for all (x,t) at which $u(x,t)$ depends solely on the initial values $u_i(x)$ in a circle

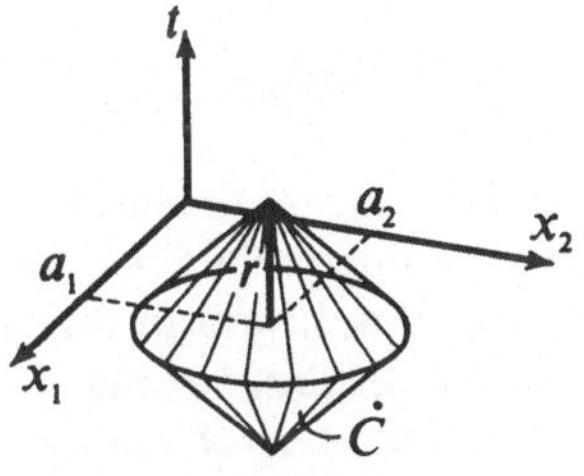

Figure 6

$\bar{C}_1\colon |x - a| \leqslant r$, then, as in Section 2.6, we obtain the *domain of determinateness*

$$\mathfrak{B}(\bar{C}_1)\colon |x - a| \leqslant r - |t|. \tag{2.48}$$

The boundary parts of $\mathfrak{B}$ are called characteristic manifolds $\dot{C}$. We have

$$\dot{C}\colon |x - a| = r - |t|. \tag{2.49}$$

These are the boundary surfaces of the double cone (2.48) which, for the wave equation in R_2, is pictured in Figure 6.

2.9 Dependence of the Wave Equation on Dimension

For the domain of dependence $\bar{\mathfrak{A}}(x_0,t_0)$ of the point x_0, t_0 in space-time a dependence on dimension arises which has the greatest consequences. Therefore we write more precisely:

$\bar{\mathcal{C}}_n(x_0,t_0)$. In Sections 2.6 and 2.8 we found that

$$\bar{\mathcal{C}}_3(x_0,t_0)\colon |x - x_0| = |t_0|, \qquad \bar{\mathcal{C}}_2(x_0,t_0)\colon |x - x_0| \leqslant |t_0|. \tag{2.50}$$

In R_2, $\bar{\mathcal{C}}_2$ consists of the disk with center x_0 and radius $|t_0|$, whereas in R_3 for $\bar{\mathcal{C}}_3$ we merely obtain the corresponding spherical surface.

Differential equations in the world point x_0, t_0 are called of Huygens' type if

$$\bar{\mathcal{C}}_n(x_0,t_0) = \dot{C} \cap \{\text{the plane } t = 0\} \tag{2.51}$$

holds, that is, if $\bar{\mathcal{C}}_n$ consists only of the intersection of $\dot{C}$ with the plane $t = 0$ in which the initial values are given. The wave equation in R_3 is of Huygens' type; in R_2 and in R_1 it is not.

Let us consider the initial-value problem for the wave equation in R_2 and R_3, respectively. Let $u_i(x) \equiv 0$ with the exception of a domain of perturbation $\mathcal{P}$, where $u_i(x) \neq 0$. How does the solution (perturbation) $u(x,t)$ behave at the point x_0, t of space-time with fixed x_0 and increasing time t? For this problem the schematic Figure 7 may serve as an

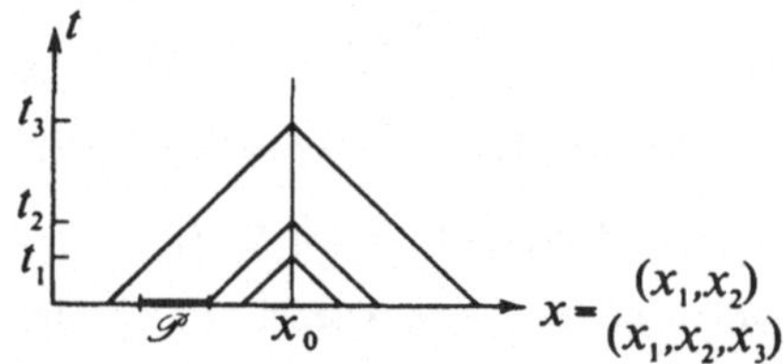

Figure 7

interpretation, where the two (three) space dimensions have been compressed into one. At x_0 there is rest at time t_1 ($u \equiv 0$, no perturbation), for in the corresponding domains of dependence $\bar{\mathcal{C}}_2$ and $\bar{\mathcal{C}}_3$, $u_i(x) \equiv 0$ so that $u \equiv 0$ in $\bar{\mathcal{B}}_2$, $\bar{\mathcal{B}}_3$ follows. At the point x_0, t_2 of space-time, $u \neq 0$. The perturbation begins, because the first parts of $\mathcal{P}$ are lying in the domains of dependence. At the point x_0, t_3 of space-time rest prevails ($u \equiv 0$, no perturbation), because the corresponding $\bar{\mathcal{C}}_3$ only consists of the surface of the sphere on which $u_i(x) \equiv 0$ already. In R_2, however, $u \neq 0$ in x_0, t_3, because $\bar{\mathcal{C}}_2$ consists of the entire disk, which also contains $\mathcal{P}$.

We see that such a perturbation in R_3 begins and vanishes abruptly; in R_2, on the other hand, it only begins abruptly and it never vanishes again for all subsequent times t. The situation for the wave equation in R_n with n odd >1 (n even >0) is analogous to the one in R_3 (R_2). These arguments become particularly clear if, for instance, in R_3 we prescribe a concrete perturbation:

$$u(x,0) = u_0(x) = 0, \qquad u_t(x,0) = u_1(x) = \begin{cases} 0 & \text{for } |x| > \epsilon, \\ 1 & \text{for } |x| \leqslant \epsilon. \end{cases}$$

Here, for this example, it does not matter that $u_1(x)$ is not in C^2. If we consider the solution of this initial-value problem at the fixed point x_0 with $|x_0| = r > \epsilon$ and all times $t \geqslant 0$, then using (2.33) we find

$$u(x_0,t) = \frac{t}{4\pi} \int_{|\nu| = 1} u_1(x_0 + \nu t)\, d\omega = \frac{1}{4\pi t} \int_{|x - x_0| = t} u_1(x)\, dS,$$

where in the last integral we put $x = x_0 + vt$. The sphere $|x - x_0| = t$ has no common points with the ball $|x| \leqslant \epsilon$, so $u(x_0,t) = 0$ since then u_1 is always zero. Common points can occur only for such t for which $r - \epsilon \leqslant t \leqslant r + \epsilon$. For those in the last formula given we then have to integrate over that part of the sphere $|x - x_0| = t$ which is contained in $|x| \leqslant \epsilon$. Using polar coordinates about the point x_0 we find

$$u(x_0,t) = \frac{1}{4\pi t} \int_0^{2\pi} \int_0^{\delta} t^2 \sin \chi \, d\chi \, d\varphi = \frac{t}{2\pi} \{1 - \cos \delta\}.$$

Here $\cos \delta$ is obtained from the elementary cosine theorem as $\cos \delta = (r^2 + t^2 - \epsilon^2)/2rt$. Thus we finally have

$$u(x_0,t) = \begin{cases} 0 \quad \text{for} \quad |t - r| > \epsilon, \\ \dfrac{1}{4\pi r} [\epsilon^2 - (r - t)^2] \quad \text{for} \quad |t - r| \leqslant \epsilon. \end{cases}$$

2.10   Continuable Initial Conditions; Determinism in Nature

We consider the initial-value problem in R_3 and, using (2.10), introduce the vector functions

$$v(x,t) = (u(x,t), u_t(x,t)), \qquad v_0(x) = (u_0(x), u_1(x)). \tag{2.52}$$

From (2.10) we find

$$v(x,t) = \mathfrak{J}(t)v_0(x) \quad \text{where} \quad \mathfrak{J}(t) = \begin{pmatrix} N_t(t), & N(t) \\ N_{tt}(t), & N_t(t) \end{pmatrix}, \qquad N(t) = tM(t). \tag{2.53}$$

We interpret the matrix $\mathfrak{J}(t)$ as a solution operator which transforms the state v at time $t = 0$ into the state v at time t. Obviously, $\mathfrak{J}(0) = I$, the identity matrix.

This new interpretation now means that it must also be possible to obtain the state v at time t by intermediate steps: Starting from the state v at time $t = 0$, we first determine the state at time $t_1 < t$ and, after j such steps, obtain the state at time t (*requirement of J. Hadamard, determinism in nature*). Finally, that process must be possible in the opposite direction of time (retrogressing into the past), since the wave equation does not change if we replace t by $-t$. These requirements make it necessary that

$$\mathfrak{J}(t_1 + t_2) = \mathfrak{J}(t_2)\mathfrak{J}(t_1) \quad \text{with} \quad \mathfrak{J}(-t)\mathfrak{J}(t) = I, \qquad \mathfrak{J}(0) = I \tag{2.54}$$

holds, that is, the solution operators form a *group* with respect to the operation defined in (2.54). Indeed, after time t_1 we find the state $\mathfrak{J}(t_1)v_0$, and by t_2 later we find the state $\mathfrak{J}(t_2)\mathfrak{J}(t_1)v_0$. This state must also result if we progress at once by $t_1 + t_2$. For the general equations, see also E. Hille and R. Phillips[6].

It is most remarkable that the theorem from Section 2.4 is too weak to satisfy Hadamard's requirement. After the first time step we have $u \in C^2$, and consequently $u_t \in C^1$; but these values cannot be used as initial values again, since for that we would need $u \in C^3$, $u_t \in C^2$. We speak of *noncontinuable initial values* (the properties of the solution deteriorate with every time step). *Continuable initial conditions* have been given by K. O. Friedrichs and H. Lewy[7]; for these the concept of square integrability in the sense of Lebesgue is needed. Heuristically, it is easy to see that square integrability plays a decisive role by giving a physically important quantity for the wave equation which does not change its quality with the time t. This is the *total vibration energy*

$$E(t) = \int_{R_3} ((u_{x_1})^2 + (u_{x_2})^2 + (u_{x_3})^2 + (u_t)^2)\, dx, \tag{2.55}$$

where $u \in C^2$ in R_3 is a solution of the wave equation which for large $|x|$ converges to zero so rapidly that no surface integrals appear in the transformation of (2.55) by means of the Gauss integral theorem (1.7). Then, writing $\dot{E} = dE/dt$, we have

$$\dot{E}(t) = 2 \int_{R_3} \left(\sum_{i=1}^{3} u_{x_i} u_{x_i t} + u_t u_{tt} \right) dx = 2 \int_{R_3} (-\Delta_3 u + u_{tt}) u_t\, dx = 0$$

so that indeed $E(t) = $ const follows.

Problem. Determine $\mathfrak{J}(t)$ in R_1, R_2, and with the use of Section 2.1 show that in R_1 the initial values are continuable.

2.11 Wave Forms

We consider the wave equation in R_n:

$$\Delta_n u - u_{tt} = 0. \tag{2.56}$$

A solution $u(x,t) \in C^2$ of the form

$$u(x,t) = f(x)g(h(x) \pm t) \tag{2.57}$$

is called a *progressing wave*. The argument of g, that is, $h(x) \pm t$, is called the *phase* of the wave; the surfaces $h(x) = $ const are called *wave fronts*. A solution of (2.56) of the form

$$u(x,t) = f(x)g(t) \tag{2.58}$$

is called a *standing wave*. The most important example of progressing waves in R_3 is given by the spatially decreasing spherical waves: If in R_3 we look for solutions of (2.56) which depend only on $|x| = r$ and t, that is, $u = u(r,t)$, we find

$$\Delta_3 u \equiv u_{rr} + \frac{2}{r} u_r \equiv \frac{1}{r} (ru)_{rr} = u_{tt}. \tag{2.59}$$

With this, ru satisfies the wave equation in R_1 and, using (2.3), we have

$$u(r,t) = \frac{w_1(r + t)}{r} + \frac{w_2(r - t)}{r}. \tag{2.60}$$

Here the second term is an outgoing, the first an incoming spherical wave.

In R_2 we obtain rotation-symmetric standing waves, namely cylindric waves, by the product solution $u(r,t) = f(r)\, e^{i\omega t}$.[4] First, for $u(r,t)$ we obtain

$$\Delta_2 u \equiv u_{rr} + \frac{1}{r} u_r = u_{tt}. \tag{2.61}$$

If we substitute the product solution in (2.61), we obtain for $f(r)$ a *Bessel differential equation*

$$f''(r) + \frac{1}{r} f'(r) + \omega^2 f(r) = 0. \tag{2.62}$$

[4] In physical considerations, of course, only the real or the imaginary part can be used.

As is well known, all solutions of this equation are given by

$$f(r) = C_1 J_0(\omega r) + C_2 N_0(\omega r)$$
$$[\text{or} = c_1 H_0^{(1)}(\omega r) + c_2 H_0^{(2)}(\omega r)] \tag{2.63}$$

where $J_0(N_0)$ is the Bessel (Neumann) function of order zero, and $H_0^{(1)}$ ($H_0^{(2)}$) the Hankel function of the first (second) kind of order zero.

Problem. Determine standing spherical waves in R_3 of the form $u(r,t) = f(r)g(t)$.

2.12   An Initial-Boundary Value Problem in R_3

Let the following problem be given:

$$\Delta_3 u - u_{tt} = f(x,t),$$
$$u(x,0) = u_0(x), \qquad u_t(x,0) = u_1(x), \tag{2.64}$$
$$u(0, x_2, x_3, t) = f_1(x_2, x_3, t)$$

in $-\infty < x_2, x_3 < \infty$, $0 \leqslant x_1, t < \infty$, where $u_0(0, x_2, x_3) = f_1(x_2, x_3, 0)$, $u_1(0, x_2, x_3) = f_{1t}(x_2, x_3, 0)$. The treatment of this initial-boundary value problem is carried out in several steps, each of which is important by itself. We forego a precise formulation of the hypotheses.

First Step. We consider the problem

$$\Delta_3 u - u_{tt} = f(x,t),$$
$$u(x,0) = 0, \qquad u_t(x,0) = 0, \tag{2.65}$$

in $-\infty < x_1, x_2, x_3 < \infty, 0 \leqslant t < \infty$. This can be reduced to solving a pure initial-value problem with $f \equiv 0$ where the initial values are given at time $t = \tau$ and not, as has been done always so far, at time $t = 0$. This initial-value problem is of the form

$$\Delta_3 v - v_{tt} = 0,$$
$$v(x,\tau) = 0, \qquad v_t(x,\tau) = f(x,\tau). \tag{2.66}$$

Its solution appears in the form $v = v(x, t, \tau)$, and according to Section 2.4 it can be determined immediately as

$$v(x, t, \tau) = (t - \tau)M(t - \tau)f \equiv \frac{1}{4\pi}(t - \tau)\int_{|\nu|=1} f(x + \nu(t - \tau),\tau)\, d\omega. \tag{2.67}$$

When we use this solution, the solution of problem (2.65) appears in the form

$$u(x,t) = -\int_0^t v(x, t, \tau)\, d\tau. \tag{2.68}$$

We verify this. First, we have $u(x,0) = 0$, and furthermore

$$-u_t(x,t) = v(x, t, t) + \int_0^t v_t(x, t, \tau)\, d\tau = \int_0^t v_t(x, t, \tau)\, d\tau, \tag{2.69}$$

so that $u_t(x,0) = 0$ follows.   Correspondingly, we obtain

$$-u_{tt}(x,t) = v_t(x,\, t,\, t) + \int_0^t v_{tt}(x,\, t,\, \tau)\, d\tau = f(x,t) + \int_0^t v_{tt}(x,\, t,\, \tau)\, d\tau, \tag{2.70}$$

$$\Delta_3 u - u_{tt} = f(x,t) - \int_0^t (\Delta_3 v - v_{tt})\, dx = f(x,t). \tag{2.71}$$

For applications it is useful to change formula (2.68) further.   Setting $t - \tau = \tau'$, we have

$$\begin{aligned}
u(x,t) &= -\frac{1}{4\pi} \int_0^t (t - \tau) \left\{ \iint_{|\nu|=1} f(x + \nu(t - \tau),\tau)\, d\omega \right\} d\tau \\
&= -\frac{1}{4\pi} \int_0^t \tau' \left\{ \iint_{|\nu|=1} f(x + \nu\tau',\, t - \tau')\, d\omega \right\} d\tau'.
\end{aligned} \tag{2.72}$$

If we set $y = x + \nu\tau'$, then $|\nu| = 1$ corresponds to the sphere $|y - x| = \tau'$ with center x and radius τ'.   Furthermore, we then have $dS = \tau'^2\, d\omega$ so that from (2.72) we obtain

$$u(x,t) = -\frac{1}{4\pi} \int_0^t \frac{1}{\tau'} \left\{ \iint_{|y-x|=\tau'} f(y,t - \tau')\, dS \right\} d\tau' = -\frac{1}{4\pi} \int_{|y-x|\leq t} \frac{f(y,t - |y - x|)}{|y - x|}\, dy. \tag{2.73}$$

This representation of $u(x,t)$, which is used in physics, is called a "retarded potential."

Second Step.   The problem

$$\begin{aligned}
\Delta_3 u - u_{tt} &= f(x,t), \\
u(x,0) &= u_0(x), \qquad u_t(x,0) = u_1(x)
\end{aligned} \tag{2.74}$$

in $-\infty < x_1,\, x_2,\, x_3 < \infty$, $0 \leq t < \infty$, with use of Section 2.4 and (2.73), is solved by

$$u(x,t) = tM(t)u_1 + \frac{\partial}{\partial t}\left[tM(t)u_0\right] - \frac{1}{4\pi} \int_{|y-x|\leq t} \frac{f(y,t - |x - y|)}{|y - x|}\, dy. \tag{2.75}$$

Third Step.[5]   Let $u(x,t)$ be the "sufficiently good" solution of the problem (2.64).   If furthermore we put

$$\Delta_2 u \equiv u_{x_2 x_2} + u_{x_3 x_3},$$

we have

$$u_{x_1 x_1}(0,\, x_2,\, x_3) = (f + u_{tt} - \Delta_2 u)_{x_1=0} = f(0,\, x_2,\, x_3,\, t) + f_{1tt} - \Delta_2 f_1 \equiv h(x_2,\, x_3,\, t). \tag{2.76}$$

As suggested by F. John[5] we introduce a new function $v(x,t)$,

$$u(x,t) = v(x,t) + f_1(x_2,\, x_3,\, t) + \tfrac{1}{2}(x_1)^2 h(x_2,\, x_3,\, t). \tag{2.77}$$

Then $v(0,\, x_2,\, x_3,\, t) = 0$ automatically.   After an easy calculation we obtain for $v(x,t)$ the relations

$$f = \Delta_3 u - u_{tt} = \Delta_3 v - v_{tt} + f(0,\, x_2,\, x_3,\, t) + \tfrac{1}{2}(x_1)^2\{\Delta_2 h - h_{tt}\}, \tag{2.78}$$

and finally

$$\begin{aligned}
\Delta_3 v - v_{tt} &= G(x,t) \quad \text{with} \quad G = f - f(0,\, x_2,\, x_3,\, t) - \tfrac{1}{2}(x_1)^2\{\Delta_2 h - h_{tt}\}, \\
v(x,0) &= u_0(x) - f_1(x_2,\, x_3,\, 0) - \tfrac{1}{2}(x_1)^2 h(x_2,\, x_3,\, 0) \equiv V_0(x), \\
v_t(x,0) &= u_1(x) - f_{1t}(x_2,\, x_3,\, 0) - \tfrac{1}{2}(x_1)^2 h_t(x_2,\, x_3,\, 0) \equiv V_1(x)
\end{aligned} \tag{2.79}$$

where $-\infty < x_2,\, x_3 < \infty$, $0 \leq x_1,\, t < \infty$.   Now, $V_0(0,\, x_2,\, x_3) = V_1(0,\, x_2,\, x_3) = 0$.

[5] See F. John[8].

We obtain for v a problem of type (2.74) if we continue G and $V_i(x)$ as odd functions into $-\infty < x_1 \leqslant 0$. Then we have[6]

$$\begin{aligned}
\Delta_3 v - v_{tt} &= g(x,t) \quad \text{where} \quad g(x,t) = G(x,t)\ \text{sign}\ x_1, \\
v(x,0) &= v_0(x) \quad \text{where} \quad v_0(x) = V_0(x)\ \text{sign}\ x_1, \\
v_t(x,0) &= v_1(x) \quad \text{where} \quad v_1(x) = V_1(x)\ \text{sign}\ x_1
\end{aligned} \tag{2.80}$$

with $-\infty < x_1,\ x_2,\ x_3 < \infty$, $0 \leqslant t < \infty$. This problem can be solved by means of the second step. If we now reverse our way of reasoning, then (2.77) will be a solution of the problem (2.64).

[6] We define: sign $x_1 = 1$ if $x_1 > 0$; $= 0$ if $x_1 = 0$; $= -1$ if $x_1 < 0$. Note that under this definition $g(x,t)$, $v_0(x)$, $v_1(x)$ do not become discontinuous for $x_1 = 0$, for the functions $G(x,t)$, $V_0(x)$, $V_1(x)$ vanish for $x_1 = 0$.

3

THE POTENTIAL EQUATION

3.1   The Initial-Value Problem for the Potential Equation

The problem

$$\Delta_n u = 0 \quad \text{where} \quad \Delta_n u \equiv \sum_{i=1}^{n} u_{x_i x_i}, \tag{3.1}$$

$$
\begin{aligned}
u(x_1, x_2, \ldots, x_{n-1}, 0) &= u_0(x_1, x_2, \ldots, x_{n-1}), \\
u_{x_n}(x_1, x_2, \ldots, x_{n-1}, 0) &= u_1(x_1, x_2, \ldots, x_{n-1})
\end{aligned}
\tag{3.2}
$$

is called the initial-value problem for the potential equation. By way of examples, we shall show that for this initial-value problem the third requirement cannot, in general, be satisfied, and that the second requirement can in general be satisfied only if the u_i, $i = 0, 1$ are analytic functions in R_{n-1}.

The function $f(x)$ is called *analytic* in a neighborhood $|x - x^0| < R$ (with $R > 0$) of the point x^0 if $f(x)$ can be represented there by a power series, convergent in $|x - x^0| < R$:

$$f(x) = \sum_{k_1, k_2, \ldots, k_n = 0}^{\infty} \gamma_{k_1, k_2, \ldots, k_n}(x_1 - x_1^0)^{k_1}(x_2 - x_2^0)^{k_2} \ldots (x_n - x_n^0)^{k_n}.$$

Further, we say that $f(x)$ is analytic in the domain $\mathfrak{F} \subset R_n$ ($\mathfrak{F}$ not necessarily a normal domain) if for every point $x^0 \, \varepsilon \, \mathfrak{F}$ we can find a neighborhood $|x - x^0| < R$ in which $f(x)$ is analytic. Here the number R may depend on x^0: $R = R(x^0)$. If the analyticity of $f(x)$ is to be defined in $\bar{\mathfrak{F}}$, then we again have to require convergence of the power series in $|x - x^0| < R$ for every $x^0 \, \varepsilon \, \bar{\mathfrak{F}}$; the validity of the last equation, however, has to be required only in the intersection $\bar{\mathfrak{F}} \cap \{|x - x^0| < R\}$.

If $f(z)$ is a function of the complex variable $z = x_1 + ix_2$, then $f(z)$ is called *analytic* in a neighborhood $|z - z^0| < R$ of the point z^0 if $f(z)$ can be represented by a power series

$$f(z) = \sum_{k=0}^{\infty} \gamma_k(z - z^0)^k$$

convergent in $|z - z^0| < R$. Further, we say $f(z)$ is *analytic* in the domain $\mathfrak{F}$ of the z-plane if for each point $z^0 \, \varepsilon \, \mathfrak{F}$ we can find a neighborhood $|z - z^0| < R$ in which $f(z)$ is analytic. Equivalent to the last definition—but much more in use—is the following one: $f(z) = u(x_1, x_2) + iv(x_1, x_2)$ is called analytic in $\mathfrak{F}$ if the real part u and the imaginary part v satisfy the Cauchy-Riemann differential equations

$$u_{x_1} - v_{x_2} = 0, \qquad u_{x_2} + v_{x_1} = 0$$

in $\mathfrak{F}$.

Example 1. *J. Hadamard*[9]. $\Delta_2 u = 0$, $u(x_1,0) = u_0(x_1)$, $u_{x_2}(x_1,0) = u_1(x_1)$.
A solution with $u_0(x_1) = 0$, $u_1(x_1) = 0$ is given by $u(x_1,x_2) \equiv 0$.
A solution with $u_0(x_1) = 0$, $u_1(x_1) = (1/k) \sin kx_1$ is given by

$$\bar{u}(x_1,x_2) = (1/k^2) \sin kx_1 \sinh kx_2.$$

For sufficiently large k, the initial values are arbitrarily close, but the solutions are not, since $\sinh kx_2$ behaves like $\frac{1}{2} e^{kx_2}$ for large k. Thus the third requirement is violated.[7]

Example 2. $\Delta_2 u = 0$, $u(x_1,0) = 0$, $u_{x_2}(x_1,0) = u_1(x_1)$.
As is known from the elements of function theory, the solution $u(x_1,x_2)$ can be considered as the real part of an analytic function $f(z) = u + iv$ with $z = x_1 + ix_2$.
In particular, $u(x_1,x_2)$ is then analytic in the two real variables x_1, x_2. We use a reflection principle which is closely related to the *Schwarz reflection principle* from complex variable theory (see Section II-3.2), in the form given by I. I. Privalov[10]:

Let $u(x) \in C^2$ and $\Delta_2 u = 0$ in a domain $\mathfrak{F}$ of R_2. Suppose that the boundary $\dot{\mathfrak{F}}$ of $\mathfrak{F}$ contains the open analytic boundary arc $\dot{\mathfrak{F}}_1$. Further let the function $u(x)$ be continuous in $\mathfrak{F} + \dot{\mathfrak{F}}_1$ and equal to zero on $\dot{\mathfrak{F}}_1$. Then the function $u(x)$ can be continued analytically across $\dot{\mathfrak{F}}_1$, by assignment of values to it which have opposite signs at points symmetric to $\dot{\mathfrak{F}}_1$. [*Analytic boundary arc* $\dot{\mathfrak{F}}_1$ can be described by $x_1 = x_1(t)$, $x_2 = x_2(t)$ with real parameter t, where $x_1(t)$, $x_2(t)$ are analytic functions on the t interval corresponding to $\dot{\mathfrak{F}}_1$ with $(\dot{x}_1)^2 + (\dot{x}_2)^2 > 0$.]

For $\dot{\mathfrak{F}}_1$ we take the x_1-axis so that the analytic function $u(x_1,x_2)$ becomes analytic on $\dot{\mathfrak{F}}_1$. But this means that $u_1(x_1)$ also must be assumed analytic.

Thus it has been shown that because of Section 2.1 the problem (3.1) and (3.2) cannot arise in physical considerations. The first requirement (uniqueness), on the other hand, can be proved to be satisfied in the greatest generality and under the weakest assumptions. This has been done in recent times by Cl. Müller, E. Heinz, P. Hartmann, and others[11].

Therefore, as was already to be expected from Section 1.7, we consider a *boundary-value problem* for (3.1) in a normal domain $\mathfrak{D}$:

$$\Delta_n u = 0 \quad \text{in} \quad \mathfrak{D}, \qquad u(x) = \varphi(x) \quad \text{on} \quad \dot{\mathfrak{D}}. \tag{3.3}$$

The notations from Sections 1.2, 1.3, and 1.4 will be used frequently.

3.2 Singularity Functions

We shall determine solutions $u(x)$ of $\Delta_n u = 0$ in the whole of R_n which depend only on
$r = |a - x| = \sqrt{\sum_{i=1}^{n} (a_i - x_i)^2}$ where $a \in R_n$ is a fixed vector. If we put $u(x) = v(r)$, the

---

[7] The change has been made on an infinite interval. The state of affairs persists, however, if we change the initial values merely in a finite interval.

function v satisfies the equation[8]

$$v'' + \frac{n-1}{r} v' = 0, \qquad ' \equiv \frac{d}{dr}, \tag{3.4}$$

which has the solutions

$$\begin{aligned} v(r) &= C_1 + C_2 r^{2-n} \quad \text{for} \quad n > 2, \\ &= C_1 + C_2 \log r \quad \text{for} \quad n = 2, \end{aligned} \tag{3.5}$$

where log denotes the natural logarithm. We shall call

$$\begin{aligned} s(a,x) &= \frac{1}{(n-2)\omega_n} |a - x|^{2-n} \quad \text{for} \quad n > 2, \\ &= -\frac{1}{2\pi} \log |a - x| \quad \text{for} \quad n = 2, \end{aligned} \tag{3.6}$$

the *singularity function* for $\Delta_n u = 0$. The function $s(a,x)$ has the property that $s \in C^\infty$ and $\Delta_n s = 0$ for $a \neq x$. For $a = x$, s has a singularity which depends upon the dimension.

Problem.* Determine $s(a,x)$ for the equation $\Delta_3 u + \kappa u = 0$, where $\kappa = \text{const} > 0$.

3.3 The Fundamental Solution

Let a be a fixed point in $\mathfrak{D}$. A function, defined for $x \neq a$ by

$$\gamma(a,x) = s(a,x) + \Phi(x) \quad \text{where} \quad \Phi \in C^1 \quad \text{in} \quad \bar{\mathfrak{D}}, \quad \in C^2 \quad \text{in} \quad \mathfrak{D}, \quad \Delta_n \Phi = 0, \tag{3.7}$$

is called a *fundamental solution relative to* $\mathfrak{D}$. We can now prove the following theorem about fundamental solutions:

THEOREM. Let $u(x) \in C^1$ in $\bar{\mathfrak{D}}$, $\in C^2$ in $\mathfrak{D}$, and suppose that $u(x)$ is a solution of $\Delta_n u = f(x)$, where $f(x) \in C^0$ in $\bar{\mathfrak{D}}$. If $a \in \mathfrak{D}$ is an arbitrary point, then

$$u(a) = \int_{\dot{\mathfrak{D}}} \{\gamma(a,x)u_\nu(x) - u(x)\gamma_\nu(a,x)\} \, dS - \int_{\mathfrak{D}} \gamma(a,x)f(x) \, dx \tag{3.8}$$

holds.

Proof. Since $\mathfrak{D}$ is open, the ball $\bar{S}: |x - a| \leq \rho$ about any point $a \in \mathfrak{D}$ is contained in $\mathfrak{D}$ for sufficiently small $\rho > 0$. Since $\gamma(a,x)$ becomes singular as $x \to a$, we remove $\bar{S}$ from $\mathfrak{D}$ and apply formula (1.17a) for γ and u on $\mathfrak{D} - \bar{S}$. We obtain

$$\int_{\mathfrak{D}-\bar{S}} (\gamma \Delta_n u - u \Delta_n \gamma) \, dx = \int_{\dot{\mathfrak{D}}} (\gamma u_\nu - u\gamma_\nu) \, dS - \int_{|x-a|=\rho} (\gamma u_\nu - u\gamma_\nu) \, dS, \tag{3.9}$$

---

[8] The little intermediate calculation goes as follows:

$$r^2 = \sum_{i=1}^n (a_i - x_i)^2, \qquad rr_{x_i} = -(a_i - x_i), \qquad u_{x_i} = v' r_{x_i} = -v' \frac{a_i - x_i}{r},$$

$$u_{x_i x_i} = v'' \frac{(a_i - x_i)^2}{r^2} + v' \left(\frac{1}{r} - \frac{(a_i - x_i)^2}{r^3} \right),$$

$$\Delta_n u = \sum_{i=1}^n u_{x_i x_i} = v'' + \frac{n-1}{r} v'.$$

where ν in the last integral is the outer normal of $\bar{S}$. If we substitute (3.7) in the last term of (3.9), we obtain the two terms

$$T_1(a,\rho) = \int_{|x-a|=\rho} (\Phi u_\nu - u\Phi_\nu)\, dS, \qquad T_2(a,\rho) = \int_{|x-a|=\rho} (su_\nu - us_\nu)\, dS. \qquad (3.10)$$

In T_1 the integrand is continuous in $|x-a| \leqslant \rho$, so that $\lim_{\rho\to 0} T_1(a,\rho) = 0$. For the estimation of T_2 we suppose $n > 2$. $\dot{S}$ is described by the vector $x = a + \nu\rho$. Therefore, using $dS = \rho^{n-1}\, d\omega$ and (3.6) we have

$$\int_{|x-a|=\rho} su_\nu\, dS = \rho^{n-1} \int_{|\nu|=1} s(a, a + \nu\rho)u_\nu(a + \nu\rho)\, d\omega$$

$$= \frac{\rho}{(n-2)\omega_n} \int_{|\nu|=1} u_\nu(a + \nu\rho)\, d\omega. \qquad (3.11)$$

Obviously this term tends to zero for $\rho \to 0$. There now remains only[9]

$$-\int_{|x-a|=\rho} us_\nu\, dS = -\rho^{n-1} \int_{|\nu|=1} u(a + \nu\rho)s_\nu(a, a + \nu\rho)\, d\omega$$

$$= -\frac{\rho^{n-1}}{(n-2)\omega_n} \int_{|\nu|=1} u(a + \nu\rho)\rho_\nu^{2-n}\, d\omega = \frac{1}{\omega_n} \int_{|\nu|=1} u(a + \nu\rho)\, d\omega. \qquad (3.12)$$

The limit of the last expression in (3.12) as $\rho \to 0$ is $u(a)$, so that $\lim_{\rho\to 0} T_2(a,\rho) = u(a)$ is proved. The left-hand side of (3.9) as $\rho \to 0$ converges to $\int_{\mathfrak{D}} \gamma f\, dx$; thus the theorem has been proved for $n > 2$.

Problem. Prove (3.8) for $n = 2$.

3.4 Green's Function of the First Kind

If, for fixed $x \in \mathfrak{D}$, $\gamma(x,y)$ is a fundamental solution relative to y with the additional property that $\gamma(x,y) = 0$ for $y \in \dot{\mathfrak{D}}$, then $\gamma(x,y)$ is called a *Green's function of the first kind*, and is denoted by $g(x,y)$.

For $g(x,y)$, equation (3.8) takes on the form

$$u(x) = -\int_{\dot{\mathfrak{D}}} g_\nu(x,y)u(y)\, dS - \int_{\mathfrak{D}} g(x,y)f(y)\, dy. \qquad (3.13)$$

If we substitute for $u(x)$ the prescribed boundary values $\varphi(x)$ on $\dot{\mathfrak{D}}$, then the right-hand side of (3.13) is known if we know g. We may then expect that (3.13) is the solution of the boundary-value problem

$$\Delta_n u = f(x) \quad \text{in} \quad \mathfrak{D}, \qquad u = \varphi(x) \quad \text{on} \quad \dot{\mathfrak{D}}. \qquad (3.14)$$

For this, however, the existence of g has to be guaranteed, and we must prove that the $u(x)$ in (3.13) has the desired properties.

---

[9] The simple intermediate calculation is:

$$\nu_i = \frac{x_i - a_i}{\rho},$$

$$\rho_\nu^{2-n} = \sum_{i=1}^{n} (\rho^{2-n})_{x_i}\nu_i = \sum_{i=1}^{n} (\rho^{2-n})_\rho \rho_{x_i}\nu_i = \sum_{i=1}^{n} (2-n)\rho^{1-n}\frac{(x_i - a_i)^2}{\rho^2} = (2-n)\rho^{1-n}.$$

We shall now construct the Green's function for the ball $|y| \leqslant R$ of dimension $n > 2$. As suggested in a note of E. Heinz[12], for fixed x with $0 < |x| < R$, we set

$$g(x,y) = s(x,y) + \Phi(x,y) \quad \text{where} \quad \Phi(x,y) = -\frac{k}{(n-2)\omega_n}|\lambda x - y|^{2-n} \qquad (3.15)$$

and k and λ, $\lambda x \neq y$, are numbers yet to be determined. Then, according to (3.6), $\Delta_n \Phi = 0$ is satisfied in y and for $|y| = R$ we must have $g = 0$. But with (3.6), this implies

$$|x - y|^{2-n} = k|\lambda x - y|^{2-n} \quad \text{or} \quad |\lambda x - y|^2 = k^{(2/n-2)}|x - y|^2 \quad \text{for} \quad |y| = R. \qquad (3.16)$$

If this equation is written out, taking into consideration (1.4) and $|y| = R$, then

$$(1 - k^{2/(n-2)})R^2 = (k^{2/(n-2)} - \lambda^2)|x|^2 + 2(\lambda - k^{2/(n-2)})(x,y) \qquad (3.17)$$

results. If we put $k^{2/(n-2)} = \lambda$, we obtain for λ either 1 or $R^2/|x|^2$. The first value violates the requirement $\lambda x \neq y$, the second gives $k = (R/|x|)^{n-2}$. The condition $\lambda x \neq y$ is also satisfied because $R^2 > |x|^2$. Finally, for $n > 2$, we obtain

$$\begin{aligned}
g(x,y) &= \frac{1}{(n-2)\omega_n}\left\{|x - y|^{2-n} - \left(\frac{|x|}{R}\right)^{2-n}\left|\frac{R^2}{|x|^2}x - y\right|^{2-n}\right\}, \qquad x \neq 0, \\
g(0,y) &= \frac{1}{(n-2)\omega_n}\{|y|^{2-n} - R^{2-n}\}.^{10}
\end{aligned} \qquad (3.18)$$

Problem 1. Determine g for the disk $|y| \leqslant R$ $(n = 2)$ by setting

$$g(x,y) = s(x,y) + \Phi(x,y) \quad \text{where} \quad \Phi(x,y) = -\frac{1}{2\pi}\{\log k - \log|\lambda x - y|\}. \qquad (3.19)$$

We find $\lambda = R^2/|x|^2$, $k = R/|x|$; thus

$$\begin{aligned}
g(x,y) &= -\frac{1}{2\pi}\left\{\log|x - y| + \log\frac{R}{|x|} - \log\left|\frac{R^2}{|x|^2}x - y\right|\right\}, \qquad x \neq 0 \\
g(0,y) &= -\frac{1}{2\pi}\{\log|y| - \log R\}.^{10}
\end{aligned} \qquad (3.20)$$

Problem 2.* Show that in general $g(x,y) = g(y,x)$ for the domain $\mathfrak{D}$.

3.5   Poisson's Formula

In (3.13) we need the value of $g_\nu(x,y)$ for fixed $x \in \mathfrak{D}$ and all $y \in \dot{\mathfrak{D}}$. We calculate this quantity for the sphere $|y| = R$ supposing first that $n > 2$. If, for the moment, we put $r = |x - y|$, then

$$r_{y_i} = -\frac{x_i - y_i}{r}, \qquad r^{2-n}_{y_i} = r^{2-n}_r y_i = (n-2)r^{1-n}\frac{x_i - y_i}{r}.$$

Hence

$$|x - y|^{2-n}_{y_i} = \frac{n-2}{|x-y|^n}(x_i - y_i). \qquad (3.21)$$

---

[10] This expression follows immediately from the definition of $g(x,y)$.

Then we find for (3.18)

$$g_{y_i}(x,y) = \frac{1}{\omega_n}\left\{\frac{x_i - y_i}{|x - y|^n} - \left(\frac{|x|}{R}\right)^{2-n}\frac{(R^2/|x|^2)x_i - y_i}{|(R^2/|x|^2)x - y|^n}\right\}. \tag{3.22}$$

Now, $g(x,y) = 0$ for $y \in \dot{\mathfrak{D}}$ so that (3.18) also yields

$$|x - y| = \frac{|x|}{R}\left|\frac{R^2}{|x|^2}x - y\right| \quad \text{for} \quad |y| = R. \tag{3.23}$$

If we use this relation in (3.22), we obtain

$$g_{y_i}(x,y) = \frac{1}{\omega_n|x - y|^n}\left\{x_i - y_i - \left(\frac{|x|}{R}\right)^2\left(\frac{R^2}{|x|^2}x_i - y_i\right)\right\} = \frac{|x|^2 - R^2}{\omega_n|x - y|^n}\cdot\frac{y_i}{R^2}, \tag{3.24}$$

which is valid only for $|y| = R$. But the unit vector ν in the direction of the outer normal is given by $\nu_i = y_i/R$, so that using (1.15) we obtain

$$g_\nu(x,y) = \sum_{i=1}^n g_{y_i}(x,y)\nu_i = \frac{|x|^2 - R^2}{\omega_n R|x - y|^n}\sum_{i=1}^n (\nu_i)^2 = \frac{|x|^2 - R^2}{\omega_n R|x - y|^n} \quad \text{for} \quad |y| = R; \tag{3.25}$$

which concludes the calculation.

Problem.   Show that (3.25) is also valid for $n = 2$.

With this, formula (3.13) for the sphere (circle) becomes

$$u(x) = \frac{R^2 - |x|^2}{R\omega_n}\int_{|y|=R}\frac{u(y)}{|x - y|^n}\,dS - \int_{|y|\leqslant R}g(x,y)f(y)\,dy, \tag{3.26}$$

which is valid for $n \geqslant 2$.

POISSON'S THEOREM.   If $\varphi(x) \in C^0$ on $|x| = R$, then for $n \geqslant 2$

$$u(x) = \begin{cases}\dfrac{R^2 - |x|^2}{R\omega_n}\displaystyle\int_{|y|=R}\dfrac{\varphi(y)}{|x - y|^n}\,dS & \text{for} \quad |x| < R \\[2ex] \varphi(x) & \text{for} \quad |x| = R\end{cases} \tag{3.27}$$

belongs to C^0 in $|x| \leqslant R$, to C^2 in $|x| < R$, and $u(x)$ is a solution of the problem

$$\Delta_n u = 0 \quad \text{for} \quad |x| < R, \qquad u = \varphi(x) \quad \text{for} \quad |x| = R; \tag{3.28}$$

this problem satisfies the three requirements.   In $|x| < R$, $u \in C^\infty$.[11]

Note that no decisive dependence on dimension appears in this theorem.

Proof.   The proof closely follows R. Courant and D. Hilbert[13].   First we consider the second requirement.   If $|x| < R$, then $|x - y| \neq 0$ and (3.27) can be differentiated under the integral sign arbitrarily often.   We have

$$\Delta_n u(x) = \frac{1}{R\omega_n}\int_{|y|=R}\varphi(y)\Delta_n\left\{\frac{R^2 - |x|^2}{|x - y|^n}\right\}dS \tag{3.29}$$

$$\Delta_n\left\{\frac{R^2 - |x|^2}{|x - y|^n}\right\} = (R^2 - |x|^2)\Delta_n|x - y|^{-n} + 2\sum_{i=1}^n (R^2 - |x|^2)_{x_i}|x - y|^{-n}_{x_i}$$
$$+ |x - y|^{-n}\Delta_n(R^2 - |x|^2) = 0. \tag{3.30}$$

---

[11] This is arbitrarily often continuously differentiable.

The last formula is obtained after a little calculation. We have

$$|x - y|_{x_i}^{-n} = - \frac{n(x_i - y_i)}{|x - y|^{n+2}}, \ \Delta_n |x - y|^{-n} = \frac{2n}{|x - y|^{n+2}}, \ (R^2 - |x|^2)_{x_i}$$

$$= -2x_i, \ \Delta_n(R^2 - |x|^2) = -2n,$$

$$\Delta_n \left\{ \frac{R^2 - |x|^2}{|x - y|^n} \right\} = \frac{1}{|x - y|^{n+2}} \{2n(R^2 - |x|^2) + 4n(|x|^2 - (x,y)) - 2n|x - y|^2\} = 0.$$

Thus $\Delta_n u = 0$ for $|x| < R$ has been proved. The attainment of the boundary values is more difficult to verify. First, $u(x) \equiv 1$ in $\mathfrak{D}$ is a solution of $\Delta_n u = 0$ with $u \, \varepsilon \, C^2$ in $\mathfrak{D}$. If we substitute this $u(x)$ in (3.26), we obtain

$$u(x) \equiv 1 = \frac{R^2 - |x|^2}{R\omega_n} \int_{|y| = R} \frac{1}{|x - y|^n} \, dS. \tag{3.31}$$

Using (3.27), we find that

$$u(x) - \varphi(x^0) = \frac{R^2 - |x|^2}{R\omega_n} \int_{|y| = R} \frac{\varphi(y) - \varphi(x^0)}{|x - y|^n} \, dS = \frac{R^2 - |x|^2}{R\omega_n} \left\{ \int_{\dot{S}_1} + \int_{\dot{S}_2} \right\}, \tag{3.32}$$

where x^0 is an arbitrary point on $\dot{S}$: $|y| = R$ and $|x - x^0| \leqslant \rho/2$. About x^0 we construct

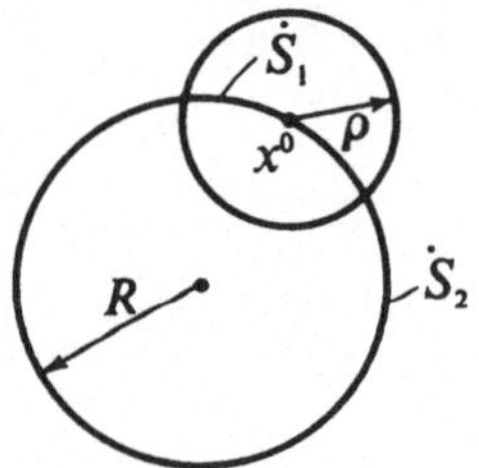

Figure 8

a sphere with radius ρ (Figure 8) and the part of $\dot{S}$ which lies in this sphere we denote by $\dot{S}_1$ ($\dot{S}_1$: $|y| = R$, $|y - x^0| \leqslant \rho$). Then $\dot{S} = \dot{S}_1 + \dot{S}_2$ with $\dot{S}_2$: $|y| = R$, $|y - x^0| > \rho$. Now we estimate the two terms in (3.32). We have

$$\left| \frac{R^2 - |x|^2}{R\omega_n} \int_{\dot{S}_1} \frac{\varphi(y) - \varphi(x^0)}{|x - y|^n} \, dS \right| \leqslant \frac{R^2 - |x|^2}{R\omega_n} \max_{y \varepsilon \dot{S}_1} |\varphi(y) - \varphi(x^0)| \int_{\dot{S}_1} \frac{dS}{|x - y|^n}$$

$$\leqslant \max_{y \varepsilon \dot{S}_1} \{|\varphi(y) - \varphi(x^0)|\} \frac{R^2 - |x|^2}{R\omega_n} \int_{|y| = R} \frac{dS}{|x - y|^n} = \max_{y \varepsilon \dot{S}_1} |\varphi(y) - \varphi(x^0)| \tag{3.33}$$

because of (3.31). Furthermore,

$$\max_{y \varepsilon \dot{S}_2} \frac{R^2 - |x|^2}{|x - y|^n} = \max_{y \varepsilon \dot{S}_2} \frac{(R - |x|)(R + |x|)}{|x - y|^n} \leqslant \frac{2R}{(\rho/2)^n} (R - |x|) \tag{3.34}$$

since

$$|x - y| = |y - x^0 + x^0 - x| \geqslant |y - x^0| - |x^0 - x| \geqslant \rho - \frac{\rho}{2} = \frac{\rho}{2}.$$

If we set $\max_{y \varepsilon \dot{S}} |\varphi(y)| = M$, for the second term in (3.32) we obtain the estimate

$$\left| \frac{R^2 - |x|^2}{R\omega_n} \int_{\dot{S}_2} \frac{\varphi(y) - \varphi(x^0)}{|x - y|^n} \, dS \right| \leqslant \frac{4M}{\omega_n(\rho/2)^n} (R - |x|) \int_{|y|=R} dS = \frac{4MR^{n-1}}{(\rho/2)^n} (R - |x|).$$

$$(3.35)$$

This concludes the evaluation of (3.32). From (3.33) and (3.35) it follows that

$$|u(x) - \varphi(x^0)| \leqslant \max_{y \varepsilon \dot{S}_1} |\varphi(y) - \varphi(x^0)| + \frac{4MR^{n-1}}{(\rho/2)^n} (R - |x|). \tag{3.36}$$

Given $\epsilon > 0$ we choose ρ such that $\max_{y \varepsilon \dot{S}_1} |\varphi(y) - \varphi(x^0)| < \epsilon/2$, which is possible because of the continuity of φ. For such a ρ we can find a $\delta > 0$ so that

$$\frac{4M}{(\rho/2)^n} R^{n-1}(R - |x|) < \frac{\epsilon}{2}$$

for all x with $|x - x^0| < \delta$, since $R - |x| = |x^0| - |x| \leqslant |x^0 - x|$. Thus

$$|u(x) - \varphi(x^0)| < \epsilon \quad \text{for} \quad |x - x^0| < \delta, \qquad \delta = \delta(\epsilon), \tag{3.37}$$

which shows that $u \in C^0$ in $|x| \leqslant R$ and that it has the correct boundary values. Thus the second requirement is satisfied.

The first requirement is dealt with in Section 3.7. The third requirement is established as follows. If $\bar{u}(x)$ is the solution of (3.28) with the boundary values $\bar{\varphi}(x)$, then using (3.27) and (3.31) we have

$$\begin{aligned}
|\bar{u}(x) - u(x)| &\leqslant \max_{|y|=R} |\bar{\varphi}(y) - \varphi(y)| \frac{R^2 - |x|^2}{R\omega_n} \int_{|y|=R} \frac{dS}{|x - y|^n} \\
&= \max_{|y|=R} |\bar{\varphi}(y) - \varphi(y)|.
\end{aligned} \tag{3.38}$$

Problem. Rewrite (3.27) for $n = 3$ in polar coordinates.

3.6   The Existence of the Green's Function in R_2

It is very remarkable that in the case $n = 2$ this question can easily be taken care of by using the following theorem.

RIEMANN MAPPING THEOREM OF FUNCTION THEORY.   Let the whole (extended) complex ($z = y_1 + iy_2$)-plane be denoted by Z. If $\mathfrak{D} \subset Z$ is a simply connected domain[12] which is different from the whole extended z-plane and from the z-plane from which a point has been deleted, then $\mathfrak{D}$ can be mapped one to one and conformally by an analytic function $w = f(z)$ onto the interior of the unit circle $|w| < 1$, and this can be done so that to an arbitrarily given point in $\mathfrak{D}$ and to an arbitrarily given direction through this point correspond the origin $w = 0$ and the direction of the positive real axis in the w-plane respectively. With this, the function $f(z)$ is determined uniquely.

---

[12] Not necessarily a normal domain.

Let $w = f(z)$ be the above mapping of $\mathfrak{D}$ onto the unit circle $|w| < 1$ which maps the point $z^1 = x_1 + ix_2$ onto the point $w = 0$ $[0 = f(z^1)]$. Then, with $x = (x_1, x_2)$ and $y = (y_1, y_2)$, the *Green's function* is given by

$$g(x,y) = g(z^1, z) = -\frac{1}{2\pi} \log |f(z)|. \tag{3.39}$$

Indeed, $f(z)$ is of the form $f(z) = (z^1 - z)\, e^{h(z)}$, since $z = z^1$ must be a simple zero of $f(z)$ (conformality of the mapping). Then we have (Re = real part)

$$g(x,y) = g(z^1, z) = -\frac{1}{2\pi} \log |z^1 - z| - \frac{1}{2\pi} \operatorname{Re} h(z).$$

So far, according to Section 3.3, this is formally a fundamental solution, because $\log |z^1 - z|$ $= \log |x - y|$, and $\Delta_2 \operatorname{Re} h = 0$ for the real part of any analytic function. Furthermore, on $\dot{\mathfrak{D}}$ we have $|f(z)| = 1$, which implies that $g(x,y) = 0$ for $y \,\varepsilon\, \dot{\mathfrak{D}}$. For the last inference, however, we also used the correspondence of the boundaries under conformal mapping[14] but this only means restriction to a sufficiently "well-behaved" boundary $\dot{\mathfrak{D}}$.

It is often useful not to assume the mapping $w = f(z)$ to be normalized by $0 = f(z^1)$. If $w = F(z)$ is an arbitrary analytic and conformal mapping of $\mathfrak{D}$ onto the unit circle, then

$$g(z^1, z) = -\frac{1}{2\pi} \log \left| \frac{F(z) - F(z^1)}{1 - \overline{F(z^1)}F(z)} \right| = -\frac{1}{2\pi} \log \left| \frac{F(z^1) - F(z)}{1 - F(z^1)\overline{F(z)}} \right|. \tag{3.40}$$

Problem 1. Compute g for the unit circle, and for the upper half-plane in R_2.

Problem 2.* Prove the estimates

$$|g_{x_i}(x,y)| \leqslant \frac{\varkappa}{|x - y|}, \qquad i = 1, 2 \quad \text{for} \quad x, y \,\varepsilon\, \mathfrak{D}, \tag{3.41}$$

and $\varkappa$ a fixed constant. As a simplification assume that $\dot{\mathfrak{D}}$ is a closed analytic curve [Then $F(z)$ is even analytic in $\bar{\mathfrak{D}}$.]

3.7   The Mean-Value and the Maximum-Minimum Properties

We shall say that $u(x)$ has the *first mean-value property* in $\mathfrak{D}$ if $u(x) \,\varepsilon\, C^0$ in $\bar{\mathfrak{D}}$, and if for every sphere $\{|y - x| \leqslant r\} \subset \mathfrak{D}$

$$u(x) = \frac{1}{\omega_n r^{n-1}} \int_{|y - x| = r} u(y)\, dS \tag{3.42a}$$

holds. If, instead of (3.42a), u satisfies

$$u(x) = \frac{n}{\omega_n r^n} \int_{|y - x| \leqslant r} u(y)\, dy, \tag{3.42b}$$

we say that $u(x)$ has the *second mean-value property*.

The two mean-value properties are equivalent so that in the future it will suffice to speak of *the* mean-value property. Indeed, from (3.42a) we have

$$u(x)\rho^{n-1} = \frac{1}{\omega_n} \int_{|y - x| = \rho} u(y)\, dS \quad \text{or} \quad u(x) \int_0^r \rho^{n-1}\, d\rho = \frac{1}{\omega_n} \int_0^r d\rho \int_{|y - x| = \rho} u(y)\, dS, \tag{3.43}$$

which completes the first step, for the last integral expression coincides with $\int_{|y-x|\leqslant r} u(y)\,dy$. From (3.42$b$) we have

$$\frac{r^n u(x)}{n} = \frac{1}{\omega_n} \int_{|y-x|\leqslant r} u(y)\,dy \equiv \frac{1}{\omega_n} \int_0^r d\rho \int_{|y-x|=\rho} u(y)\,dS, \tag{3.44}$$

and differentiation with respect to r gives (3.42a).

THEOREM 1.   MAXIMUM-MINIMUM PRINCIPLE.   If $u(x)$ has the mean-value property in $\mathfrak{D}$, then the maximum and the minimum of $u(x)$ occur on $\dot{\mathfrak{D}}$.

Proof.   Since $u(x) \in C^0$ in $\bar{\mathfrak{D}}$, there is at least one point $x^0 \in \bar{\mathfrak{D}}$ at which $u(x)$ attains its maximum; let $u(x^0) = M$.

1. If $u(x) \equiv M$, every point is a maximum and a minimum point and the theorem is proved.

2. If $u(x) \not\equiv M$, we suppose that $x^0 \in \mathfrak{D}$ is a maximum point.   Then we consider the set $\mathfrak{M}$ of all maximum points $x \in \mathfrak{D}$:

$$\mathfrak{M}: x \in \mathfrak{D} \quad \text{with} \quad u(x) = M.$$

Because of the hypothesis, $\mathfrak{M}$ is not empty; furthermore, $\mathfrak{M} \subset \mathfrak{D}$.   We show later that $\mathfrak{M}$ is both an open and a closed set relative to $\mathfrak{D}$.[13]   But then, $\mathfrak{M} \equiv \mathfrak{D}$ and this means $u(x) \equiv M$ in $\mathfrak{D}$.   It follows immediately from the continuity of u that $u(x) \equiv M$ in $\bar{\mathfrak{D}}$, and this contradicts $u(x) \not\equiv M$.

(a) $\mathfrak{M}$ is open in $\mathfrak{D}$.   If $\bar{x} \in \mathfrak{M}$ and $\bar{S}: |x - \bar{x}| \leqslant r$ is contained in $\mathfrak{D}$, then $u = M$ on $\bar{S}$. If this is not the case, there is an $x \in \bar{S}$ at which $u(x) < M$.   But from the continuity of u and (3.42b) we have

$$M = u(\bar{x}) = \frac{n}{\omega_n r^n} \int_{|x-\bar{x}|\leqslant r} u(x)\,dx < M\,\frac{n}{\omega_n r^n} \int_{|x-\bar{x}|\leqslant r} dx = M, \tag{3.45}$$

which is an obvious contradiction.

(b) $\mathfrak{M}$ is closed in $\mathfrak{D}$.   If $\{x^i\}$ is a convergent sequence of maximum points $[u(x^i) = M]$ such that $\lim_{i \to \infty} x^i = x \in \mathfrak{D}$, then from the continuity of u, we obtain $u(x) = \lim_{i \to \infty} u(x^i) = M$ which implies $x \in \mathfrak{M}$.

For the minimum we argue analogously, which completes the proof.

The function $u(x)$ is called *harmonic* in $\mathfrak{D}$ if

$$u(x) \in C^0 \text{ in } \bar{\mathfrak{D}}, \quad \in C^2 \text{ in } \mathfrak{D}, \quad \text{and} \quad \Delta_n u = 0 \text{ in } \mathfrak{D}. \tag{3.46}$$

THEOREM 2. $u(x)$ is harmonic in $\mathfrak{D}$ if and only if it possesses the mean-value property.

[13] By definition $\mathfrak{D}$ is an open point set in R_n; however, $\mathfrak{D}$ is open as well as closed relative to $\mathfrak{D}$.

Proof. 1. Let $u(x)$ be harmonic in $\mathfrak{D}$. First, we suppose that the origin is contained in $\mathfrak{D}$. Then, from (3.27), we have

$$u(x) = \frac{r^2 - |x|^2}{r\omega_n} \int_{|y|=r} \frac{u(y)}{|x-y|^n}\, dS \quad \text{and} \quad u(0) = \frac{1}{\omega_n r^{n-1}} \int_{|y|=r} u(y)\, dS. \tag{3.47}$$

If in the last formula we translate the origin $y = 0$ to the point x, we obtain

$$u(x) = \frac{1}{\omega_n r^{n-1}} \int_{|y-x|=r} u(y)\, dS \tag{3.48}$$

and $u(x)$ has the mean-value property.

2. Let $u(x)$ have the mean-value property. Let $S \subset \mathfrak{D}$ be an arbitrary sphere. Suppose $v(x)$ is a solution of the boundary-value problem $\Delta_n v = 0$ in S, $v = u$ on $\dot{S}$. The fact that the problem can be solved has been established in Section 3.5 by means of Poisson's theorem under the assumption that the center of S lies at the origin. A simple translation of the origin frees us from this assumption. Thus $v(x)$ is harmonic in S. Then, because of the first part of the proof, $v(x)$ has the mean-value property in S and with it so does $w(x) = u(x) - v(x)$. Now $w(x) = 0$ on $\dot{S}$. Following Theorem 1, the maxima and minima of $w(x)$ must lie on $\dot{S}$; thus $w \equiv 0$ in $\bar{S}$ or $u = v$ in $\bar{S}$. Thus it has been shown that $u(x)$ is harmonic in any sphere in $\mathfrak{D}$ and since $u \, \varepsilon \, C^0$ in $\mathfrak{D}$, it is harmonic in $\mathfrak{D}$.

The following are simple corollaries.

THEOREM 3. FIRST REQUIREMENT. The boundary-value problem

$$\Delta_n u = f(x) \quad \text{in} \quad \mathfrak{D}, \qquad u = \varphi(x) \quad \text{on} \quad \dot{\mathfrak{D}}, \qquad f(x) \, \varepsilon \, C^0 \quad \text{in} \quad \mathfrak{D},$$
$$\varphi(x) \, \varepsilon \, C^0 \quad \text{on} \quad \dot{\mathfrak{D}} \tag{3.49}$$

has at most one solution $u(x) \, \varepsilon \, C^0$ in $\bar{\mathfrak{D}}$, εC^2 in $\mathfrak{D}$.

Proof. Suppose $u^1(x)$, $u^2(x)$ are two solutions. Then for $v = u^1 - u^2$,

$$\Delta_n v = 0 \quad \text{in} \quad \mathfrak{D}, \qquad v = 0 \quad \text{on} \quad \dot{\mathfrak{D}}. \tag{3.50}$$

Hence $v(x)$ is harmonic, has the mean-value property because of Theorem 2, and by Theorem 1 the maxima and minima of v lie on $\dot{\mathfrak{D}}$. Thus $v \equiv 0$ in $\mathfrak{D}$.

THEOREM 4. THIRD REQUIREMENT. If $u^i(x)$, $i = 1, 2$, are solutions of

$$\Delta_n u^i = f(x) \quad \text{in} \quad \mathfrak{D}, \qquad u^i = \varphi^i(x) \quad \text{on} \quad \dot{\mathfrak{D}} \quad \text{where} \quad \max_{x \varepsilon \dot{\mathfrak{D}}} |\varphi^1(x) - \varphi^2(x)| = \epsilon,$$
$$\tag{3.51}$$

then

$$|u^1(x) - u^2(x)| \leqslant \epsilon \quad \text{in} \quad \mathfrak{D}. \tag{3.52}$$

Proof. $v = u^1 - u^2$ satisfies the problem $\Delta_n v = 0$, $v = \varphi^1(x) - \varphi^2(x)$ on $\dot{\mathfrak{D}}$. Thus v is harmonic and the conclusion follows easily from Theorems 1 and 2.

THEOREM 5.   FIRST HARNACK THEOREM.   Let $u^j(x)$, $j = 1, 2, 3, \ldots$, be a sequence of harmonic functions in $\mathfrak{D}$ which converge uniformly in $\bar{\mathfrak{D}}$ to a limit function u.   Then $u(x)$ is harmonic in $\mathfrak{D}$.

Proof.   By Theorem 2, $u^j(x)$ has the mean-value property; in particular,

$$u^j(x) = \frac{1}{\omega_n r^{n-1}} \int_{|y-x|=r} u^j(y)\, dS. \tag{3.53}$$

Because of the uniformity of the convergence, the passage to the limit $j \to \infty$ may be carried out under the integral sign and (3.42a) follows for $u(x)$.   Since further $u(x) \in C^0$ in $\bar{\mathfrak{D}}$, it has the mean-value property and, by Theorem 2, is harmonic.

Problem 1.*   Let $u(x) \in C^2$ for $x \in R_n$ with $u(x) \not\equiv 0$ and $\Delta_n u = 0$ in the whole space. Then $\int_{R_n} u^2(x)\, dx$ does not exist.

Problem 2.*   If $u(x)$ is harmonic, $\in C^3$ in $\mathfrak{D}$, and $\in C^1$ in $\bar{\mathfrak{D}}$, then the maximum of $|\mathrm{grad}\ u(x)|$ lies on $\mathfrak{D}$.

3.8   Harnack's Inequality

THEOREM 1.   Let $u(x)$ be harmonic for $|x| < R$ [that is, $u(x) \in C^0$ in $|x| \leqslant R$, $\in C^2$ in $|x| < R$, and $\Delta_n u = 0$].   Furthermore, suppose that $u(x) \geqslant 0$.   Then for $|x| < R$ the following inequality holds:

$$\frac{1 - |x|/R}{(1 + |x|/R)^{n-1}}\, u(0) \leqslant u(x) \leqslant \frac{1 + |x|/R}{(1 - |x|/R)^{n-1}}\, u(0). \tag{3.54}$$

Proof.   From (3.27) we have the representation

$$u(x) = \frac{1}{R\omega_n} \int_{|y|=R} \frac{|y|^2 - |x|^2}{|x - y|^n}\, u(y)\, dS \quad \text{for} \quad |x| < R. \tag{3.55}$$

For $|y| = R$ and $|x| < R$ we have the inequality

$$\frac{R^2 - |x|^2}{(R + |x|)^n} \leqslant \frac{|y|^2 - |x|^2}{|x - y|^n} \leqslant \frac{R^2 - |x|^2}{(R - |x|)^n}. \tag{3.56}$$

Since $u(x) \geqslant 0$, inequality (3.56) may be multiplied by $(1/R\omega_n)u(y)$.   By integration we find

$$\frac{R^2 - |x|^2}{(R + |x|)^n} \frac{1}{R\omega_n} \int_{|y|=R} u(y)\, dS \leqslant u(x) \leqslant \frac{R^2 - |x|^2}{(R - |x|)^n} \frac{1}{R\omega_n} \int_{|y|=R} u(y)\, dS. \tag{3.57}$$

Since $u(x)$ is harmonic, it has the mean-value property

$$u(0) = \frac{1}{R^{n-1}\omega_n} \int_{|y|=R} u(y)\, dS.$$

If we use this on the right- and left-hand sides of (3.57), the theorem follows after an easy computation.

We shall now use this inequality to prove a theorem which is closely related to Liouville's theorem in function theory.

THEOREM 2.   Let $u(x)$ be harmonic in the whole of R_n; that is, $u(x) \varepsilon C^2$ and $\Delta_n u = 0$ for $x \varepsilon R_n$.   Further, suppose that $u(x)$ is bounded from above: $u(x) \leqslant C$. Then $u(x)$ is a constant.

Proof.   The function $U(x) = C - u(x)$ is harmonic in every sphere $|x| < R$, and $U(x) \geqslant 0$.   Because of Harnack's inequality (3.54) we then have

$$\frac{1 - |x|/R}{(1 + |x|/R)^{n-1}}\, U(0) \leqslant U(x) \leqslant \frac{1 + |x|/R}{(1 - |x|/R)^{n-1}}\, U(0).$$

Now R may be taken arbitrarily large.   If we keep x fixed and let R tend to ∞ then $U(x) = U(0)$ follows and $u(x) = u(0)$.   This concludes the proof.

3.9   H. Weyl's Lemma in the Simplest Case

Let $u(x)$ be harmonic in $\mathfrak{D}$ and let $v(x) \varepsilon C^2$ in $\mathfrak{D}$ be a function which vanishes sufficiently close to $\dot{\mathfrak{D}}$.   The second Green's formula then gives

$$\int_{\mathfrak{D}} u(y)\, \Delta_n v(y)\, dy = 0. \tag{3.58}$$

Weyl's lemma is a converse to these considerations and from the validity of (3.58) for all such $v(x)$ it concludes that the function $u(x)$ must be harmonic.

LEMMA.   Let $u(x) \varepsilon C^0$ in $\bar{D}$.   Further let relation (3.58) hold for all $v(x) \varepsilon C^2$ in $\mathfrak{D}$ which vanish in a suitable neighborhood of $\dot{\mathfrak{D}}$.   Then $u(x)$ is harmonic in $\mathfrak{D}$.

Proof.   The proof is given only in R_3 since later we prove a considerably more general version of Weyl's lemma.   Let $x \varepsilon \mathfrak{D}$ be an arbitrary point and let S_ϵ: $|y - x| < \epsilon$ be a ball with center x and radius $\epsilon > 0$ such that $\bar{S}_\epsilon \subset \mathfrak{D}$.   We set

$$v(y) = \begin{cases} (|y - x|^2 - \epsilon^2)^3 & \text{for} \quad |y - x| \leqslant \epsilon, \\ 0 & \text{for} \quad |y - x| \geqslant \epsilon. \end{cases} \tag{3.59}$$

This $v(y)$ has the required properties.   If we form $\Delta_3 v(y)$, we find

$$\Delta_3 v(y) = \begin{cases} 6(7|y - x|^4 - 10\epsilon^2|y - x|^2 + 3\epsilon^4) & \text{for} \quad |y - x| \leqslant \epsilon, \\ 0 & \text{for} \quad |y - x| \geqslant \epsilon. \end{cases} \tag{3.60}$$

For this v, relation (3.58) then gives

$$0 = \int_{|y-x|\leqslant\epsilon} u(y)\{7|y - x|^4 - 10\epsilon^2|y - x|^2 + 3\epsilon^4\}\, dy \equiv \int_0^\epsilon d\rho \int_{|y-x|=\rho} u(y)\{\cdot\cdot\cdot\}\, dS. \tag{3.61}$$

Differentiation with respect to ϵ gives (after omitting a common factor)

$$0 = \int_{|y-x| \leqslant \epsilon} u(y)\{-5|y-x|^2 + 3\epsilon^2\}\, dy \equiv \int_0^\epsilon d\rho \int_{|y-x|=\rho} u(y)\{\cdots\}\, dS, \qquad (3.62)$$

and further differentiation with respect to ϵ and omission of a common factor yields

$$\epsilon \int_{|y-x|=\epsilon} u(y)\, dS = 3 \int_{|y-x| \leqslant \epsilon} u(y)\, dy. \qquad (3.63)$$

This can also be written in the form

$$\frac{1}{4\pi\epsilon^2} \int_{|y-x|=\epsilon} u(y)\, dS = \frac{3}{4\pi\epsilon^3} \int_{|y-x| \leqslant \epsilon} u(y)\, dy. \qquad (3.64)$$

Finally, further differentiation with respect to ϵ gives

$$\frac{d}{d\epsilon}\left(\frac{1}{4\pi\epsilon^2} \int_{|y-x|=\epsilon} u(y)\, dS \right) = 0. \qquad (3.65)$$

Thus the expression in large parentheses does not depend on ϵ but is a constant. For $\epsilon \to 0$, the value of this constant turns out to be $u(x)$. Thus

$$u(x) = \frac{1}{4\pi\epsilon^2} \int_{|y-x|=\epsilon} u(y)\, dS. \qquad (3.66)$$

But this means that $u(x)$ has the mean-value property. From Theorem 2 in Section 3.7 it follows that $u(x)$ is harmonic in $\mathfrak{D}$. The proof followed W. I. Smirnow [15].

3.10   Mean-Value Properties for $\Delta_n u = f$ and $\Delta_3 u + k^2 u = f$

THEOREM 1.   Suppose $u(x) \in C^1$ in $\bar{\mathfrak{D}}$, $\in C^2$ in $\mathfrak{D}$ is a solution of $\Delta_n u = f$ with $f \in C^0$ in $\bar{\mathfrak{D}}$. Then for every sphere $\{|y-x| \leqslant r\} \subset D$ with arbitrary center $x \in \mathfrak{D}$ and radius r the mean-value relation

$$u(x) = \frac{1}{\omega_n r^{n-1}} \int_{|y-x|=r} u(y)\, dS - \int_{|y-x| \leqslant r} \Gamma_n(x,y) f(y)\, dy \qquad (3.67)$$

holds, where

$$\Gamma_n(x,y) = \begin{cases} \dfrac{1}{(n-2)\omega_n}\,(|x-y|^{2-n} - r^{2-n}) & \text{for} \quad n > 2 \\[2ex] -\dfrac{1}{2\pi}\,(\log|x-y| - \log r) & \text{for} \quad n = 2. \end{cases}$$

Remark.   For $f \equiv 0$, Theorem 1 is contained in Theorem 2 of Section 3.7.

Proof.   For $n \geqslant 2$, $\Gamma_n(x,y)$ is (according to Section 3.3) a fundamental solution relative to y for fixed $x \in \mathfrak{D}$ with the additional property that $\Gamma_n(x,y) = 0$ for all y on the sphere $|y-x| = r$. The theorem of Section 3.3 then gives

$$u(x) = -\int_{|y-x|=r} (\Gamma_n(x,y))_\nu u(y)\, dS - \int_{|y-x| \leqslant r} \Gamma_n(x,y) f(y)\, dy. \qquad (3.68)$$

Since $(\Gamma_n)_\nu = -(1/\omega_n) r^{1-n}$ on $|y-x| = r$, the proof is complete.

Frequently mean-value relations for the equation $\Delta_n u + k^2 u = f$, where k is a nonzero constant, are of importance. We restrict ourselves to the case $n = 3$ although the case of an arbitrary dimension can be treated quite analogously.

In Section 3.2 we determined the singularity function for the equation $\Delta_3 u + k^2 u = 0$. For fixed $a \in \mathfrak{D}$ it was given by

$$s(a,x) = \frac{1}{4\pi} \frac{\cos k|a - x|}{|a - x|}. \tag{3.69}$$

Correspondingly, $\sin k|a - x|/|a - x|$ is a solution of our equation.

In analogy to Section 3.3, the function (defined for $x \neq a$)

$$\gamma(a,x) = s(a,x) + \Phi(x) \quad \text{where} \quad \Phi \in C^1 \text{ in } \bar{\mathfrak{D}}, \quad \in C^2 \text{ in } \mathfrak{D}, \quad \text{and}$$
$$\Delta_3 \Phi + k^2 \Phi = 0 \quad \text{in} \quad \mathfrak{D} \tag{3.70}$$

is called a fundamental solution relative to $\mathfrak{D}$. For this we have Theorem 2.

THEOREM 2. Let $u(x) \in C^1$ in $\bar{\mathfrak{D}}$, $\in C^2$ in $\mathfrak{D}$ be a solution of $\Delta_3 u + k^2 u = f$ with $f(x) \in C^0$ in $\bar{\mathfrak{D}}$. If $a \in \mathfrak{D}$, then

$$u(a) = \int_{\dot{\mathfrak{D}}} \{\gamma(a,x)u_\nu(x) - u(x)\gamma_\nu(a,x)\} \, dS - \int_{\mathfrak{D}} \gamma(a,x)f(x) \, dx. \tag{3.71}$$

The proof is completely analogous to that in Section 3.3. We only have to use the formula

$$\int_{\mathfrak{D}-\bar{s}} (\gamma \, \Delta_3 u - u \, \Delta_3 \gamma) \, dx = \int_{\mathfrak{D}-\bar{s}} \{\gamma(\Delta_3 u + k^2 u) - u(\Delta_3 \gamma + k^2 \gamma)\} \, dx. \tag{3.72}$$

THEOREM 3. Let $u(x) \in C^1$ in $\bar{\mathfrak{D}}$, $\in C^2$ in $\mathfrak{D}$ be a solution of $\Delta_3 u + k^2 u = f$ where $f(x) \in C^0$ in $\bar{\mathfrak{D}}$. Then for every sphere $\{|y - x| \leqslant r\} \subset \mathfrak{D}$ with arbitrary center $x \in \mathfrak{D}$ and radius r the mean-value relation

$$u(x) \frac{\sin kr}{kr} = \frac{1}{4\pi r^2} \int_{|y-x|=r} u(y) \, dS - \int_{|y-x| \leqslant r} \Gamma(x,y)f(y) \, dy \tag{3.73}$$

holds where

$$\Gamma(x,y) = \frac{1}{4\pi kr} \left(\sin kr \frac{\cos k|x - y|}{|x - y|} - \cos kr \frac{\sin k|x - y|}{|x - y|} \right).$$

Proof. For every fixed x, the function

$$\tilde{\Gamma}(x,y) = \frac{1}{4\pi} \frac{\cos k|x - y|}{|x - y|} - \frac{1}{4\pi} \frac{\cos kr}{\sin kr} \frac{\sin k|x - y|}{|x - y|} \quad \text{relative} \quad y \in \mathfrak{D}$$

is a fundamental solution of the equation $\Delta_3 u + k^2 u = 0$ with the additional property that $\tilde{\Gamma}(x,y) = 0$ for all y on the sphere $|y - x| = r$. For this, note that $(\sin k|x - y|)/|x - y|$ does not have a singularity for $x = y$, and exclude such r for which $\sin kr = 0$. Calculation of $\tilde{\Gamma}_\nu(x,y)$ on $|y - x| = r$ gives $-(1/4\pi r)(k/\sin kr)$. If we substitute this $\tilde{\Gamma}$ in (3.71) for γ, the desired conclusion follows if we multiply (3.71) by $(\sin kr)/kr$. Again it holds for arbitrary r.

Problem. Find the mean-value relation for $\Delta_2 u + k^2 u = f$.

4

THE HEAT EQUATION

4.1   The Existence Theorem for the Initial-Value Problem

According to Section 1.7, the initial-value problem for the heat equation will have the form

$$\Delta_n u - u_t = 0 \quad \text{and} \quad u(x,0) = u_0(x) \tag{4.1}$$

for a suitable time scale in (1.38).   Here x stands for $(x_1, \ldots, x_n)$, t is time, and $u = u(x,t)$.   We shall consider (4.1) for $-\infty < x_1, \ldots, x_n < \infty$ and $t \geqslant 0$.   The values $u_0(x)$ represent the initial condition to be prescribed arbitrarily at time $t = 0$.

The function

$$s(x,t) = \frac{1}{(4\pi t)^{n/2}} e^{-|x|^2/4t} \tag{4.2}$$

is called a *singularity function;* it has the property of satisfying equation (1.1) for $t > 0$ while having a singularity at $t = 0$ which depends on the dimension.   The easy calculation goes as follows:

$$s_{x_i} = -\frac{x_i}{2t}\frac{1}{(4\pi t)^{n/2}} e^{-|x|^2/4t}, \qquad s_{x_i x_i} = \frac{1}{(4\pi t)^{n/2}}\left(\frac{(x_i)^2}{4t^2} - \frac{1}{2t}\right)e^{-|x|^2/4t},$$

$$\Delta_n s = \sum_{i=1}^{n} s_{x_i x_i} = \frac{1}{(4\pi t)^{n/2}}\left(\frac{|x|^2}{4t^2} - \frac{n}{2t}\right)e^{-|x|^2/4t} = s_t.$$

As in the case of the potential equation, and in contrast to the wave equation, there is no decisive dependence on dimension for the heat equation.   Therefore we restrict ourselves to R_1 and x will simply be a real variable.

Posing of the Problem.   We consider the heat equation

$$u_{xx} - u_t = 0 \quad \text{in} \quad -\infty < x < \infty, \qquad 0 < t \leqslant T \tag{4.3}$$

with initial conditions

$$u(x,0) = u_0(x) \quad \text{in} \quad -\infty < x < \infty. \tag{4.4}$$

Suppose that $u_0(x) \in C^0$ in $-\infty < x < \infty$ and that it satisfies an estimate $|u_0(x)| \leqslant M\, e^{Ax^2}$ with constants $M, A \geqslant 0$.

We look for functions $u(x,t) \in C^0$ in $-\infty < x < \infty, 0 \leqslant t \leqslant T$ for which $u_{xx}, u_t \in C^0$ in $-\infty < x < \infty, 0 < t \leqslant T$ and which satisfy equation (4.3) and the initial conditions

(4.4). Further, we require that $u(x,t)$ satisfy an estimate of the form $|u(x,t)| \leqslant M_1 e^{A_1 x^2}$ in $-\infty < x < \infty$, $0 \leqslant t \leqslant T$, with constants M_1, $A_1 \geqslant 0$. Such $u(x,t)$ will be called solutions. For these the following theorem holds.

THEOREM.[14]   Let $u_0(x) \; \varepsilon \; C^0$ in R_1 and $|u_0(x)| \leqslant M \, e^{Ax^2}$ with constants M, $A \geqslant 0$. Then

$$u(x,t) = \begin{cases} \int_{-\infty}^{+\infty} s(x-y,\,t)u_0(y)\,dy \equiv \dfrac{1}{\sqrt{4\pi t}} \int_{-\infty}^{\infty} e^{-(x-y)^2/4t}\, u_0(y)\,dy \quad \text{for} \\ \qquad\qquad\qquad\qquad\qquad\qquad\qquad\qquad\qquad\qquad 0 < t \leqslant T, \quad (4.5) \\ u_0(x) \quad \text{for} \quad t = 0, \end{cases}$$

where $T < 1/4A$, is a solution of the problem (4.3) and (4.4) with the formerly defined properties. Moreover, $u \, \varepsilon \, C^\infty$ in $-\infty < x < \infty$, $0 < t \leqslant T$. The problem (4.3) and (4.4), with the foregoing notion of solution, satisfies the three requirements.

Before proving the theorem we want to obtain the solution (4.5) in a heuristic way. We represent the initial values $u_0(x)$ by using the Fourier integral theorem

$$u_0(x) = \int_0^\infty \{a(\mu) \cos \mu x + b(\mu) \sin \mu x\}\, d\mu$$

where

$$a(\mu) = \frac{1}{\pi} \int_{-\infty}^{\infty} u_0(y) \cos \mu y\, dy, \qquad b(\mu) = \frac{1}{\pi} \int_{-\infty}^{\infty} u_0(y) \sin \mu y\, dy.$$

Now for the solution we try

$$u(x,t) = \int_0^\infty \{a(\mu) \cos \mu x + b(\mu) \sin \mu x\} f(t,\mu)\, d\mu.$$

To give this $u(x,t)$ the correct initial values, $\lim\limits_{t\to 0} u(x,t) = u_0(x)$, it is plausible to require that $\lim\limits_{t\to 0} f(t,\mu) = 1$. If we substitute this $u(x,t)$ in the heat equation and carry out the differentiation under the integral sign, we obtain for f the further condition $f_t + \mu^2 f = 0$ with the solution $f(t,\mu) = e^{-\mu^2 t}$. If we substitute the formulas for $a(\mu)$, $b(\mu)$ and this value of f in our formula for $u(x,t)$ and then interchange the order of integration we obtain

$$u(x,t) = \int_{-\infty}^{+\infty} u_0(y) \left\{ \frac{1}{\pi} \int_0^\infty \cos \mu(x-y)\, e^{-\mu^2 t}\, d\mu \right\} dy.$$

If we further take note of the elementary formula

$$\int_0^\infty e^{-z^2} \cos 2\alpha z\, dz = \frac{\sqrt{\pi}}{2} e^{-\alpha^2},$$

we find that $\{\cdot \; \cdot \; \cdot\} = s(x-y,\,t)$. Of course these considerations do not prove anything, because the use of the Fourier integral theorem is not allowed for the $u_0(x)$ that we have admitted.

---

[14] See A. N. Tychonow[16]. The presentation in Sections 4.1, 4.2, and 4.3 is based on lectures by F. John given at New York University, 1954–55.

The smaller we choose A, the better the theorem becomes; for then T becomes larger. We take this into consideration by arguing as follows for a function $f(x)$ continuous in $-\infty < x < \infty$ which satisfies

$$|f(x)| \leqslant M \, e^{Ax^2}, \qquad -\infty < x < \infty. \tag{4.6}$$

We consider the set $\mathcal{C}$ of all constants A for which there exist corresponding constants M such that (4.6) holds. If in particular $\bar{A} \in \mathcal{C}$, then also $\bar{A} \in \mathcal{C}$ whenever $\bar{A} \geqslant \tilde{A}$. The greatest lower bound of $\mathcal{C}$ is denoted by $\gamma(f(x))$.

LEMMA. Let $f(x)$, $g(x)$ be continuous functions on $-\infty < x < \infty$, where $f(x)$ satisfies (4.6) and where $g(x)$ satisfies the estimate $|g(x)| \leqslant \bar{M} \, e^{\bar{A}x^2}$. If

$$\gamma(f(x)) + \gamma(g(x)) < 0,$$

then

$$h(x) = \int_{-\infty}^{\infty} g(x-y)f(y) \, dy \tag{4.7}$$

is continuous in $-\infty < x < \infty$ and satisfies an estimate $|h(x)| \leqslant \tilde{M} \, e^{\tilde{A}x^2}$ with

$$\gamma(h(x)) \leqslant \frac{\gamma(f(x)) \cdot \gamma(g(x))}{\gamma(f(x)) + \gamma(g(x))}. \tag{4.8}$$

Proof. First we have

$$|h(x)| \leqslant M\bar{M} \int_{-\infty}^{\infty} \exp\left[Ay^2 + \bar{A}(x-y)^2\right] dy$$

$$= M\bar{M} \exp\left(\frac{A\bar{A}}{A+\bar{A}} x^2\right) \int_{-\infty}^{\infty} \exp\left[(A+\bar{A})\left(y - \frac{\bar{A}}{A+\bar{A}} x\right)^2\right] dy, \tag{4.9}$$

which is easily verified. By hypothesis, we may assume $A + \bar{A} < 0$. The last integral can be evaluated explicitly by using the well-known formula $\int_{-\infty}^{\infty} e^{-x^2} \, dx = \sqrt{\pi}$. Therefore we have

$$|h(x)| \leqslant M\bar{M} \sqrt{\frac{-\pi}{A+\bar{A}}} \exp\left(\frac{A\bar{A}}{A+\bar{A}} x^2\right), \tag{4.10}$$

which is equivalent to (4.8). The integral (4.7) converges uniformly on any finite x-interval because of (4.9) and (4.10); thus $h(x)$ is continuous.

Proof of the Theorem. Second Requirement. According to the lemma, the integral in (4.5) is certainly convergent whenever $\gamma(s(x,t)) + \gamma(u_0(x)) < 0$. Now, from (4.5) we have $\gamma(s(x,t)) = -1/4t$ so that we obtain

$$-\frac{1}{4t} + \gamma(u_0(x)) < 0. \tag{4.11}$$

This implies the restriction

$$0 \leqslant t \leqslant T \quad \text{where} \quad T < \begin{cases} \dfrac{1}{4\gamma(u_0(x))} & \text{if} \quad \gamma(u_0(x)) > 0, \\ \infty & \text{if} \quad \gamma(u_0(x)) \leqslant 0, \end{cases} \tag{4.12}$$

for T. Thus the bounds mentioned in the theorem have been found. Now it is easy to show that the integrals

$$\int_{-\infty}^{\infty} |s_t(x - y, t)u_0(y)| \, dy \quad \text{and} \quad \int_{-\infty}^{\infty} |s_{xx}(x - y, t)u_0(y)| \, dy \tag{4.13}$$

are uniformly convergent in (x,t) for $0 < T_1 \leqslant t \leqslant T$ and any finite x-interval $X_0 \leqslant x \leqslant X_1$. Thus differentiation with respect to x and t may be carried out under the integral sign in (4.5) and, since $s_{xx} - s_t = 0$, this leads to the desired result. For the same reason we may differentiate under the integral sign arbitrarily often. It remains to prove that $u(x,t)$ is continuous in $0 \leqslant t$. For this we must show that the two-dimensional limit

$$\lim_{(x,t) \to (x^0,0)} u(x,t) = u_0(x^0) \tag{4.14}$$

exists for $t > 0$, $-\infty < x^0 < \infty$ and that it has the given value. Without loss of generality we may suppose $x^0 = 0$. Then we have

$$u_0(x) = u_0(0) + v_0(x) \quad \text{where} \quad v_0(x) = u_0(x) - u_0(0). \tag{4.15}$$

Now because of (4.5) we have

$$u(x,t) = \int_{-\infty}^{\infty} s(x - y, t)u_0(0) \, dy + \int_{-\infty}^{\infty} s(x - y, t)v_0(y) \, dy, \tag{4.16}$$

provided that both integrals converge. The first integral can be evaluated explicitly by using the formula $\int_{-\infty}^{\infty} e^{-x^2} \, dx = \sqrt{\pi}$. We find

$$\int_{-\infty}^{\infty} s(x - y, t)u_0(0) \, dy = u_0(0). \tag{4.17}$$

For the second integral in (4.16) we note that $v_0(x)$ is continuous and that $v_0(0) = 0$. Therefore, to a given $\epsilon > 0$ there is a $\delta(\epsilon)$ such that $|v_0(x)| < \epsilon$ for $|x| < \delta(\epsilon)$. By hypothesis, $v_0(x)$ satisfies an estimate of the form $|v_0(x)| \leqslant \tilde{M} e^{\tilde{A}x^2}$. Now $\tilde{A}$ can be chosen positive and sufficiently large so that $\tilde{A} = B(\epsilon)$ for $|x| \geqslant \delta(\epsilon)$

$$|v_0(x)| \leqslant \epsilon \, e^{B(\epsilon)x^2}. \tag{4.18}$$

Because $e^{B(\epsilon)x^2} \geqslant 1$, both estimates can be combined in

$$|v_0(x)| \leqslant \epsilon \, e^{B(\epsilon)x^2} \quad \text{in} \quad -\infty < x < \infty. \tag{4.19}$$

Now in (4.16) we obtain

$$\left| \int_{-\infty}^{\infty} s(x - y, t)v_0(y) \, dy \right| \leqslant \frac{\epsilon}{\sqrt{4\pi t}} \int_{-\infty}^{\infty} e^{-(x-y)^2/4t} \, e^{By^2} \, dy = \frac{\epsilon}{\sqrt{1 - 4Bt}} \, e^{Bx^2/(1-4Bt)} \tag{4.20}$$

where, of course, $t < 1/4B$ has been assumed. Obviously, from (4.20) we obtain

$$\lim_{(x,t) \to (0,0)} \left| \int_{-\infty}^{\infty} s(x - y, t)v_0(y) \, dy \right| \leqslant \epsilon. \tag{4.21}$$

Since $\epsilon > 0$ was arbitrary, this limit must be zero, and from (4.16) follows

$$\lim_{(x,t)\to(0,0)} u(x,t) = u_0(0). \tag{4.22}$$

To prove that the problem satisfies the second requirement it merely remains to show that $u(x,t)$ satisfies the desired estimate. From (4.5) and (4.8) follows the estimate

$$\gamma(u(x,t)) \leqslant \frac{\gamma(s(x,t))\gamma(u_0(x))}{\gamma(s(x,t)) + \gamma(u_0(x))} = \frac{-(1/4t)\gamma(u_0(x))}{\gamma(u_0(x)) - 1/4t} = \frac{\gamma(u_0(x))}{1 - 4t\gamma(u_0(x))}. \tag{4.23}$$

Therefore if $0 \leqslant t \leqslant T$, then $|u(x,t)| \leqslant M_1 e^{A_1 x^2}$ for suitable constants M_1, A_1.

4.2 The Uniqueness Theorem for the Initial-Value Problem

In Section 3.7 the first requirement for the boundary-value problem of the potential equation had been settled by means of the maximum-minimum principle. We also need such a tool for the heat equation.

THEOREM. MAXIMUM-MINIMUM PRINCIPLE. In $\mathfrak{D}$ (Figure 9) we consider the equation

$$u_{xx} - u_t = 0. \tag{4.24}$$

Here $\dot{\mathfrak{D}}_1$ is open and $\dot{\mathfrak{D}}_2$, under inclusion of the points A and B, is closed. Let $u(x,t) \in C^0$ in $\mathfrak{D}$ with u_{xx}, $u_t \in C^0$ in $\mathfrak{D} + \dot{\mathfrak{D}}_1$ be a solution of (4.24) in $\mathfrak{D} + \dot{\mathfrak{D}}_1$. Then $u(x,t)$ assumes its maximum and minimum on $\dot{\mathfrak{D}}_2$.

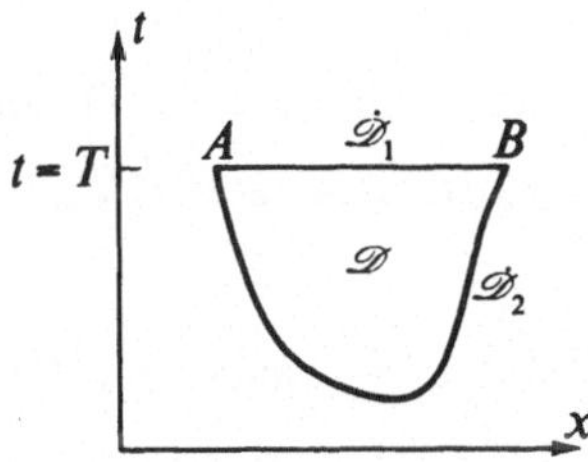

Figure 9

Proof. For arbitrary $\epsilon > 0$ we consider $v(x,t) = u(x,t) - \epsilon t$, and v is continuous in $\mathfrak{D}$. Hence v must assume its maximum and minimum in $\mathfrak{D}$. Suppose the maximum of v is attained at the point $(x_0,t_0) \in \mathfrak{D} + \dot{\mathfrak{D}}_1$. Then, for sufficiently small $k > 0$ the points $x_0 - k \leqslant x \leqslant x_0 + k$, $t = t_0$ lie in $\mathfrak{D} + \dot{\mathfrak{D}}_1$. Now at (x_0,t_0) we have $v_{xx}(x_0,t_0) \leqslant 0$, and furthermore

$$v_{xx} - v_t = u_{xx} - u_t + \epsilon = \epsilon, \quad \text{or} \quad v_t(x_0,t_0) \leqslant -\epsilon. \tag{4.25}$$

Now $h > 0$ can be chosen so small that $v_t(x_0,t) \leqslant -\epsilon/2$ for $t_0 - h \leqslant t \leqslant t_0$. Then we have

$$v(x_0,t_0) - v(x_0, t_0 - h) = \int_{t_0-h}^{t_0} v_t(x_0,t)\, dt \leqslant -\frac{\epsilon}{2} h < 0, \tag{4.26}$$

or $v(x_0,t_0) < v(x_0, t_0 - h)$, which contradicts our assumption. Hence the maximum of v lies on $\dot{\mathfrak{D}}_2$. Since $\epsilon > 0$ can be made arbitrarily small $u(x,t)$ must assume its maximum on $\dot{\mathfrak{D}}_2$. For the minimum we argue in an analogous fashion.

Proof of the Theorem in Section 4.1. *First Requirement.* Let $u^1(x,t)$, $u^2(x,t)$ be two solutions of the initial-value problem (4.3) and (4.4) with the properties required there. Then $u(x,t) = u^1(x,t) - u^2(x,t)$ is a solution of the problem (4.3) and (4.4) with $u_0(x) \equiv 0$. Thus u has the desired properties; in particular it satisfies an estimate of the form

$$|u(x,t)| \leqslant M_1 \, e^{A_1 x^2}$$

in $-\infty < x < \infty, 0 \leqslant t \leqslant T$ with suitable constants $M_1, A_1 \geqslant 0$. We have to show that $u \equiv 0$ in $-\infty < x < \infty, 0 \leqslant t \leqslant T$.

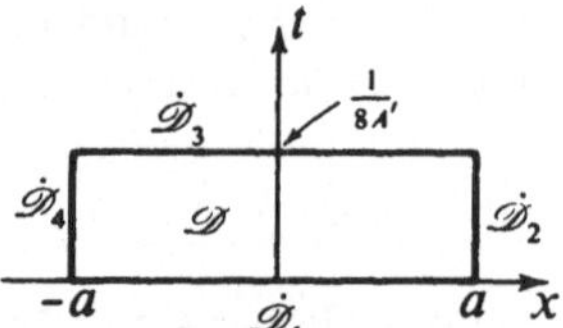

Figure 10

For this demonstration we fix an interval $-b \leqslant x \leqslant b$ with $b > 0$ and prescribe an $\epsilon > 0$ with $\epsilon < M_1$. Furthermore, let $A' > A_1$ be a constant, and let

$$a \geqslant \left(\frac{\log M_1/\epsilon}{A' - A_1}\right)^{1/2} \tag{4.27}$$

be another constant which we choose so that $a > b$. Then it follows that

$$|u(\pm a,t)| \leqslant M_1 \, e^{A_1 a^2} \leqslant \epsilon \, e^{A' a^2}. \tag{4.28}$$

The easy argument for this is

$$(A' - A_1)a^2 \geqslant \log \frac{M_1}{\epsilon} \quad \text{and} \quad M_1 \, e^{A_1 a^2} \leqslant \epsilon \, e^{A' a^2}.$$

Furthermore, for $\mathfrak{D}$ we consider the rectangle with vertices $(-a,0)$, $(a,0)$, $(a,1/8A')$, and $(-a,1/8A')$. As a solution of the heat equation in $\mathfrak{D}$ we consider the function

$$v(x,t) = \frac{\epsilon}{\sqrt{1 - 4A't}} \exp\left(\frac{A'x^2}{1 - 4A't}\right), \quad \text{where} \quad 0 \leqslant t \leqslant \frac{1}{8A'}. \tag{4.29}$$

On $\dot{\mathfrak{D}}_2$, $\dot{\mathfrak{D}}_4$ (Figure 10) we have

$$v(\pm a,t) = \frac{\epsilon}{\sqrt{1 - 4A't}} \exp\left(\frac{A'a^2}{1 - 4A't}\right) \geqslant \epsilon \, e^{A' a^2}. \tag{4.30}$$

On $\dot{\mathfrak{D}}_1$ we have

$$v(x,0) \geqslant \epsilon > 0. \tag{4.31}$$

Because of (4.28) and the hypothesis $u(x,0) = 0$, on $\dot{\mathfrak{D}}_4 + \dot{\mathfrak{D}}_1 + \dot{\mathfrak{D}}_2$ the estimate

$$|u(x,t)| \leqslant v(x,t), \quad \text{or} \quad u + v \geqslant 0 \quad \text{and} \quad u - v \leqslant 0, \tag{4.32}$$

holds. The functions $w = u \pm v$ satisfies the heat equation and, according to the theorem, assumes its maximum and minimum on $\dot{\mathfrak{D}}_4 + \dot{\mathfrak{D}}_1 + \dot{\mathfrak{D}}_2$. Hence (4.32) holds in $\mathfrak{D}$. In particular, (4.29) yields the estimate

$$|u(x,t)| \leqslant \frac{\epsilon}{\sqrt{\frac{1}{2}}} e^{A'b^2/4} \quad \text{in} \quad -b \leqslant x \leqslant b, \qquad 0 \leqslant t \leqslant \frac{1}{8A'}.$$

Since $\epsilon > 0$ was arbitrary, $u(x,t) \equiv 0$ in $-b \leqslant x \leqslant b, 0 \leqslant t \leqslant 1/8A'$. Since b was chosen arbitrarily finally

$$u(x,t) \equiv 0 \quad \text{in} \quad -\infty < x < \infty, \qquad 0 \leqslant t \leqslant \frac{1}{8A'}. \tag{4.33}$$

By repetition of the argument for parallel t-strips, the statement follows immediately.

Proof of the Theorem in Section 4.1. *Third requirement.* We shall show that, if $u^i(x,t)$, $i = 1, 2$ are two solutions of the problem (4.3) and (4.4), satisfying

$$u^i(x,0) = u_0^i(x), \qquad \max_{-\infty < x < \infty} |u_0^1(x) - u_0^2(x)| = \eta, \tag{4.34}$$

then

$$|u^1(x,t) - u^2(x,t)| \leqslant \eta \quad \text{in} \quad 0 \leqslant t \leqslant T, \qquad -\infty < x < \infty. \tag{4.35}$$

From (4.5) we have

$$|u^1(x,t) - u^2(x,t)| = \left| \int_{-\infty}^{\infty} s(x - y, t)[u_0^1(y) - u_0^2(y)] \, dy \right| \leqslant \eta \int_{-\infty}^{\infty} s(x - y, t) \, dy = \eta, \tag{4.36}$$

because the last integral is the solution of equation (4.3) with initial value 1. Hence the permitted use of the existence and uniqueness theorem gives

$$\int_{-\infty}^{\infty} s(x - y, t) \, dy \equiv 1.$$

Thus all parts of the theorem in Section 4.1 have been proved.

It is clear that similar considerations (which follow immediately from the maximum-minimum principle) show that the problem

$$u_{xx} - u_t = 0 \quad \text{in} \quad \mathfrak{D}(\text{Figure } 9), \qquad u = u_0 \quad \text{on} \quad \dot{\mathfrak{D}}_2(\text{Figure } 9), \tag{4.37}$$

satisfies the first and third requirements also.

4.3   Counterexamples

We show that an improvement of the theorem in Section 4.1 is essentially impossible.

Example 1.   The function $u(x,t) = s_x(x,t)$ is a solution of $u_{xx} - u_t = 0$ in

$$0 < t < \infty, \qquad -\infty < x < \infty.$$

Furthermore, $\lim_{t \to 0} s_x(x,t) = 0$ for fixed x. Nevertheless, $u(x,t)$ is not continuous in $-\infty <

$x < \infty$, $0 \leqslant t < \infty$; if we approach the point $(0,0)$ on the curve $x^2 = t$, then

$$s_x = -\frac{x}{4\sqrt{\pi}\,t^{3/2}}\,e^{-x^2/4t} \quad \text{with} \quad s_x(\sqrt{t},t) = -\frac{1}{4\sqrt{\pi}\,t}\,e^{-1/4} \tag{4.38}$$

is not even bounded for $t \to 0$. Thus, because of the first requirement, the condition $u \,\varepsilon\, C^0$ in $-\infty < x < \infty$, $0 \leqslant t$ cannot be dispensed with.

Example 2. *A. N. Tychonow*[17]. A solution of the equation $u_{xx} - u_t = 0$ is given in the form

$$u(x,t) = \sum_{k=0}^{\infty} \frac{d^k f(t)}{dt^k}\,\frac{x^{2k}}{(2k)!}, \qquad -\infty < x,t < \infty, \tag{4.39}$$

provided the convergence is sufficiently good, because then we have, using $j = k - 1$,

$$u_{xx} = \sum_{k=1}^{\infty} \frac{d^k f(t)}{dt^k}\,\frac{x^{2k-2}}{(2k-2)!} = \sum_{j=0}^{\infty} \frac{d^{j+1} f(t)}{dt^{j+1}}\,\frac{x^{2j}}{(2j)!}, \tag{4.40}$$

$$u_t = \sum_{k=0}^{\infty} \frac{d^{k+1} f(t)}{dt^{k+1}}\,\frac{x^{2k}}{(2k)!}, \tag{4.41}$$

which shows the agreement. The series (4.39) certainly converges if $f(t)$ is analytic for all t. Tychonow, however, sets

$$f(t) = \begin{cases} e^{-1/t^2} & \text{for} \quad t \neq 0, \\ 0 & \text{for} \quad t = 0. \end{cases}$$

As is well known, this defines a function which for all t, $-\infty < t < \infty$, is arbitrarily often continuously differentiable but which is not analytic for $t = 0$. Therefore the convergence of (4.39) has to be investigated separately. If we denote the kth derivative of $f(t)$ by $f^{(k)}(t)$, we have to find an estimate for $f^{(k)}(t)$. For this we use (from function theory) the theorem below.

THEOREM OF CAUCHY. Let the complex function $f(z)$ with $z = x + iy$ be analytic in $\mathfrak{D}$. Let $\mathfrak{F} \subset \mathfrak{D}$ be a circle with the point z in its interior. Then the following formulas hold:

$$f^{(k)}(z) = \frac{k!}{2\pi i} \int_{\mathfrak{F}} \frac{f(\zeta)}{(\zeta - z)^{k+1}}\,d\zeta, \qquad k = 1, 2, \ldots. \tag{4.42}$$

First we restrict ourselves to the case $0 < t < \infty$. For $\mathfrak{F}$ we use, in the complex ζ-plane, the circle $|\zeta - t| \leqslant \frac{1}{2}t$ with positive t, which can also be described by $\zeta = t + (t/2)\,e^{i\theta}$, $0 \leqslant \theta < 2\pi$.

For $f(\zeta)$ we may use e^{-1/ζ^2}, since $f(\zeta)$ is analytic on the circle and in its interior. By an easy calculation we see that $1/\zeta$ describes a circle with center $4/3t$ and radius $4/6t$ as θ changes from 0 to 2π. Thus

$$\frac{1}{\zeta} = \frac{4}{3t}\left(1 + \tfrac{1}{2}e^{i\varphi}\right) = \frac{4}{3t}\left\{(1 + \tfrac{1}{2}\cos\varphi) + \frac{i}{2}\sin\varphi\right\}, \qquad 0 \leqslant \varphi < 2\pi, \tag{4.43}$$

$$\frac{1}{\zeta^2} = \frac{16}{9t^2}\left\{(1 + \tfrac{1}{2}\cos\varphi)^2 - \tfrac{1}{4}\sin^2\varphi + i(\cdot\,\cdot\,\cdot)\right\}, \tag{4.44}$$

$$\mathrm{Re}\,\frac{1}{\zeta^2} = \frac{16}{9t^2}\left\{\tfrac{1}{2}(1 + \cos\varphi)^2 + \tfrac{1}{4}\right\} \geqslant \frac{4}{9t^2}. \tag{4.45}$$

Then from (4.42) we have the estimate

$$|f^{(k)}(t)| \leqslant \frac{k!}{2\pi} \int_{|\zeta - t| = \frac{1}{2}t} \frac{e^{-4/9t^2}}{|\zeta - t|^{k+1}} \, d\zeta = \frac{2^k k!}{t^k} e^{-4/9t^2}. \tag{4.46}$$

Now, by putting $f(t) = e^{-1/t^2}$, from (4.39) we obtain

$$|u(x,t)| \leqslant \sum_{k=0}^{\infty} \frac{2^k k!}{(2k)!} \frac{x^{2k}}{t^k} e^{-4/9t^2} < \sum_{k=0}^{\infty} \frac{1}{k!} \left(\frac{x^2}{t}\right)^k e^{-4/9t^2} = e^{x^2/t} e^{-4/9t^2}. \tag{4.47}$$

From this it follows, however, that $\lim_{t \to 0} u(x,t) = 0$ uniformly on every interval $X_0 \leqslant x \leqslant X_1$ which justifies (4.39), and (4.41) follows from an analogous argument. Nevertheless, although the $u(x,t)$ constructed belongs to C^0 in $-\infty < x < \infty$, $0 \leqslant t < \infty$, the first requirement is not satisfied, because $\bar{u}(x_1 t) \equiv 0$ is a further solution with the same initial values. Therefore, the hypothesis $|u(x,t)| \leqslant M_1 e^{A_1 x^2}$ in the theorem of Section 4.1 is obviously violated. Thus, because of the first requirement, the condition $|u(x_1 t)| \leqslant M_1 e^{A_1 x^2}$ cannot be dispensed with.

The consequences of A. N. Tychonow's example go considerably further. Whereas in the case of the wave equation we could determine the solution of the initial-value problem for the past, that is, for $t < 0$ (the wave equation did not change under substitution for t of $-t$), such a plan is forbidden in (4.3) for two reasons at the start.

1. If we replace t by $-t$, equation (4.3) changes to the equation $u_{xx} + u_t = 0$.

2. For a negative t expression (4.5), which we had found as a solution to the initial-value problem, does not make sense because of $\sqrt{4\pi t}$.

In Tychonow's example, on the other hand, values of $t < 0$ may be considered.[15] It shows that the solution of the initial-value problem is uniquely determined neither for the future $(t > 0)$ nor for the past $(t < 0)$ without an estimate of the form $|u(x,t)| \leqslant M_1 e^{A_1 x^2}$.

The following example is also of interest.

Example 3. E. Rothe[18]. $u_{xx} - u_t = 0$, $u(x,0) = u_0(x)$, $-\infty < x < \infty$. If $u_0(x) \equiv 0$, then $u(x,t) \equiv 0$ is a solution. If $u_0(x) = \lambda \sin (x/\lambda)$, $0 < \lambda$, then $u(x,t) = \lambda e^{-t/\lambda^2} \sin (x/\lambda)$ is a solution. For sufficiently small $\lambda > 0$ the initial values come arbitrarily close to zero, but the solutions do so only for $t \geqslant 0$. If $t < 0$, the third requirement is violated. For the past $(t < 0)$, the third requirement cannot be expected to be satisfied even if the second requirement is.

4.4   Remarks

As in Section 2.3, it makes more sense physically to consider an *initial-boundary value problem*. The simplest one would be of the form

$$u_{xx} - u_t = 0, \qquad u(x,0) = u_0(x), \qquad u(0,t) = u(l,t) = 0, \qquad 0 \leqslant x \leqslant l. \tag{4.48}$$

It can be solved, as in Section 2.3, by a suitable extension of $u_0(x)$ to $-\infty < x < \infty$. We must not, however, be silent about the fact that by this method we can do justice only to the very simple problems. In applications we have to deal with a much greater variety of boundary conditions, so that it will later be advantageous to use stronger methods. Nevertheless we discuss a few more complicated problems in Section 4.5.

---

[15] If $t < 0$, we consider the circle $|\zeta - t| \leqslant \frac{1}{2}|t|$ and obtain $|u(x,t)| \leqslant \exp\left(|x|^2/|t| - 4/9t^2\right)$.

The considerations of Section 2.10 can be carried over immediately. If we introduce a suitable solution operation $\mathfrak{I}(t)$ which transforms the state at time $t = 0$ into the state at time t, then, using Section 4.1, we find

$$u(x,t) = \mathfrak{I}(t)u_0 \quad \text{where} \quad \mathfrak{I}(t) = \int_{-\infty}^{\infty} s(x - y, t)\{\cdot \cdot \cdot\}\, dy. \tag{4.49}$$

As in Section 2.10, *Hadamard's requirement* is given by

$$\mathfrak{I}(t_1 + t_2) = \mathfrak{I}(t_2)\mathfrak{I}(t_1). \tag{4.50}$$

However, $\mathfrak{I}(-t)$ does not make sense now so that the solution operators only form a semi-group[16] under the composition defined in (4.50). Somewhat superficially we may say that the solution operators of *reversible* processes in nature form groups and those of *irreversible* processes in nature form semigroups.

The theorem in Section 4.1 shows that the initial values there are *continuable* ones. If we consider the solution (4.5) at the point (x_0,t_0) in space-time, it becomes obvious from (4.5) that $u(x_0,t_0)$ depends on all initial values. Therefore we have

$$\bar{\mathfrak{a}}(x_0,t_0): -\infty < x < \infty. \tag{4.51}$$

A consequence of this is as follows. If $u_0(x) \equiv 0$ with the exception of a domain of perturbation $\mathcal{P}: X_0 < x < X_1$, in which $u_0(x) \neq 0$, then all points (x_0,t_0) in space-time with $t_0 > 0$ "receive information" from this perturbation. Hence the perturbation propagates itself with infinite velocity. Since such a statement is physically senseless, it has been shown that the heat equation describes the physical state of affairs imperfectly.

Perhaps the most beautiful theorem about the initial-value problem is due to D. V. Widder[19]. It shows that the initial-value problem is solved by (4.5) under the mere assumption of continuity of the initial values whenever $u(x,t)$ is bounded from below: $u(x,t) \geqslant c$. Physically this is always the case, since the temperature is bounded from below by absolute zero.

4.5 Initial-Boundary Value Problems

We shall give the formal solutions of a few initial-boundary value problems. We consider the problem

$$u_{xx} - u_t = 0 \quad \text{in} \quad 0 < x,t < \infty; \qquad u(x,0) = u_0(x), \qquad 0 < x < \infty \tag{4.52}$$

with one of the boundary conditions

$$u(0,t) = 0, \qquad 0 < t < \infty; \tag{4.53a}$$
$$u_x(0,t) = 0, \qquad 0 < t < \infty; \tag{4.53b}$$
$$u_x(0,t) + \alpha u(0,t) = 0, \qquad 0 < t < \infty. \tag{4.53c}$$

In all cases we shall attempt to extend $u_0(x)$ for $x < 0$ in such a way that the function $u(x,t)$, which is given by (4.5), automatically has the correct boundary values. It is then self-evident that $u(x,t)$ satisfies the heat equation and that it has the correct boundary values for $x > 0$. Our considerations are analogous to those in Section 2.3. We assume

---

[16] The definition of a semigroup differs from that of a group merely by the absence of the postulate about the inverse element.

here that $u_0(x)$ is bounded and $u_0(x) \in C^0$ in $0 \leqslant x < \infty$, and in addition for problem (4.53a) that $u_0(0) = 0$.

If in problem (4.53a) we extend $u_0(x)$ as an odd function: $u_0(x) = -u_0(-x)$ for $x < 0$, then $u_0(x) \in C^0$ for $-\infty < x < \infty$. Then we obtain the solution of the first problem by

$$u(x,t) = \int_{-\infty}^{\infty} s(x-y,\,t)u_0(y)\,dy = \int_0^{\infty} s(x-y,\,t)u_0(y)\,dy$$
$$- \int_{-\infty}^0 s(x-y,\,t)u_0(-y)\,dy$$
$$= \int_0^{\infty} \{s(x-y,\,t) - s(x+y,\,t)\}u_0(y)\,dy. \tag{4.54}$$

We see immediately that $s(0-y,\,t) - s(0+y,\,t) = 0$, so that it is plausible that the boundary condition $u(0,t) = 0$ is satisfied. We would merely have to prove that the passage to the limit $x \to 0$ may be carried out under the improper integral. But this is permissible since the integral is uniformly convergent in x and t for $0 < t_0 \leqslant t \leqslant t_1$.

Problem (4.53b) is settled analogously by an even extension of $u_0(x)$: $u_0(x) = u_0(-x)$ for $x < 0$. Then from (4.54) we obtain

$$u(x,t) = \int_0^{\infty} \{s(x-y,\,t) + s(x+y,\,t)\}u_0(y)\,dy \tag{4.55}$$

as a solution. Here we see immediately that $s_x(0-y,\,t) + s_x(0+y,\,t) = 0$ so that the validity of the second boundary condition is plausible. We settle the third problem by putting

$$u(x,t) = \int_{-\infty}^{\infty} s(x-y,\,t)u_0(y)\,dy = \int_0^{\infty} s(x-y,\,t)u_0(y)\,dy$$
$$+ \int_{-\infty}^0 s(x-y,\,t)u_0(y)\,dy \tag{4.56}$$
$$= \int_0^{\infty} \{s(x-y,\,t)u_0(y) + s(x+y,\,t)u_0(-y)\}\,dy.$$

Here $u_0(x)$ is given for $x > 0$, and the continuation of $u_0(x)$ is to be determined for $x < 0$ so that $u_0(x)$ is continuous for all x. Using (4.56) and (4.5) we find formally that

$$\sqrt{4\pi t}\,u(0,t) = \int_0^{\infty} \{u_0(y) + u_0(-y)\}\,e^{-y^2/4t}\,dy. \tag{4.57}$$

Correspondingly, after an easy calculation, we find

$$\sqrt{4\pi t}\,u_x(0,t) = \int_0^{\infty} \{u_0(y) - u_0(-y)\}\,\frac{y}{2t}\,e^{-y^2/4t}\,dy \tag{4.58}$$
$$= -\int_0^{\infty} \{u_0(y) - u_0(-y)\}\,\frac{\partial}{\partial y}\,e^{-y^2/4t}\,dy.$$

Under consideration of the requirement of continuity

$$\lim_{\substack{x \to 0 \\ x > 0}} u_0(x) = \lim_{\substack{x \to 0 \\ x > 0}} u_0(-x),$$

integration by parts gives

$$\sqrt{4\pi t}\,u_x(0,t) = \int_0^{\infty} \frac{d}{dy}\{u_0(y) - u_0(-y)\}\,e^{-y^2/4t}\,dy. \tag{4.59}$$

We try now to achieve the validity of the boundary condition for $t > 0$ by requiring validity of the boundary condition for the corresponding integrand. Writing $u' = du/dy$, we obtain the requirement,

$$u_0'(-y) - \alpha u_0(-y) = u_0'(y) + \alpha u_0(y). \tag{4.60}$$

The values $u_0(y)$ are known for $y \geqslant 0$. Hence (4.60) is an ordinary differential equation for the function $u_0(-y)$. If we set $-y = z$ it has the form

$$\frac{d}{dz} u_0(z) + \alpha u_0(z) = f(-z) \quad \text{where} \quad f(-z) = \frac{d}{dz} u_0(-z) - \alpha u_0(-z), \tag{4.61}$$

with the solution

$$u_0(z) = e^{-\alpha z} \left\{ u_0(0) + \int_0^z f(-\eta) \, e^{\alpha \eta} \, d\eta \right\}. \tag{4.62}$$

If we substitute the expression for $f(-z)$ and carry out the integration by parts we finally obtain, by the substitution $z = -y$,

$$u_0(-y) = u_0(y) + 2\alpha \, e^{\alpha y} \int_0^y e^{-\alpha \xi} \, u_0(\xi) \, d\xi \tag{4.63}$$

for $y \geqslant 0$. Thus $u_0(x)$ has been determined for $-\infty < x < \infty$, and it is plausible that (4.56) is the solution of problem (4.53c). Note that, for $\alpha = 0$, (4.63) gives the correct continuation for the second problem.

Among the important initial-boundary value problems another one, of a somewhat different kind, is of technical significance. Suppose an infinite heat conductor $(-\infty < x < \infty)$ consists of two distinct materials in the intervals $-\infty < x < 0$ and $0 < x < \infty$. At the transition point $x = 0$, the temperatures as well as the heat flow in either direction must coincide. We denote the temperature in the left-hand part by $u(x,t)$, in the right-hand part by $U(x,t)$. Using (1.38), we arrive at the problem

$$\begin{aligned}
\gamma u_{xx} - u_t &= 0, & -\infty &< x < 0, & \Gamma U_{xx} - U_t &= 0, & 0 &< x < \infty, \\
u(x,0) &= u_0(x), & -\infty &< x < 0, & U(x,0) &= U_0(x), & 0 &< x < \infty, \\
u(0,t) &= U(0,t), & \omega u_x(0,t) &= \Omega U_x(0,t), & 0 &< t < \infty.
\end{aligned} \tag{4.64}$$

Here ω, Ω denote the conductivities and γ, Γ the diffusion coefficients. Furthermore we suppose that $u_0(0) = U_0(0)$.

By a suitable continuation of $u_0(x)$, this problem can be reduced to the simple problem

$$\gamma u_{xx} - u_t = 0, \quad -\infty < x < \infty; \quad u(x,0) \text{ given in } -\infty < x < \infty. \tag{4.65}$$

To begin with, we assume that the problem (4.64) has been solved and that $u(x,t)$ and $U(x,t)$ are the solutions. Then, as suggested by F. John, we define a function $u(x,t)$ for $x > 0$ by

$$u(x,t) = au(-x,t) + bU(\sqrt{\Gamma/\gamma} \, x,t) \tag{4.66}$$

where a and b are arbitrary constants. By a simple verification we confirm that this $u(x,t)$ in $0 < x < \infty$ satisfies the equation $\gamma u_{xx} - u_t = 0$. Thus a function $u(x,t)$ has been defined in $-\infty < x < \infty$, $x \neq 0$, $0 < t < \infty$ which satisfies the equation $\gamma u_{xx} - u_t = 0$ and which assumes the values $u_0(x)$ for $t = 0$, $-\infty < x < 0$. In addition we require that $u(x,t)$ and $u_x(x,t)$ behave continuously at $x = 0$. From the two last conditions in (4.64), the following requirements result:

$$\begin{aligned}
u(0,t) &= au(0,t) + bu(0,t), \\
\Omega u_x(0,t) &= -a\Omega u_x(0,t) + b\omega\sqrt{\Gamma/\gamma}\,u_x(0,t),
\end{aligned} \tag{4.67}$$

or

$$1 = a + b, \qquad \Omega = -a\Omega + b\omega\sqrt{\Gamma/\gamma}, \tag{4.68}$$

which determine the values a and b. From (4.66), then, the values of $u_0(x)$ for $x > 0$ follow, namely

$$u_0(x) = au_0(-x) + bU_0(\sqrt{\Gamma/\gamma}\,x) \quad \text{for} \quad 0 < x < \infty. \tag{4.69}$$

With this, we finally have

$$\gamma u_{xx} - u_t = 0, \qquad -\infty < x < \infty, \qquad x \neq 0;$$
$$u(x,0) = \begin{cases} u_0(x) & \text{for} \quad -\infty < x < 0, \\ au_0(-x) + bU_0(\sqrt{\Gamma/\gamma}\,x) & \text{for} \quad 0 \leqslant x < \infty; \end{cases} \tag{4.70}$$

for $u(x,t)$. Now we drop the assumption that (4.64) has been solved and solve (4.70) directly. From (4.5) we obtain, by introducing a new x-coordinate, $\bar{x} = x/\sqrt{\gamma}$, and then reversing this transformation,

$$u(x,t) = \int_{-\infty}^{\infty} \frac{1}{\sqrt{4\pi t}}\,\exp\left[-\frac{1}{4t}\left(\frac{x}{\sqrt{\gamma}} - y\right)^2\right] u(\sqrt{\gamma}\,y,0)\,dy. \tag{4.71}$$

Using (4.66), we are now able to compute $U(x,t)$. Thus the problem (4.64) has been solved, as we see immediately, if we perform the separate steps in inverse order.

REFERENCES

1. O. D. KELLOGG, *Foundations of Potential Theory* (Berlin: Springer Verlag), 1929, Chap. 4 [or (New York: Dover Publications), 1953—Translator's note].

2. CL. MÜLLER, *Grundprobleme der mathematischen Theorie elektromagnetischer Schwingungen* (Berlin: Springer Verlag), 1957.

3. O. D. KELLOGG, *Foundations of Potential Theory* (Berlin: Springer Verlag), 1929, p. 119.

4. R. SAUER, *Einführung in die theoretische Gasdynamik* (Berlin: Springer Verlag), 1951.

5. F. JOHN, *Plane Waves and Spherical Means Applied to Partial Differential Equations* (New York: Interscience Publishers, Inc.), 1955.

6. E. HILLE and R. PHILLIPS, *Functional Analysis and Semi-Groups*, 2nd ed. (Providence: American Mathematical Society), 1957, pp. 617–618 [or pp. 388ff. in 1st ed., 1948].

7. K. O. FRIEDRICHS and H. LEWY, *Nachr. Wiss. Ges. Göttingen* **26**, 135–143 (1932).

8. F. JOHN, *Commun. Pure Appl. Math.* **7**, 245–269 (1954).

9. J. HADAMARD, *Lectures on Cauchy's Problem in Linear Partial Differential Equations* (New York: Dover Publications), 1952.

10. I. I. PRIWALOW, *Einführung in die Funktionentheorie*, Part III (Leipzig: B. G. Teubner), 1959, p. 27 [or I. I. PRIVALOV, *Vvedeniye v teoriyu funkciĭ komplexnogo peremennogo*, 10th ed. (Moscow: Fizmatgiz), 1960, p. 350; also L. V. AHLFORS, *Complex Analysis* (New York: McGraw-Hill Book Company, Inc.), 1953, pp. 189–193—Translator's note].

11. Cl. Müller, *Commun. Pure Appl. Math.* **7**, 505–515 (1954). E. Heinz, *Nachr. Akad. Wiss. Göttingen, Math.-Phys. Kl.* **IIa,** No. 1 (1955). P. Hartmann, *Transactions of the Symposium on Partial Differential Equations* (Berkeley, California, 1955) (New York: Interscience Publishers, Inc.), 1956 [reprinted from *Commun. Pure Appl. Math.* **9,** No. 3 (1956)], pp. 137–147.

12. E. Heinz, "Elliptische Differentialgleichungen" (lecture notes), Math. Inst. Univ. Göttingen, 1954–55.

13. R. Courant and D. Hilbert, *Methoden der mathematischen Physik*, Part II (Berlin: Springer Verlag), 1937, p. 244 [or R. Courant and D. Hilbert, *Methods of Mathematical Physics*, Vol. II (New York: Interscience Publishers, Inc.), 1962—Translator's note].

14. A. Hurwitz and R. Courant, *Allgemeine Funktionentheorie* . . . (Berlin: Springer Verlag), 1929, p. 400.

15. W. I. Smirnow, *Lehrgang der höheren Mathematik*, Part IV (Berlin: VEB Deutscher Verlag der Wissenschaften), 1958, p. 409 [or V. I. Smirnov, Course of Higher Mathematics, Vol. IV, 5th ed. (Moscow: Fizmatgiz), 1958, p. 481 (Russian)—Translator's note].

16. A. N. Tychonow, *Moscow Univ. Bull. Sér. intern., Section A, Math. Méchan.* **1,** No. 9, 1–14 (1938).

17. A. N. Tychonow, *Akad. Nauk SSSR Mat. Sb.* **42,** 199–215 (1935).

18. E. Rothe, *Math. Z.* **27,** 76–86 (1928).

19. D. V. Widder, *Trans. Am. Math. Soc.* **55,** 85–94 (1944).

Part II. Classification into Types, Theory of Characteristics, and Normal Form

1

DIFFERENTIAL EQUATIONS OF THE SECOND ORDER

1.1 Classification into Types

The domains $\mathfrak{D}$ or $\bar{\mathfrak{D}}$ which appear in this chapter need not necessarily be normal domains. In particular, we shall always admit the whole of R_n for $\mathfrak{D}$ or $\bar{\mathfrak{D}}$. As a special case of the general equation (I-1.1) we consider

$$Du \equiv Au + h = 0 \quad \text{where} \quad Au \equiv \sum_{i,k=1}^{n} a_{ik} u_{x_i x_k}. \tag{1.1}$$

Here $u = u(x) = u(x_1, x_2, \ldots, x_n)$ is the unknown function.[1]

(a) Equation (1.1) is said to be *quasi-linear* if the a_{ik} and h are functions of the $2n + 1$ variables $x_1, x_2, \ldots, x_n, u, u_{x_1}, \ldots, u_{x_n}$.

(b) Equation (1.1) is said to be *almost linear* if the a_{ik} are functions of x alone and h a function as in (a).

(c) Equation (1.1) is said to be *linear* if $h = \sum_{i=1}^{n} a_i u_{x_i} + au + f$ where a_{ik}, a_i, a, f are functions of x alone.

Here (1.1) is considered in the $(2n + 1)$-dimensional domain: $x \in \bar{\mathfrak{D}}$; $|u| \leqslant C$, $|u_{x_i}| \leqslant C$. To the equation (1.1) we assign the quadratic form

$$Q(y) = \sum_{i,k=1}^{n} a_{ik} y_i y_k. \tag{1.2}$$

In case (b) the coefficients a_{ik} are functions of $x \in \bar{\mathfrak{D}}$ only. In case (a) we assume that a solution[2] of (1.1) has been substituted in the a_{ik} so that *with respect to this solution* the coefficients are then functions of x only. If we fix $x \in \bar{\mathfrak{D}}$, then the a_{ik} are numbers and (1.2) can, by a linear *coordinate transformation*

$$z = z(y): z_i = z_i(y_1, y_2, \ldots, y_n), \quad i = 1, 2, \ldots, n, \tag{1.3}$$

with nonvanishing functional determinant, be transformed to *principal axes*:

$$\tilde{Q}(z) = \sum_{i=1}^{n} \kappa_i (z_i)^2. \tag{1.4}$$

---

[1] In equations of the form (1.1) we shall assume, here and in the future, that $a_{ik} = a_{ki}$. This is done without loss of generality because $u_{x_i x_k} = u_{x_k x_i}$ for $u(x) \in C^2$.

[2] The question if and how such a solution can be found remains open here.

In algebra, the number of negative κ_i is called the index of inertia T, and the number of vanishing κ_i the defect D of the quadratic form (1.2), since, as is well known, T and D are independent of the choice of such a linear coordinate transformation. Hence a classification of (1.2) by means of these *invariants* is possible.

Classification into Types. At the fixed point $x \,\varepsilon\, \mathfrak{D}$, (1.1) is said to be

of *elliptic type* if $D = 0$, $T = 0$, or if $D = 0$, $T = n$;
of *hyperbolic type* if $D = 0$, $T = 1$, or if $D = 0$, $T = n - 1$;
of *parabolic type* if $D > 0$;
of *ultrahyperbolic type* if $D = 0$, $1 < T < n - 1$.

This is a *complete classification* relative to the fixed point $x \,\varepsilon\, \mathfrak{D}$. In case (a) it also depends on the selected solution $u = u(x)$.

The expression Du is said to be of *elliptic type in* $\mathfrak{D}$ if Du has this property for every $x \,\varepsilon\, \mathfrak{D}$. We adopt analogous terminology for the other types. Relative to $\mathfrak{D}$ such a classification is no longer complete. We also have to consider equations of *mixed type* in $\mathfrak{D}$. In the case $n = 2$ and $a_{ik} \,\varepsilon\, C^0(\mathfrak{D})$, equations of *elliptic-parabolic, hyperbolic-parabolic,* and *hyperbolic-parabolic-elliptic types* may occur.

If Du is of elliptic type in $\mathfrak{D}$, then obviously this fact can also be described as follows: The quadratic form $\sum\limits_{i,k=1}^{n} a_{ik} y_i y_k$ is definite. Since (1.1) may be multiplied by -1, if necessary, this form may even be assumed *positive definite*.

Important representatives of the first three types are the *potential, wave,* and *heat equations*. Here we have to interpret the point (x,t) in space-time (with $x \,\varepsilon\, R_n$) as a point x in R_{n+1}. An equation of ultrahyperbolic type would be, for instance,

$$u_{x_1 x_1} + u_{x_2 x_2} - u_{x_3 x_3} - u_{x_4 x_4} = 0.$$

The purely algebraic point of view—maintained so far—does not do justice to the theory of partial differential equations. We suppose that $a_{ik} \,\varepsilon\, C^1(\mathfrak{D})$ and assign to the quadratic form (1.2) the *partial differential equation of the first order:*

$$\sum\limits_{i,k=1}^{n} a_{ik}\varphi_{x_i}\varphi_{x_k} = 0. \tag{1.5}$$

We look for real solutions $\varphi(x) \,\varepsilon\, C^1$. For each such $\varphi(x)$ we define a point set by

$$\varphi(x) \equiv \varphi(x_1, \ldots, x_n) = 0 \quad \text{with} \quad \sum\limits_{i=1}^{n} (\varphi_{x_i})^2 > 0 \tag{1.6}$$

and close this set by adding its accumulation points. We call this closed set of points—if it is not empty—a *characteristic manifold* of the differential equation (1.1). Every such characteristic manifold will be denoted by $\dot{C}$. It is clear that for elliptic equations (1.1) such $\dot{C}$ do not exist.

In Section I-2, in the case of the wave equation, we have already attached the name "characteristic manifolds" to certain point sets. In general, the foregoing definition will be attached to these larger point sets. For the wave equation in R_1, in the terminology of Section I-2.2, the straight lines $|x - a| = r - |t|$ (for arbitrary a and $r > 0$) were called characteristic manifolds $\dot{C}$. They formed the boundary of the square of Figure 3. According to the new definition we must first consider equation (1.5). For the wave equation in R_1, in the notation of Section I-2.2, (1.5) is of the form $(\varphi_x)^2 - (\varphi_t)^2 = 0$. If we look for solutions of the form (1.6): $\varphi(x,t) = 0$, then $\varphi_t \neq 0$ because $\varphi_t = 0$ would also imply $\varphi_x = 0$. Therefore they may be brought into the form $t = t(x)$. From $\varphi_x + \varphi_t(dt/dx) = 0$ it follows that $(\varphi_x/\varphi_t)^2 = (dt/dx)^2 = 1$. Thus we have $\varphi = t \pm x + \text{const}$, and for $\dot C: t \pm x + \text{const} = 0$. It is clear that this set of points, for arbitrary constants, contains the straight line segments $|x - a| = r - |t|$. In R_2 we denoted the lateral surface of the double cone in Figure 6, Section I-2.8 by $\dot C$. By setting $x_3 = t$ we find for the wave equation in R_2

$$(\varphi_{x_1})^2 + (\varphi_{x_2})^2 - (\varphi_t)^2 = 0 \quad \text{with} \quad \varphi(x_1, x_2, t) = (t - \delta) \pm \sqrt{\sum_{i=1}^{2} (x_i - \gamma_i)^2}$$

as solutions with arbitrary constants γ_i, δ. These can be verified by substitution or, better, found systematically as solutions of a first-order partial differential equation.[3] The corresponding closed point sets

$$\dot C: (t - \delta) \pm \sqrt{\sum_{i=1}^{2} (x_i - \gamma_i)^2} = 0 \tag{1.7}$$

for $\gamma_i = a_i$, $\delta = r$, and the plus sign (or $\gamma_i = a_i$, $\delta = -r$ and the minus sign) contain the upper (or the lower) half of the lateral surface of the double cone in Figure 6,[4] but otherwise they give more.

For a hyperbolic equation (1.1) the $\dot C$ defined here have, in general, the form of *conoids*, that is, conelike surfaces.

If we consider the *heat equation* (I-4.1) in one space dimension $x = x_1$ and put $x_2 = t$ in (1.5), we obtain $(\varphi_{x_1})^2 = 0$. We look for curves of the form $\varphi(x_1,t) = 0$, and since $d\varphi/dx_1 = \varphi_{x_1} + \varphi_t(dt/dx_1) = 0$, we have $dt/dx_1 = 0$. From this it follows that for the heat equation $\dot C$ is given by $t = \text{const}$, and it becomes understandable from the mathematical point of view why in Section I-4.1 only the initial values $u(x,0) = u_0(x)$ were prescribed. Problem 1 in Section I-2.2 shows that in the case of the wave equation only u may be prescribed if the prescription is made on $\dot C$.

1.2   Invariance Properties of $\dot C$

We shall show that the equation (1.5) of the characteristic manifolds $\dot C$ is *invariant* under coordinate transformations so that the $\dot C$'s have an invariant meaning, as they must if they are to be the boundaries of the domains of determinateness $\mathfrak{B}$ in the hyperbolic case. We only consider the *almost-linear case* of (1.1). Let

$$y = y(x): y_i = y_i(x_1, x_2, \ldots, x_n), \qquad i = 1, \ldots, n,$$

---

[3] The theory of a partial differential equation of the first order can be reduced completely to that of ordinary differential equations so that these things will not be discussed here.

[4] Note that without taking the closure, because $\sum_{i=1}^{n} (\varphi_{x_i})^2 > 0$, the vertices (a_1, a_2, r) and $(a_1, a_2, -r)$ of the double cone in Figure 6 would not be included.

be the coordinate transformation, $u(x) = v(y)$ the transformed solution of (1.1), and $\varphi(x) = \psi(y) = 0$ the transformed $\dot{C}$. Then we have

$$u_{x_i} = \sum_{j=1}^{n} v_{y_j} y_{jx_i}, \qquad u_{x_i x_k} = \sum_{j,l=1}^{n} v_{y_j y_l} y_{jx_i} y_{lx_k} + \cdots, \tag{1.8}$$

$$\sum_{i,k=1}^{n} a_{ik} u_{x_i x_k} = \sum_{i,k=1}^{n} \alpha_{ik} v_{y_i y_k} + \cdots, \qquad \alpha_{ik} = \sum_{j,l=1}^{n} a_{jl} y_{jx_i} y_{lx_k}. \tag{1.9}$$

According to (1.8), relation (1.5) transforms to

$$\sum_{i,k=1}^{n} a_{ik}\varphi_{x_i}\varphi_{x_k} = \sum_{i,k,j,l=1}^{n} a_{ik}\psi_{y_j} y_{jx_i}\psi_{y_l} y_{lx_k} = \sum_{i,k=1}^{n} \alpha_{ik}\psi_{y_i}\psi_{y_k}. \tag{1.10}$$

But the last expression is equation (1.5) for the transformed equation (1.1).

1.3   Characteristic Directions

For $n = 2$ we consider the *quasi-linear equation* (1.1) in a domain $\mathfrak{D}$ in R_2. To begin with, we suppose $a_{ik} \in C^0$ and $a_{11} > 0$ in $\mathfrak{D}$. Then (1.2) becomes

$$Q(y) = a_{11}\left[\left(y_1 + \frac{a_{12}}{a_{11}}y_2\right)^2 + \frac{d}{a_{11}^2}y_2^2\right] \quad \text{where} \quad d = a_{11}a_{22} - a_{12}^2. \tag{1.11}$$

According to the considerations in Section 1.1, (1.1) is of *elliptic, hyperbolic,* or *parabolic* type in $\mathfrak{D}$ depending on whether $d > 0$, $d < 0$, or $d = 0$ there.[5]

Example. Equation (I-1.33). This is a second-order quasi-linear partial differential equation with

$$d = [a^2 - (u_{x_1})^2][a^2 - (u_r)^2] - (u_{x_1})^2(u_r)^2 = a^2(a^2 - |v|^2) \tag{1.11a}$$

where $v = \operatorname{grad} u$. Equation (I-1.33) is hyperbolic, parabolic, or elliptic depending on whether the *velocity of flow* $|v|$ is greater than, equal to, or less than the local *sound velocity a*.

Equation (1.5) becomes

$$\sum_{i,k=1}^{2} a_{ik}\varphi_{x_i}\varphi_{x_k} = \frac{1}{a_{11}}\{a_{11}\varphi_{x_1} + (a_{12} + \sqrt{-d})\varphi_{x_2}\}\{a_{11}\varphi_{x_1} + (a_{12} - \sqrt{-d})\varphi_{x_2}\} = 0. \tag{1.12}$$

If we attempt to find characteristic manifolds $\varphi(x_1,x_2) = 0$, then for $x_2 = x_2(x_1)$ we have $dx_2/dx_1 = -\varphi_{x_1}/\varphi_{x_2}$. If we divide (1.12) by $(\varphi_{x_2})^2$, we find two equations for dx_2/dx_1. They are

$$\frac{dx_2}{dx_1} = \frac{1}{a_{11}}\{a_{12} \pm \sqrt{-d}\}. \tag{1.13}$$

The directions given by (1.13) are called *characteristic directions*. In the elliptic, hyperbolic, or parabolic case there is (are) at each point of $\mathfrak{D}$ either no, precisely two, or precisely one characteristic direction(s). If $u(x) \in C^1(\mathfrak{D})$ for $n = 2$ and if $\alpha = (\alpha_1(x), \alpha_2(x))$ is a vector

---

[5] As coordinate transformation we can use $z_1 = y_1 + (a_{12}/a_{11})\,y_2$, $z_2 = y_2$.

field in $\mathfrak{D}$, where $|\alpha| > 0$ and $\alpha_i \in C^0$, then, along with equation (I-1.15),

$$u_\alpha(x) = \sum_{i=1}^{2} \alpha_i(x)u_{x_i}(x) \tag{1.14}$$

is called a *directional derivative* at the point x in the direction of the vector α. u_α and u_β are said to be *linearly independent in* $\mathfrak{D}$ if α and β are linearly independent at every point of $\mathfrak{D}$. In particular, for $n = 2$ any arbitrary directional derivative u_γ can then be written as a linear combination of u_α and u_β.

Problem. Show that instead of the known identity $u_{x_1 x_2} = u_{x_2 x_1}$, for directional derivatives we have the relation

$$u_{\alpha\beta} - qu_\alpha = u_{\beta\alpha} - \bar{q}u_\beta. \tag{1.15}$$

Determine $q, \bar{q}$.

In the hyperbolic case, two such vector fields in $\mathfrak{D}$ are given by (1.13), namely,

$$\alpha = (Ma_{11}, M(a_{12} - \sqrt{-d})), \qquad \beta = (Na_{11}, N(a_{12} + \sqrt{-d})) \quad \text{where}$$
$$\alpha_1\beta_2 - \alpha_2\beta_1 > 0; \tag{1.16}$$

here a common factor $M > 0$, $N > 0$ always remains free.

1.4   Normal Form in the Hyperbolic Case for $n = 2$

By introducing the *characteristic directional derivatives* given in (1.16), we can simplify considerably equation (1.1) in the almost linear case; if with arbitrary α, β we set

$$u_{\alpha\beta} \equiv \beta_1(\alpha_1 u_{x_1} + \alpha_2 u_{x_2})_{x_1} + \beta_2(\alpha_1 u_{x_1} + \alpha_2 u_{x_2})_{x_2} = \sum_{i,k=1}^{2} a_{ik}u_{x_i x_k} + \cdots, \tag{1.17}$$

where the dots stand for suitable terms which contain only first derivatives of u, then we see that in general (1.17) is satisfied if and only if

$$\alpha_1\beta_1 = a_{11}, \qquad \alpha_1\beta_2 + \alpha_2\beta_1 = 2a_{12}, \qquad \alpha_2\beta_2 = a_{22} \tag{1.18}$$

holds. By elementary calculations we find with the notation for d as in (1.11) that

$$2\sqrt{-d} = \alpha_1\beta_2 - \alpha_2\beta_1, \qquad 2a_{12} = \alpha_1\beta_2 + \alpha_2\beta_1. \tag{1.19}$$

Addition and subtraction, under consideration of (1.18), give

$$\frac{\alpha_2}{\alpha_1} = \frac{a_{12} - \sqrt{-d}}{a_{11}}, \qquad \frac{\beta_2}{\beta_1} = \frac{a_{12} + \sqrt{-d}}{a_{11}}, \tag{1.20}$$

and with this we obtain (1.16) for α and β, where M and N satisfy $MN = 1/a_{11}$ because of the relations (1.18). Since furthermore u_{x_1}, u_{x_2} can be written as linear combinations of u_α, u_β, we obtain the *normal form* in the *almost-linear hyperbolic case:*

$$u_{\alpha\beta} + h_1(x_1, x_2, u, u_\alpha, u_\beta) = 0 \quad \text{or} \quad u_{\beta\alpha} + h_2(x_1, x_2, u, u_\alpha, u_\beta) = 0. \tag{1.21}$$

Its construction requires only the most elementary tools. Furthermore we can achieve the outcome—not by elementary means—that the directional derivatives go over into partial derivatives. If

$$\xi_1 = \varphi_1(x_1, x_2), \qquad \xi_2 = \varphi_2(x_1, x_2) \quad \text{where} \quad \delta = \frac{\partial(\varphi_1, \varphi_2)}{\partial(x_1, x_2)} \neq 0 \quad \text{in} \quad \mathfrak{D} \qquad (1.22)$$

is a one-to-one coordinate transformation, then we require that for any two functions $v(x_1, x_2)$ and $w(\xi_1, \xi_2)$ with $v(x_1, x_2) = w(\xi_1, \xi_2)$ the two relations

$$v_\alpha = w_{\xi_1}, \qquad v_\beta = w_{\xi_2} \qquad (1.23)$$

hold. This implies

$$v_{x_1} = w_{\xi_1}\varphi_{1x_1} + w_{\xi_2}\varphi_{2x_1}, \qquad v_{x_2} = w_{\xi_1}\varphi_{1x_2} + w_{\xi_2}\varphi_{2x_2} \qquad (1.24)$$

with the solution

$$w_{\xi_1} = \frac{1}{\delta}(v_{x_1}\varphi_{2x_2} - v_{x_2}\varphi_{2x_1}), \qquad w_{\xi_2} = \frac{1}{\delta}(v_{x_2}\varphi_{1x_1} - v_{x_1}\varphi_{1x_2}). \qquad (1.25)$$

Using (1.25) and (1.16), we can bring requirement (1.23) into the form

$$\frac{\alpha_2}{\alpha_1} = \frac{a_{12} - \sqrt{-d}}{a_{11}} = -\frac{\varphi_{2x_1}}{\varphi_{2x_2}}, \qquad \frac{\beta_2}{\beta_1} = \frac{a_{12} + \sqrt{-d}}{a_{11}} = -\frac{\varphi_{1x_1}}{\varphi_{1x_2}}. \qquad (1.26)$$

This implies existence of the relations

$$a_{11}\varphi_{1x_1} + (a_{12} + \sqrt{-d})\varphi_{1x_2} = 0, \qquad a_{11}\varphi_{2x_1} + (a_{12} - \sqrt{-d})\varphi_{2x_2} = 0. \qquad (1.27)$$

By comparison with (1.12) we see that the *characteristic manifolds* $\dot{C}$ (*characteristic curves*) have to be introduced as coordinates in order to obtain the *integrable normal form* in the almost-linear hyperbolic case:

$$u_{\xi_1\xi_2} + h_3(\xi_1, \xi_2, u, u_{\xi_1}, u_{\xi_2}) = 0. \qquad (1.28)$$

The characteristic manifold $\dot{C}$ is found in this case, more simply as a solution of (1.13). If we now set $\eta_1 = \frac{1}{2}(\xi_1 + \xi_2)$, $\eta_2 = \frac{1}{2}(\xi_1 - \xi_2)$, then (1.28) becomes

$$u_{\eta_1\eta_1} - u_{\eta_2\eta_2} + h_4(\eta_1, \eta_2, u, u_{\eta_1}, u_{\eta_2}) = 0, \qquad (1.29)$$

an equation whose principal part consists precisely of the terms of the *wave equation*.

Problem. Determine $\dot{C}$ for $u_{x_1x_1} - (x_1)^p u_{x_2x_2} = 0$, $(x_1)^p u_{x_1x_1} - u_{x_2x_2} = 0$ for $x_1 \geqslant 1$, and determine the normal forms (1.28), (1.29). Here p is a positive integer.

1.5   Normal Form in the Elliptic Case for $n = 2$

We consider the almost-linear equation (1.1) and suppose that it is elliptic in $\mathfrak{D}$ ($d > 0$). If we repeat the considerations in 1.4, the coordinates ξ_1, ξ_2 become complex conjugates of one another. On the other hand, using $\eta_1 = \frac{1}{2}(\xi_1 + \xi_2)$, $\eta_2 = (1/2i)(\xi_1 - \xi_2)$ where $i = \sqrt{-1}$, we can pass to real coordinates in (1.28) and thus obtain the *integrable normal form*

$$u_{\eta_1\eta_1} + u_{\eta_2\eta_2} + h_5(\eta_1, \eta_2, u, u_{\eta_1}, u_{\eta_2}) = 0. \qquad (1.30)$$

Here the principal part in (1.30), if it is set equal to zero, is just the *potential equation*. This procedure via the complex plane necessarily requires that the a_{ik} be analytic in $\mathfrak{D}$; equations

(1.13), for instance, have to be integrated in the complex plane. Hence we reject this method as unsuitable.

Far better but more difficult are the following considerations in which we suppose $a_{ik} \in C^1(\mathfrak{D})$. We try to determine two directional derivatives $u_\alpha = \sum_{i=1}^{2} \alpha_i u_{x_i}$, $u_\beta = \sum_{i=1}^{2} \beta_i u_{x_i}$ so that

$$u_{\alpha\alpha} + u_{\beta\beta} = \sum_{i,k=1}^{2} a_{ik} u_{x_i x_k} + \cdots, \tag{1.31}$$

where the dots stand for terms which contain only first derivatives of u. Now we have

$$\begin{aligned} u_{\alpha\alpha} &= \alpha_1(\alpha_1 u_{x_1} + \alpha_2 u_{x_2})_{x_1} + \alpha_2(\alpha_1 u_{x_1} + \alpha_2 u_{x_2})_{x_2}, \\ u_{\beta\beta} &= \beta_1(\beta_1 u_{x_1} + \beta_2 u_{x_2})_{x_1} + \beta_2(\beta_1 u_{x_1} + \beta_2 u_{x_2})_{x_2} \end{aligned} \tag{1.32}$$

so that (1.31) implies the requirements

$$(\alpha_1)^2 + (\beta_1)^2 = a_{11}, \qquad \alpha_1\alpha_2 + \beta_1\beta_2 = a_{12}, \qquad (\alpha_2)^2 + (\beta_2)^2 = a_{22}. \tag{1.33}$$

By assuming $\alpha_1\beta_2 - \alpha_2\beta_1 > 0$, with the notation d in (1.11) we find

$$\sqrt{d} = \alpha_1\beta_2 - \alpha_2\beta_1. \tag{1.34}$$

Using (1.33) and (1.34) we have

$$\begin{aligned} a_{12}\alpha_2 + \sqrt{d}\,\beta_2 &= a_{22}\alpha_1, & a_{12}\beta_2 - \sqrt{d}\,\alpha_2 &= a_{22}\beta_1, \\ a_{12}\alpha_1 - \sqrt{d}\,\beta_1 &= a_{12}\alpha_2, & a_{12}\beta_1 + \sqrt{d}\,\alpha_1 &= a_{11}\beta_2, \end{aligned} \tag{1.35}$$

or

$$\begin{aligned} \alpha_1 &= \frac{1}{\sqrt{d}} (a_{11}\beta_2 - a_{12}\beta_1), & \beta_1 &= \frac{1}{\sqrt{d}} (a_{12}\alpha_1 - a_{11}\alpha_2), \\ \alpha_2 &= \frac{1}{\sqrt{d}} (a_{12}\beta_2 - a_{22}\beta_1), & \beta_2 &= \frac{1}{\sqrt{d}} (a_{22}\alpha_1 - a_{12}\alpha_2). \end{aligned} \tag{1.36}$$

Thus we obtain the normal form

$$u_{\alpha\alpha} + u_{\beta\beta} + h_6(x_1, x_2, u, u_\alpha, u_\beta) = 0, \tag{1.37}$$

which again can be constructed by elementary means. Here one vector field can be described arbitrarily. The second is then computed by using (1.36).

If we carry out the coordinate transformation (1.22) and require (1.23), then, from the formulas in (1.25), we obtain immediately

$$\alpha_1 = \frac{\varphi_{2x_2}}{\delta}, \qquad \alpha_2 = -\frac{\varphi_{2x_1}}{\delta}, \qquad \beta_1 = -\frac{\varphi_{1x_2}}{\delta}, \qquad \beta_2 = \frac{\varphi_{1x_1}}{\delta} \tag{1.38}$$

so that (1.36) can be brought into the form

$$\begin{cases} \varphi_{1x_1} = \dfrac{1}{\sqrt{d}} (a_{12}\varphi_{2x_1} + a_{22}\varphi_{2x_2}), \\[2ex] \varphi_{1x_2} = -\dfrac{1}{\sqrt{d}} (a_{11}\varphi_{2x_1} + a_{12}\varphi_{2x_2}), \end{cases} \qquad \begin{cases} \varphi_{2x_1} = -\dfrac{1}{\sqrt{d}} (a_{12}\varphi_{1x_1} + a_{22}\varphi_{1x_2}), \\[2ex] \varphi_{2x_2} = \dfrac{1}{\sqrt{d}} (a_{11}\varphi_{1x_1} + a_{12}\varphi_{1x_2}). \end{cases} \tag{1.39}$$

Each one of these is a system of two first-order partial differential equations for two unknown functions. It suffices to consider one such system. It is called *Beltrami's system*. If we have found an arbitrary *pair of solutions* $\varphi_1(x_1,x_2)$, $\varphi_2(x_1,x_2) \in C^2$ in $\mathfrak{D}$ which map $\mathfrak{D}$ one-to-one onto a domain of the (ξ_1,ξ_2)-plane and for which $\delta \equiv \varphi_{1x_1}\varphi_{2x_2} - \varphi_{1x_2}\varphi_{2x_1} \neq 0$, then the *integrable normal form*

$$u_{\xi_1\xi_1} + u_{\xi_2\xi_2} + h_7(\xi_1,\ \xi_2,\ u,\ u_{\xi_1},\ u_{\xi_2}) = 0 \tag{1.40}$$

can be achieved in $\mathfrak{D}$.

For $\varphi_i(x_1,x_2)$, $i = 1, 2$ by means of the commutation relation $\varphi_{x_1x_2} = \varphi_{x_2x_1}$, there arises the second-order Beltrami differential equation

$$\left(\frac{a_{11}\varphi_{x_1} + a_{12}\varphi_{x_2}}{\sqrt{d}}\right)_{x_1} + \left(\frac{a_{12}\varphi_{x_1} + a_{22}\varphi_{x_2}}{\sqrt{d}}\right)_{x_2} = 0. \tag{1.41}$$

To obtain the integrable normal form in $\mathfrak{D}$, it is necessary to guarantee the existence of a solution $\varphi_1(x_1,x_2) \in C^2$ of (1.41) which is not constant.[6]   Using (1.39) we easily determine another solution $\varphi_2(x_1,x_2)$ so that $\delta \neq 0$, because we have

$$\varphi_2(x_1,x_2) = \int (\varphi_{2x_1}\, dx_1 + \varphi_{2x_2}\, dx_2)$$
$$= \int \left[-\frac{1}{\sqrt{d}}\, (a_{12}\varphi_{1x_1} + a_{22}\varphi_{1x_2})\, dx_1 + \frac{1}{\sqrt{d}}\, (a_{11}\varphi_{1x_1} + a_{12}\varphi_{1x_2})\, dx_2 \right]. \tag{1.42}$$

With this, the mapping is guaranteed to be one-to-one only locally; globally this property must be an *ad hoc* assumption.

This kind of problem arises in various places in mathematics; for instance, in differential geometry in the case of *conformal mapping* of a *singularity-free, simply connected, nonanalytic surface (piece of a surface)* S onto a domain in the plane.   Let S be described by $y = f(x_1,x_2)$ where f is a vector field $f = (f_1,\ f_2,\ f_3) \in C^1$ for $x_1,\ x_2 \in \mathfrak{D}$.   The line element on S then becomes

$$\begin{aligned} (ds)^2 &= E(dx_1)^2 + 2F\, dx_1\, dx_2 + G(dx_2)^2; \\ E &= (f_{x_1},f_{x_1}), \qquad F = (f_{x_1},f_{x_2}), \qquad G = (f_{x_2},f_{x_2}), \end{aligned} \tag{1.43}$$

and we suppose $EG - F^2 > 0$ and $E > 0$.[7]   We look for a one-to-one map $u_1 = \varphi_1(x_1,x_2)$, $u_2 = \varphi_2(x_1,x_2) \in C^1$ which transforms $\mathfrak{D}$ into a domain $\mathfrak{D}^*$ of the $u_1,\ u_2$-plane such that

$$(ds)^2 = \lambda(u_1,u_2)\{(du_1)^2 + (du_2)^2\} \quad \text{with} \quad \lambda > 0 \quad \text{in} \quad \mathfrak{D}^*. \tag{1.44}$$

This mapping problem is equivalent to finding a pair of solutions of the *Beltrami system*

$$\varphi_{1x_1} = \frac{1}{\sqrt{W}}\, (-F\varphi_{2x_1} + E\varphi_{2x_2}),$$
$$\varphi_{1x_2} = \frac{1}{\sqrt{W}}\, (-G\varphi_{2x_1} + F\varphi_{2x_2}), \tag{1.45}$$
$$W = EG - F^2, \qquad \frac{\partial(\varphi_1,\varphi_2)}{\partial(x_1,x_2)} \neq 0,$$

which give a one-to-one mapping of $\mathfrak{D}$ onto $\mathfrak{D}^*$.   The equivalence is to be understood as follows: Any pair of solutions of the first problem is also a pair of solutions to the second

---

[6] This is done in Part IV.   There, however, we obtain this result as a special case of a much more general existence theorem.   For equation (1.41), also see L. Lichtenstein[1].

[7] Our notation here deviates from the usual one of differential geometry.   Here S is given in parametric form, and the parameters are x_1, x_2, which in differential geometry are usually denoted by u, v.

problem; to every pair of solutions of the second problem, there exists a $\lambda(u_1,u_2) > 0$ for which it is a pair of solutions of the first problem.

Problem. Confirm the foregoing statements.

Problem. Bring $u_{x_1 x_1} + (x_1)^p u_{x_2 x_2} = 0$, for $x_1 > 0$, into the normal form (1.30). Since $(x_1)^p$ is analytic use the complex procedure. Here p is a positive integer.

1.6 Normal Form in the Parabolic Case for $n = 2$

According to (1.11), $d = 0$, and we can show that

$$u_{\alpha\alpha} = \alpha_1(\alpha_1 u_{x_1} + \alpha_2 u_{x_2})_{x_1} + \alpha_2(\alpha_1 u_{x_1} + \alpha_2 u_{x_2})_{x_2} = \sum_{i,k=1}^{2} a_{ik} u_{x_i x_k} + \cdots. \qquad (1.46)$$

This implies the requirements

$$a_{11} = (\alpha_1)^2, \qquad a_{12} = \alpha_1\alpha_2, \qquad a_{22} = (\alpha_2)^2, \quad \text{or} \quad \frac{\alpha_2}{\alpha_1} = \frac{a_{12}}{a_{11}} \quad \text{because of}$$

$$a_{11}a_{22} - (a_{12})^2 = 0. \qquad (1.47)$$

Finally we find $\alpha = (\sqrt{a_{11}}, (1/\sqrt{a_{11}})a_{12})$. If further we take an *arbitrary vector field* $\beta = (\beta_1,\beta_2)$ so that $\alpha_1\beta_2 - \alpha_2\beta_1 \neq 0$, we obtain the *normal form*

$$u_{\alpha\alpha} + h_8(x_1, x_2, u, u_\alpha, u_\beta) = 0. \qquad (1.48)$$

If we proceed as in Section 1.4, then, with the requirements in (1.23), the *integrable normal form*

$$u_{\xi_1 \xi_1} + h_9(\zeta_1, \zeta_2, u, u_{\xi_1}, u_{\xi_2}) = 0 \qquad (1.49)$$

results. The characteristic manifolds $\dot{C}$ are described by

$$\varphi(x_1,x_2) = 0$$

where $\varphi(x_1,x_2)$ satisfies equation (1.12) with $d = 0$:

$$a_{11}\varphi_{x_1} + a_{12}\varphi_{x_2} = 0 \quad \text{or} \quad \frac{dx_2}{dx_1} = \frac{a_{12}}{a_{11}}. \qquad (1.50)$$

Finally; if the original equation (1.1) was linear, then so is equation (1.49); it may then be assumed in the form

$$u_{\xi_1 \xi_1} + B_1 u_{\xi_1} + B_2 u_{\xi_2} + Cu + F = 0. \qquad (1.51)$$

In case $B_2(\xi_1,\xi_2)$ does not vanish, (1.51) can be simplified considerably.

By the transformation $\bar{\xi}_1 = \chi(\xi_1,\xi_2)$, $\bar{\xi}_2 = \xi_2$ where

$$\chi = \int^{\xi_1} \sqrt{|B_2(\tau,\xi_2)|} \, d\tau,$$

an equation of the form (1.51) results but the term u_{ξ_2} has the coefficient ± 1. Hence we may assume $B_2 = \pm 1$ in (1.51) right away. By $u = vw$ where

$$\log w = - \int^{\xi_1} \frac{B_1(\tau,\xi_2)}{2} \, d\tau,$$

we introduce a new unknown function v. Then an equation in v of type (1.51), with $B_2 = \pm 1$ and $B_1 = 0$, results so that we can finally bring the normal form into

$$v_{\xi_1\xi_1} \pm v_{\xi_2} + \bar{C}(\xi_1,\xi_2)v + \bar{F}(\xi_1,\xi_2) = 0. \tag{1.52}$$

Here the principal part, if set equal to zero, can for $\xi_2 \geqslant 0$ be considered as the heat equation for the future $(-)$, or for the past $(+)$.

1.7   Differential Equations of Mixed Type for $n = 2$

Whenever an equation (1.1) is not of a unique type in $\mathfrak{D}$, it is called of *mixed type*. Simple examples for an equation (1) of *elliptic-parabolic*, (2) of *hyperbolic-parabolic*, and (3) of *hyperbolic-parabolic-elliptic* type are

$$x_2 u_{x_1 x_1} + u_{x_2 x_2} = 0 \quad \text{where} \quad -\infty < x_1 < \infty \tag{1.53}$$

and   (1) $x_2 \geqslant 0$,   (2) $x_2 \leqslant 0$,   (3) $-\infty < x_2 < \infty$.

From now on we consider the almost-linear equation (1.1) for $n = 2$ and suppose it is of *mixed type* in $\mathfrak{D}$. Then some new notions arise.

The point set $k \subset \mathfrak{D}$ on which (1.1) is of parabolic type $[a_{11}a_{22} - (a_{12})^2 = 0]$ is called *parabolic curves*. Here we suppose that k is of two-dimensional measure zero relative to $\mathfrak{D}$ and that it only consists of sufficiently smooth arcs which can be described by the parametric representations $x_1 = x_1(t)$, $x_2 = x_2(t)$ where $x_i(t) \in C^1$ and $(\dot{x}_1)^2 + (\dot{x}_2)^2 > 0$, all in suitable intervals.

For k we have to distinguish between precisely *three cases:*

(a) k is *characteristic*, that is, according to (1.12) and (1.13), the equation

$$a_{11}(\dot{x}_2)^2 - 2a_{12}\dot{x}_1\dot{x}_2 + a_{22}(\dot{x}_1)^2 = 0 \tag{1.54}$$

holds along k,

(b) k is *noncharacteristic*, that is, (1.54) is true at no point of k,

(c) k is *partly characteristic* and partly *noncharacteristic*.

Examples. For equation (1.53) with $0 \leqslant x_1, x_2 < \infty$, $k: x_2 = 0$ is of type (b). For the equation $u_{x_1 x_1} + x_2 u_{x_2 x_2} = 0$, where $0 \leqslant x_1, x_2 < \infty$, $k: x_2 = 0$ is of type (a).

We shall not enter into the question of normal forms for equations of mixed type. It has been investigated by F. Tricomi[2] and M. Cibrario[3].

2

SYSTEMS OF DIFFERENTIAL EQUATIONS OF THE FIRST ORDER

2.1 Hyperbolic Systems in Two Independent Variables

Every quasi-linear system of first-order partial differential equations in two independent variables x_k, $k = 1, 2$, and n unknown functions $u_j = u_j(x)$ with $x = (x_1, x_2)$, $j = 1, 2, \ldots, n$ is of the form

$$l^i \equiv \sum_{j=1}^{n} \sum_{k=1}^{2} a_{jk}^i u_{jx_k} + f^i = 0, \qquad i = 1, 2, \ldots, n. \tag{2.1}$$

We consider (2.1) in the domain $\mathfrak{D}$ of the x_1, x_2-plane and suppose $a_{jk}^i \varepsilon C^0$ in $\mathfrak{D}$.

(a) The system (2.1) is called a *quasi-linear system* if a_{jk}^i, f^i are functions of x and $u = (u_1(x), u_2(x), \ldots, u_n(x))$.

(b) It is called an *almost-linear system* if the a_{jk}^i are functions depending only on x and the f^i are functions as in (a).

(c) It is called a *linear system* if $f^i = \sum_{j=1}^{n} b_j^i u_j + c^i$ and if the a_{jk}^i, b_j^i, c^i are functions of x only.

Let $v(x) \varepsilon C^1$ in $\mathfrak{D}$ and let α be a vector field there with $|\alpha| > 0$ and with components

$$\alpha_k(x) \varepsilon C^0, \qquad k = 1, 2.$$

Then as in (1.14)

$$v_\alpha = \sum_{k=1}^{2} \alpha_k(x) v_{x_k} \tag{2.2}$$

is called the *directional derivative* of $v(x)$ in direction of the vector α. We see that in general (2.1) contains n^2 such directional derivatives.

If $\lambda_i(x, u)$ are suitable functions, yet to be determined, then by a suitable linear combination of (2.1),

$$\sum_{i=1}^{n} \lambda_i l^i \equiv \sum_{i,j=1}^{n} \lambda_i \sum_{k=1}^{2} a_{jk}^i u_{jx_k} + \sum_{i=1}^{n} \lambda_i f^i = 0, \tag{2.3}$$

we try to find a system of equations, equivalent to (2.1), which contains exactly one directional derivative $u_\tau = \sum_{k=1}^{2} \overset{\tau}{\alpha}_k u_{x_k}$, $\tau = 1, \ldots, n$ in every equation where the $\overset{\tau}{\alpha}$ are vectors which are real and pairwise linearly independent at every point $x \varepsilon \mathfrak{D}$. (See R. Courant and K. O. Friedrichs[4], I. G. Petrovski[5], W. Haack and G. Hellwig[6], and G. Hellwig[7]).

To construct the desired system we obviously need n linearly independent vectors λ^r with components λ_i^r. If we imagine these substituted in (2.3), then according to the foregoing requirements the resulting system should have the form

$$l^r \equiv \sum_{i=1}^{n} \lambda_i^r l^i = \sum_{j=1}^{n} g_j^r u_{j\alpha}^r + \sum_{j=1}^{n} \lambda_j^r f^j = 0, \qquad r = 1, \ldots, n. \tag{2.4}$$

Setting (2.3) and (2.4) equal to each other gives

$$\sum_{i=1}^{n} \lambda_i^r a_{j1}^i = g_j^r \overset{r}{\alpha}_1, \qquad \sum_{i=1}^{n} \lambda_i^r a_{j2}^i = g_j^r \overset{r}{\alpha}_2. \tag{2.5}$$

If we suppose $\overset{r}{\alpha}_1 \neq 0$ in $\mathfrak{D}$, which can be achieved by a rotation of the coordinate system and by a restriction of the domain $\mathfrak{D}$, then, with $\zeta^r = \overset{r}{\alpha}_2 / \overset{r}{\alpha}_1$, from (2.5) follows

$$\sum_{i=1}^{n} \lambda_i^r (a_{j1}^i \zeta^r - a_{j2}^i) = 0. \tag{2.6}$$

If we require that the nth-degree algebraic equation[8] in ζ,

$$|a_{j1}^i \zeta - a_{j2}^i| = 0, \tag{2.7}$$

possesses n distinct real roots ζ^r in $\mathfrak{D}$, then we are given n field vectors $\overset{r}{\alpha}$. Based on the algebraic equation (2.7), a *classification into types* becomes possible for the general system (2.1). In the quasi-linear case we again have to suppose that a fixed system of solutions $u = (u_1(x), \ldots, u_n(x))$ is known and has been substituted so that in $\mathfrak{D}$ the a_{jk}^i are functions of x only.

Classification into types. At the fixed point $x \in \mathfrak{D}$, (2.1) is said to be

of *elliptic type*, if (2.7) possesses no real root ζ,
of *hyperbolic type*, if (2.7) possesses precisely n distinct real roots ζ,
of *parabolic type*, if (2.7) possesses precisely ν distinct real roots ζ,

with $1 \leqslant \nu \leqslant n - 1$.

Equation (2.1) is said to be of hyperbolic type in $\mathfrak{D}$ if it has this property at every $x \in \mathfrak{D}$, and so forth. From now on we assume (2.1) to be of hyperbolic type in $\mathfrak{D}$. By means of the pairwise linearly independent vectors $\overset{r}{\alpha}$ we find n linearly independent vectors λ^r from (2.6). Thus such systems can always be assumed in the form (2.4), which we call the *invariant form* of (2.1). In the quasi-linear case, g_j^r and λ_j^r turn out to be functions of x and u. Furthermore, the coefficients in (2.4) have the same differentiability properties in $\mathfrak{D}$ as those in (2.1). Besides, the determinant $|g_j^r| \neq 0$, which follows from the requirements about the vectors $\overset{r}{\alpha}$. We can also omit the proof of this and include $|g_j^r| \neq 0$ in the requirements for the definition of hyperbolic type.

---

[8] The left-hand side of Equation (2.7) indicates the determinant where the term in the ith row and jth column is displayed.

2.2   Characteristic Manifolds; Normal Form

By *characteristic manifolds* $\dot{C}$ of (2.1) we understand the families of curves in $\mathfrak{D}$ which are given as solutions of

$$\dot{x}_1 = \overset{\tau}{\alpha}_1, \qquad \dot{x}_2 = \overset{\tau}{\alpha}_2; \qquad \tau = 1, 2, \ldots, n, \qquad \dot{x}_i = \frac{dx_i}{dt}; \tag{2.8}$$

they can be described by $\varphi^\tau(x_1,x_2) = 0$, where $(\varphi^\tau_{x_1})^2 + (\varphi^\tau_{x_2})^2 > 0$. We assume $\overset{\tau}{\alpha}_k \ \varepsilon \ C^1$ in $\mathfrak{D}$, for which $a^i_{j_k} \ \varepsilon \ C^1$ in $\mathfrak{D}$ is sufficient. In the hyperbolic case they form an *n-web* in the sense of Blaschke-Bol[8] which is called the *characteristic n-web*.[9] Furthermore we may assume $(\overset{\tau}{\alpha}_1)^2 + (\overset{\tau}{\alpha}_2)^2 = 1$, which fixes the differentiation parameter in (2.8) as the arc length s. More precisely, $\underset{\tau}{s}$ is the arc length on the τth *characteristic family*.

Thus for the characteristic manifolds $\dot{C}$, $\varphi^\tau_{x_1} dx_1 + \varphi^\tau_{x_2} dx_2 = 0$ holds. Using (2.8) we find for ζ^τ the representation $\zeta^\tau = -\varphi^\tau_{x_1}/\varphi^\tau_{x_2}$. If we substitute this representation in (2.7), we are led to the first-order partial differential equation

$$\left| \sum_{k=1}^{2} a^i_{j_k} \varphi_{x_k} \right| = 0 \tag{2.8a}$$

which has the solutions $\varphi(x) = \varphi^\tau(x), \tau = 1, \ldots, n$. If, conversely, we look for solutions of (2.8a) and if in R_2 we define point sets by

$$\varphi(x) = 0 \quad \text{where} \quad \sum_{k=1}^{2} (\varphi_{x_k})^2 > 0$$

and close them by adding their accumulation points, we obtain the characteristic manifolds $\dot{C}$. Equation (2.8a) is the appropriate extension of the equation (1.5) to our systems.

Often it is necessary to adjust the differentiation parameter in (2.8) more closely to the given circumstances. This can be best achieved by means of suitable *linear forms*.

To a pair of vectors $\overset{\tau}{\alpha}$, say $\overset{\rho}{\alpha}, \overset{\sigma}{\alpha}$, there is a pair of *contragradient vectors* $\underset{\rho}{\alpha}, \underset{\sigma}{\alpha}$ with components $\underset{\rho}{\alpha}_k, \underset{\sigma}{\alpha}_k$ which are uniquely determined by the requirements

$$\sum_{k=1}^{2} \overset{\rho}{\alpha}_k \underset{\rho}{\alpha}_k = 1, \qquad \sum_{k=1}^{2} \overset{\sigma}{\alpha}_k \underset{\sigma}{\alpha}_k = 1,$$
$$\sum_{k=1}^{2} \overset{\sigma}{\alpha}_k \underset{\rho}{\alpha}_k = 0, \qquad \sum_{k=1}^{2} \overset{\rho}{\alpha}_k \underset{\sigma}{\alpha}_k = 0; \tag{2.9}$$

they generate the *linear forms*

$$\underset{\rho}{\omega} = \sum_{k=1}^{2} \underset{\rho}{\alpha}_k \, dx_k, \qquad \underset{\sigma}{\omega} = \sum_{k=1}^{2} \underset{\sigma}{\alpha}_k \, dx_k. \tag{2.10}$$

[9] Note that in $\varphi^\tau(x_1,x_2) = 0$ there is contained an arbitrary constant. We define: The n families of curves $\varphi^\tau(x_1,x_2) = 0, \tau = 1, \ldots, n$ are called an n-web in $\mathfrak{D}$ if $\varphi^\tau \ \varepsilon \ C^1$ for $x \ \varepsilon \ \mathfrak{D}$ and if none of the functional determinants $\partial(\varphi^\rho,\varphi^\sigma)/\partial(x_1,x_2)$ with $\rho, \sigma = 1, \ldots, n, \rho \neq \sigma$ has a zero in $\mathfrak{D}$. The geometric meaning of the nonvanishing of the functional determinants is that the curves of the ρth and of the σth family have different tangents at an arbitrary point $x \ \varepsilon \ \mathfrak{D}$.

In particular on the ρth *characteristic family of curves* we have, from (2.8),

$$\underset{\rho}{\omega} = \sum_{k=1}^{2} \underset{\rho}{\alpha_k} \frac{dx_k}{\underset{\rho}{ds}} \underset{\rho}{ds} = \sum_{k=1}^{2} \underset{\rho}{\alpha_k} \overset{\rho}{\alpha_k} \underset{\rho}{ds} = \underset{\rho}{ds}, \tag{2.10a}$$

and thus $\int_{\rho} \underset{\rho}{\omega} = s$, while, on the other hand, on the ρth characteristic family we have $\underset{\sigma}{\omega} = 0$. Correspondingly, on the σth characteristic family we find $\int_{\sigma} \underset{\sigma}{\omega} = s, \underset{\rho}{\omega} = 0$.

In the *almost-linear hyperbolic case* (2.1), the invariant form (2.4) of (2.1) can be simplified considerably. If we introduce new unknown functions by

$$U^\tau = \sum_{j=1}^{n} g_j^\tau u_j \quad \text{where} \quad u_j = \sum_{l=1}^{n} h_{jl} U^l \quad \text{and} \quad \sum_{j=1}^{n} g_j^\tau h_{jl} = \delta_l^\tau = \begin{cases} 1, & \tau = l, \\ 0, & \tau \neq l, \end{cases} \tag{2.11}$$

then, because

$$\sum_{j=1}^{n} g_j^\tau \underset{\alpha}{u_j^\tau} = \underset{\tau}{U_\tau^\tau} - \sum_{j=1}^{n} (g_j^\tau)_\tau \underset{\alpha}{u_j}, \tag{2.12}$$

(2.4) appears in the form

$$\underset{\alpha}{U_\tau^\tau} + F^\tau(x,U) = 0, \qquad \tau = 1, \ldots, n. \tag{2.13}$$

Equation (2.13) is called the *normal form* of (2.1). Here $F^\tau \in C^0$ in $\mathfrak{D}$ if $a_{jk}^i \in C^1$ and $f^i \in C^0$ for $x \in \mathfrak{D}$ and $-\infty < u_j < \infty$.

The normal form plays a distinguished role, since (1) it makes it possible to pass to the *Volterra integral equations* immediately, and (2) it offers a suitable difference method for *numerical calculation* of the solutions for various initial and boundary conditions.

2.3   Normal Form for the Quasi-Linear Case

In I-1.6 it has been shown that the equation of gas dynamics can be transformed into a quasi-linear second-order partial differential equation. If we consider only rotation-symmetric flows, they are described by (I-1.33). This equation can be transformed into a quasi-linear system, as we shall see presently. For this reason it would be desirable to obtain a normal form for the quasi-linear case. However, even in the case of two independent variables, these efforts can meet with only partial success.

To begin with, in the quasi-linear hyperbolic system (2.1) the invariant form (2.4) can always be obtained. However, the normal form (2.13) cannot be obtained immediately, since the transformation (2.12) does not remain valid because

$$g_j^\tau = g_j^\tau(x,u).$$

Instead of (2.11) we introduce the new unknown functions by[10]

$$U^\tau = \Phi^\tau(x,u) = \Phi^\tau(x_1, x_2; u_1, u_2, \ldots, u_n). \tag{2.14}$$

If we multiply (2.4) by the nonvanishing factors $M^\tau(x,u)$, we obtain

$$M^\tau l^\tau \equiv \sum_{j=1}^{n} M^\tau g_j^\tau \underset{\alpha}{u_{j\tau}} + \sum_{j=1}^{n} M^\tau \lambda_j^\tau f^j = 0. \tag{2.15}$$

[10] See the remark of J. Nitsche in W. Haack and G. Hellwig[9].

We make the tentative requirement that (2.15) can be put into the form (2.13). First, using (2.14) we have

$$U^\tau_\tau = \sum_{\substack{j=1 \\ \alpha}}^{n} \Phi^\tau_{u_j} u_{j\tau} + \sum_{k=1}^{2} \Phi^\tau_{z_k} \alpha_k. \tag{2.16}$$

Consequently this requirement implies

$$M^\tau l^\tau = \sum_{\substack{j=1 \\ \alpha}}^{n} \Phi^\tau_{u_j} u_{j\tau} + \sum_{k=1}^{2} \Phi^\tau_{z_k} \alpha_k + F^\tau, \tag{2.17}$$

where the two last terms may be combined to one $\tilde{F}^\tau$. Comparison with (2.15) shows that our requirement is equivalent to the validity of the relations

$$M^\tau g^\tau_j = \Phi^\tau_{u_j}. \tag{2.18}$$

This means that the linear forms $\theta^\tau = \sum_{j=1}^{n} g^\tau_j \, du_j$ can be made integrable in the space of the u_j by means of the multipliers M^τ; that is, that they can be transformed into total differentials. (The x appearing in g^τ_j, M^τ merely play the role of parameters.) As is well known, for this it is necessary and sufficient that the commutation relations

$$\Phi^\tau_{u_j u_i} = (M^\tau g^\tau_j)_{u_i} = (M^\tau g^\tau_i)_{u_j} = \Phi^\tau_{u_i u_j}, \qquad i, j = 1, \ldots, n, \tag{2.19}$$

hold. These are $\binom{n}{2}$ conditions for every M^τ. But now every M^τ can be determined so that it satisfies only one such condition, that is, $1 = \binom{n}{2}$, hence $n = 2$. Therefore, in the quasi-linear case the normal form (2.13) can in general be achieved for $n = 2$ only. Of course, the vector fields $\overset{\tau}{a}$ are functions of the arguments x and U.

Example. The Rotation-Symmetric, Stationary, Vortex-Free Flow of an Ideal Gas. According to I-1.6, the velocity potential $u = u(x_1, r)$, which had been introduced in the velocity vector v of the flow by $v = \text{grad } u$, satisfies equation (I-1.33):

$$[a^2 - (u_x)^2]u_{xx} + [a^2 - (u_r)^2]u_{rr} - 2u_x u_r u_{xr} + \frac{a^2}{r} u_r = 0. \tag{2.20}$$

Here we wrote x instead of x_1. As we see from Section 1.1, this is a second-order quasi-linear partial differential equation which is hyperbolic for $|v| > a$. The function $a = a(|v|)$ represents the local sound velocity where

$$|v| = |\text{grad } u| = \sqrt{(u_x)^2 + (u_r)^2}.$$

By putting $u_x = u_1$, $u_r = u_2$ we obtain the quasi-linear system

$$[a^2 - (u_1)^2]u_{1x} + [a^2 - (u_2)^2]u_{2r} - u_1 u_2(u_{1r} + u_{2x}) + \frac{a^2}{r} u_2 = 0,$$
$$u_{1r} - u_{2x} = 0, \tag{2.21}$$

which in the hyperbolic case is to be transformed into the normal form. An immediate application of the theory is not advisable. Moreover, from the physical point of view it is useful first to rewrite (2.21) with physically distinguished directional derivatives. In polar coordinates we have

$$u_1 = u_x = |v| \cos \vartheta, \qquad u_2 = u_r = |v| \sin \vartheta. \tag{2.22}$$

Then

$$\begin{aligned}
\beta &= (|v| \cos \vartheta, |v| \sin \vartheta), \\
\bar{\beta} &= (-|v| \tan \gamma \sin \vartheta, |v| \tan \gamma \cos \vartheta)
\end{aligned} \tag{2.23}$$

are two vector fields, of which β coincides with the velocity vector v and $\bar{\beta}$ is orthogonal to it (Figure 11). Here γ is the Mach angle with $\sin \gamma = a/|v|$, and we must now assume

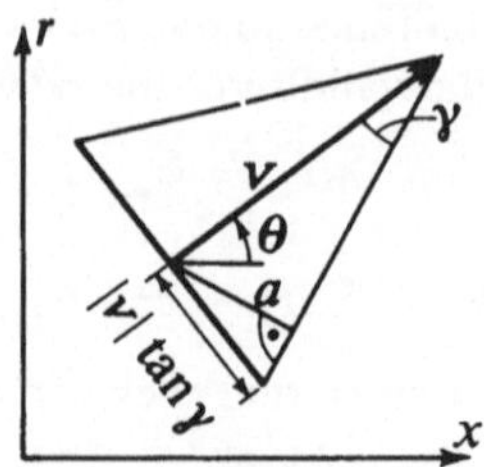

Figure 11

$a < |v|$. Using $|v|^2 = (u_1)^2 + (u_2)^2$ we then have

$$\begin{aligned}
|v||v|_\beta &= u_1 u_{1\beta} + u_2 u_{2\beta} = u_1(u_1 u_{1x} + u_2 u_{1r}) + u_2(u_1 u_{2x} + u_2 u_{2r}) \\
&= (u_1)^2 u_{1x} + (u_2)^2 u_{2r} + u_1 u_2(u_{1r} + u_{2x}), \\
u_{1x} + u_{2r} &= |v|_x \cos \vartheta - |v| \sin \vartheta \vartheta_x + |v|_r \sin \vartheta + |v| \cos \vartheta \vartheta_r
\end{aligned} \tag{2.24}$$

$$= \frac{1}{|v|}|v|_\beta + \frac{1}{\tan \gamma} \vartheta_{\bar{\beta}}. \tag{2.25}$$

Thus, after division by a^2 and use of (2.24) and (2.25), for the first equation in (2.21) we obtain

$$\frac{1}{|v|}|v|_\beta + \frac{1}{\tan \gamma} \vartheta_{\bar{\beta}} - \frac{|v||v|_\beta}{a^2} + \frac{u_2}{r} = 0. \tag{2.26}$$

Taking into account $1 - |v|^2/a^2 = -\cot^2 \gamma$ and (2.22), (2.26) immediately becomes

$$\cot \gamma \cdot |v|_\beta - |v|\vartheta_{\bar{\beta}} = \frac{|v|^2 \sin \vartheta \tan \gamma}{r}. \tag{2.27}$$

An analogous rewriting of the second equation in (2.21) gives

$$\cot \gamma \cdot |v|_{\bar{\beta}} - |v|\vartheta_\beta = 0. \tag{2.28}$$

The system (2.27), (2.28) is equivalent to (2.21).

As a side remark, let us state that the *natural equations of gas dynamics* can be obtained in this way without difficulty. If we denote the arc length by s, and the radius of curvature of the stream lines (curves which fit the vector field v) by ρ_s, then using (2.23) we obtain

$$\vartheta_\beta = |v| \frac{d\vartheta}{ds} = \frac{1}{\rho_s}|v| \qquad \left(\text{because } \frac{d\vartheta}{ds} = \frac{1}{\rho_s} \right).$$

If we now denote the arc length and the radius of curvature for the orthogonal trajectories of the stream lines by n and ρ_n, respectively, then from (2.23) we obtain

$$\vartheta_{\bar{\beta}} = |v| \, \tan \gamma \, \frac{d\vartheta}{dn} = \frac{|v| \, \tan \gamma}{\rho_n}.$$

After substitution in (2.27) and (2.28) we obtain the natural equations

$$\frac{d \log |v|}{ds} = \tan^2 \gamma \left(\frac{1}{\rho_n} + \frac{\sin \vartheta}{r} \right), \qquad \frac{d \log |v|}{dn} = \frac{1}{\rho_s}.$$

If for an arbitrary function $S(x,r) \, \varepsilon \, C^1$ we put $S_\alpha = S_\beta + S_{\bar{\beta}}$, $S_{\bar{\alpha}} = S_\beta - S_{\bar{\beta}}$, where

$$\alpha = \left(\frac{|v|}{\cos \gamma} \cos (\gamma + \vartheta), \frac{|v|}{\cos \gamma} \sin (\gamma + \vartheta) \right),$$

$$\bar{\alpha} = \left(\frac{|v|}{\cos \gamma} \cos (\vartheta - \gamma), \frac{|v|}{\cos \gamma} \sin (\vartheta - \gamma) \right),$$

$$\tag{2.29}$$

then from (2.27) and (2.28) we obtain the following system, equivalent to (2.21):

$$\frac{\cot \gamma}{|v|} |v|_\alpha - \vartheta_\alpha = \frac{|v|}{r} \sin \vartheta \, \tan \gamma, \qquad \frac{\cot \gamma}{|v|} |v|_{\bar{\alpha}} + \vartheta_{\bar{\alpha}} = \frac{|v|}{r} \sin \vartheta \, \tan \gamma. \tag{2.30}$$

According to Section 2.2, this is the invariant form of the system (2.21) and, therefore, (2.29) determines the characteristic 2-web.

Furthermore, because $a = a(|v|)$, the expression $\cot \gamma / |v| = f(|v|)$ is a function of $|v|$ only. If we put

$$U^1 = \int_{a(|v|)}^{|v|} f(\omega) \, d\omega - \vartheta, \qquad U^2 = \int_{a(|v|)}^{|v|} f(\omega) \, d\omega + \vartheta, \tag{2.31}$$

then, because $f(a) = 0$, (2.30) becomes

$$U^1_\alpha = \frac{|v|}{r} \sin \vartheta \, \tan \gamma, \qquad U^2_{\bar{\alpha}} = \frac{|v|}{r} \sin \vartheta \, \tan \gamma. \tag{2.32}$$

This is the desired normal form of the system (2.21). At the same time it has been proved (again) that (2.21) is hyperbolic for $a < |v|$. The presentation here is based on that in W. Haack and G. Hellwig[9]. A detailed account of the connections between gas dynamics and partial differential equations can be found in R. Sauer[10].

2.4 Theory of Characteristics for General Systems

A greater generality will make the connections between Sections 1 and 2 clearer. To begin with, we consider the most general quasi-linear system of m first-order partial differential equations in n independent variables $x = (x_1, x_2, \ldots, x_n)$ and m unknown functions $u(x) = (u_1(x), \ldots, u_m(x))$:

$$l^i \equiv \sum_{j=1}^{m} \sum_{k=1}^{n} a^i_{jk} u_{jx_k} + f^i = 0, \qquad i = 1, 2, \ldots, m, \tag{2.33}$$

where the a^i_{jk} and f^i are functions of x and u. We imagine a known system of solutions substituted for u so that under this substitution the a^i_{jk} and f^i are functions of x alone.

Characteristic manifolds $\dot{C}$ will be of the form (1.6).   As in Section 2.1, we use $\lambda_i(x,u)$, which are yet to be determined, and form

$$\sum_{i=1}^{m} \lambda_i l^i \equiv \sum_{i,j=1}^{m} \lambda_i \sum_{k=1}^{n} a_{jk}^i u_{jx_k} + \sum_{i=1}^{m} \lambda_i f^i = 0; \tag{2.34}$$

we hope to obtain a simplification in this way.   What kind of simplification this should be we have to find out from an analysis of the considerations in Sections 2.1 and 2.2.   If we suppose that the characteristic manifolds $\dot{C}$, which are given by (2.8), are described by $\varphi^\tau(x) = 0$, then the *invariant form* (2.4) of (2.1) contains only *inner differentiations* relative to $\dot{C}$.   This means that if $\overset{\tau}{\nu}$ is the normal vector for $\varphi^\tau(x) = 0$, then $(\overset{\tau}{\nu}, \overset{\tau}{\alpha}) = 0, \tau = 1, \ldots, n$. This formulation is suitable for the general case.

In (2.34), $u_j(x)$ is differentiated in the direction α (which is dependent on j) where α is given by

$$\alpha = (\sum_{i=1}^{m} \lambda_i a_{j1}^i, \sum_{i=1}^{m} \lambda_i a_{j2}^i, \ldots, \sum_{i=1}^{m} \lambda_i a_{jm}^i). \tag{2.35}$$

We now consider nonempty point sets in R_n which are described by $\varphi(x) = 0$ where $\varphi(x) \in C^1$ and $\sum_{i=1}^{n} (\varphi_{x_i})^2 > 0$.   If for such a set it is possible to choose the $\lambda_i(x,u)$ so that differentiation in the direction α becomes an inner differentiation relative to this point set, then this set—after addition of its points of accumulation—is called a *characteristic manifold* $\dot{C}$.   Then it must be possible to choose the $\lambda_i = \lambda_i(x,u)$ so that

$$0 = (\nu, \alpha) = \sum_{k=1}^{n} \sum_{i=1}^{m} \lambda_i a_{jk}^i \nu_k, \qquad j = 1, 2, \ldots, m. \tag{2.36}$$

Here

$$\nu = (\nu_1, \nu_2, \ldots, \nu_n) \quad \text{where} \quad \nu_k = \kappa \varphi_{x_k}(x) \quad \text{and} \quad \kappa = \frac{1}{[\sum_{i=1}^{n} (\varphi_{x_i})^2]^{\frac{1}{2}}} \tag{2.37}$$

is the normal vector to $\dot{C}$.   Relation (2.36) gives m homogeneous equations for the unknown λ_i not all of which can be zero, so that in accordance with (2.37) it has to be required that the determinant

$$\left| \sum_{k=1}^{n} a_{jk}^i \varphi_{x_k} \right| = 0. \tag{2.38}$$

This is the first-order partial differential equation for the characteristic manifolds $\dot{C}$; it is the correct generalization of (2.8a) and (1.5).   If there are m such $\varphi(x)$, namely $\varphi^\tau(x)$ with $\tau = 1, \ldots, m$ so that the corresponding $\lambda^\tau(x,u)$ form m linearly independent m-vectors with components λ_i^τ, then our system (2.33) is called *hyperbolic*, and the *invariant form*

$$l^\tau \equiv \sum_{i=1}^{m} \lambda_i^\tau l^i, \qquad \tau = 1, \ldots, m, \tag{2.39}$$

can be obtained; it contains only *inner directional derivatives* relative to $\dot{C}$.   For $n > 2$, however, this does not give any essential simplification.   The presentation followed that of R. Courant and K. O. Friedrichs (see Section 2.1).

The algebraic point of view represented in Section 1.1 can be adopted immediately. For this purpose we have to consider the *algebraic form*

$$Q(y) = \left| \sum_{k=1}^{n} a_{jk}^i y_k \right| = 0, \qquad y = (y_1, \ldots, y_n), \tag{2.40}$$

of degree m and to classify it algebraically.

Problem. Determine $\dot{C}$ for the Maxwell equations (I-1.20).

2.5 Classification into Types for Simple Systems

As in Section 1.3, we want to make possible a quick decision about the types for simple systems. We consider a *quasi-linear system* in two unknown functions $u = u(x)$, $v = v(x)$ with $x = (x_1, x_2)$ and write it in the form

$$\begin{aligned}
l &\equiv \sum_{k=1}^{2} \{a_k u_{x_k} + b_k v_{x_k}\} + f = 0, \\
\bar{l} &\equiv \sum_{k=1}^{2} \{\bar{a}_k u_{x_k} + \bar{b}_k v_{x_k}\} + \bar{f} = 0,
\end{aligned} \tag{2.41}$$

where the $a_k, \ldots, f$ are functions of x, u, v. The form (2.40) then becomes a quadratic form. We find

$$Q(y) = \sum_{i,k=1}^{2} \begin{vmatrix} a_i & b_k \\ \bar{a}_i & \bar{b}_k \end{vmatrix} y_i y_k, \tag{2.42}$$

with the discriminant

$$d = AC - \left(\frac{B}{2}\right)^2 \quad \text{where} \quad \begin{aligned} A &= \begin{vmatrix} a_1 & b_1 \\ \bar{a}_1 & \bar{b}_1 \end{vmatrix}, \qquad C = \begin{vmatrix} a_2 & b_2 \\ \bar{a}_2 & \bar{b}_2 \end{vmatrix}, \\ B &= \left\{ \begin{vmatrix} a_1 & b_2 \\ \bar{a}_1 & \bar{b}_2 \end{vmatrix} + \begin{vmatrix} a_2 & b_1 \\ \bar{a}_2 & \bar{b}_1 \end{vmatrix} \right\}. \end{aligned} \tag{2.43}$$

Therefore (2.41) is of *elliptic, hyperbolic*, or *parabolic type* in $\mathfrak{D}$ depending on whether $d > 0$, $d < 0$, or $d = 0$ there. The *characteristic manifolds* $\dot{C}$: $\varphi(x_1, x_2) = 0$ are determined from

$$A(\varphi_{x_1})^2 + B\varphi_{x_1}\varphi_{x_2} + C(\varphi_{x_2})^2 = 0. \tag{2.44}$$

Problem. Show that $u_{x_1} - u_{x_2} - v_{x_2} = 0$, $au_{x_2} - v_{x_1} + v_{x_2} + f(x_1 + x_2) = 0$ is hyperbolic for $a > 0$, parabolic for $a = 0$, elliptic for $a < 0$. Furthermore, determine $\dot{C}$.

2.6 Normal Form for Elliptic Systems[11]

We suppose that (2.41) is almost linear and of *elliptic type* in $\mathfrak{D}$ ($d > 0$). First we bring (2.44) into a somewhat different form. For $\dot{C}$ we have $d\varphi = \varphi_{x_1}\, dx^1 + \varphi_{x_2}\, dx^2 = 0$ or

---

[11] There is a change of notation in this particular section; see the Translator's Note after the Preface.

$dx^2/dx^1 = -\varphi_{x^1}/\varphi_{x^2}$. If we take this into consideration in (2.44), we obtain the *definite form*

$$Q \equiv C(dx^1)^2 - B\,dx^1\,dx^2 + A(dx^2)^2 \tag{2.45}$$

in $\mathfrak{D}$. If for abbreviation we introduce the expressions

$$q = \frac{1}{\sqrt{d}}\begin{vmatrix} a^1 & a^2 \\ \bar{a}^1 & \bar{a}^2 \end{vmatrix}, \qquad p = \frac{1}{2\sqrt{d}}\left\{\begin{vmatrix} a^1 & b^2 \\ \bar{a}^1 & \bar{b}^2 \end{vmatrix} - \begin{vmatrix} a^2 & b^1 \\ \bar{a}^2 & \bar{b}^1 \end{vmatrix}\right\}, \qquad r = \frac{1}{\sqrt{d}}\begin{vmatrix} b^1 & b^2 \\ \bar{b}^1 & \bar{b}^2 \end{vmatrix}, \tag{2.46}$$

then, because of $d > 0$ and (2.43), $1 = rq - p^2$ holds. Hence $q \neq 0$, $r \neq 0$ in $\mathfrak{D}$. We may suppose Q in (2.45) to be always positive definite and thus interpret Q/d as a quadratic fundamental form (metric) of a Riemannian geometry in the x^1, x^2-plane. Writing

$$g_{11} = \frac{C}{d}, \qquad g_{12} = g_{21} = -\frac{B}{2d}, \qquad g_{22} = \frac{A}{d}, \tag{2.47}$$

we obtain the element of arc length $ds = \sqrt{g_{ik}\,dx^i\,dx^k}$. Here we have to take the sum from 1 to 2 over those terms in which upper and lower indices are the same.[12] We now introduce two fields of unit vectors α, $\bar{\alpha}$ with components α^i, $\bar{\alpha}^i$ which are orthogonal relative to our metric, and which are determined by

$$g_{ik}\alpha^i\alpha^k = 1, \qquad g_{ik}\alpha^i\bar{\alpha}^k = 0, \qquad g_{ik}\bar{\alpha}^i\bar{\alpha}^k = 1. \tag{2.48}$$

Because of the connection with the original system (2.41) arising from (2.47), these and the *linear forms*

$$\omega_1 = \alpha_i\,dx^i, \qquad \omega_2 = \bar{\alpha}_i\,dx^i \quad \text{where} \quad \alpha_i = g_{ik}\alpha^k, \qquad \bar{\alpha}_i = g_{ik}\bar{\alpha}^k \tag{2.49}$$

are of particular significance in the elliptic case. Then from

$$g_{ik}g^{kl} = \delta_i^l = \begin{cases} 1 & \text{for} \quad i = l, \\ 0 & \text{for} \quad i \neq l, \end{cases} \tag{2.50}$$

we have the relations

$$g^{ik}\alpha_i\alpha_k = 1, \qquad g^{ik}\alpha_i\bar{\alpha}_k = 0, \qquad g^{ik}\bar{\alpha}_i\bar{\alpha}_k = 1, \qquad \begin{vmatrix} \alpha_1 & \alpha_2 \\ \bar{\alpha}_1 & \bar{\alpha}_2 \end{vmatrix} = \frac{1}{\sqrt{d}}. \tag{2.51}$$

If $S(x^1,x^2) \in C^1$ in $\mathfrak{D}$ we define $S_\alpha = \sum_{i=1}^{2}\alpha^i S_{x^i}$, $S_{\bar{\alpha}} = \sum_{i=1}^{2}\bar{\alpha}^i S_{x^i}$ with the inversion

$$S_{x^i} = \alpha_i S_\alpha + \bar{\alpha}_i S_{\bar{\alpha}}. \tag{2.52}$$

Using this, (2.41) can be rewritten with these derivatives. We find

$$\begin{aligned} l &\equiv a^k(\alpha_k u_\alpha + \bar{\alpha}_k u_{\bar{\alpha}}) + b^k(\alpha_k v_\alpha + \bar{\alpha}_k v_{\bar{\alpha}}) + f = 0, \\ \bar{l} &\equiv \bar{a}^k(\alpha_k u_\alpha + \bar{\alpha}_k u_{\bar{\alpha}}) + \bar{b}^k(\alpha_k v_\alpha + \bar{\alpha}_k v_{\bar{\alpha}}) + \bar{f} = 0. \end{aligned} \tag{2.53}$$

---

[12] This is the Einstein convention. For example, we have with this convention

$$\sum_{i,k=1}^{2} g_{ik}dx^i dx^k \equiv g_{ik}dx^i dx^k, \quad \sum_{k=1}^{2} g_{ik}g^{kl} \equiv g_{ik}g^{kl}, \quad \sum_{k=1}^{2} a^k\alpha_k \equiv a^k\alpha_k.$$

It is our aim to obtain a form of (2.41) which is as simple as possible, by using suitable linear combinations of the foregoing equations. For this purpose we form

$$l = \bar{b}^k \alpha_k l - b^k \alpha_k \bar{l}, \qquad \bar{\bar{l}} = \bar{b}^k \bar{\alpha}_k l - b^k \bar{\alpha}_k \bar{l}. \tag{2.54}$$

Then immediately we find

$$
\begin{aligned}
l &\equiv \begin{vmatrix} a^k \alpha_k & b^k \alpha_k \\ \bar{a}^k \alpha_k & \bar{b}^k \alpha_k \end{vmatrix} u_\alpha + \begin{vmatrix} a^k \bar{\alpha}_k & b^k \alpha_k \\ \bar{a}^k \bar{\alpha}_k & \bar{b}^k \alpha_k \end{vmatrix} u_{\bar\alpha} + \begin{vmatrix} b^k \bar{\alpha}_k & b^k \alpha_k \\ \bar{b}^k \bar{\alpha}_k & \bar{b}^k \alpha_k \end{vmatrix} v_{\bar\alpha} + \tilde{f} = 0, \\[4pt]
\bar{\bar{l}} &\equiv \begin{vmatrix} a^k \alpha_k & b^k \bar{\alpha}_k \\ \bar{a}^k \alpha_k & \bar{b}^k \bar{\alpha}_k \end{vmatrix} u_\alpha + \begin{vmatrix} a^k \bar{\alpha}_k & b^k \bar{\alpha}_k \\ \bar{a}^k \bar{\alpha}_k & \bar{b}^k \bar{\alpha}_k \end{vmatrix} u_{\bar\alpha} + \begin{vmatrix} b^k \alpha_k & b^k \bar{\alpha}_k \\ \bar{b}^k \alpha_k & \bar{b}^k \bar{\alpha}_k \end{vmatrix} v_\alpha + \bar{\tilde{f}} = 0.
\end{aligned}
\tag{2.55}
$$

The determinants appearing here can easily be calculated from the relations mentioned above. We have

$$\begin{vmatrix} a^k \bar{\alpha}_k & b^k \bar{\alpha}_k \\ \bar{a}^k \bar{\alpha}_k & \bar{b}^k \bar{\alpha}_k \end{vmatrix} = \begin{vmatrix} a^k \alpha_k & b^k \alpha_k \\ \bar{a}^k \alpha_k & \bar{b}^k \alpha_k \end{vmatrix} = 1, \tag{2.56}$$

$$\begin{vmatrix} b^k \alpha_k & b^k \bar{\alpha}_k \\ \bar{b}^k \alpha_k & \bar{b}^k \bar{\alpha}_k \end{vmatrix} = - \begin{vmatrix} b^k \bar{\alpha}_k & b^k \alpha_k \\ \bar{b}^k \bar{\alpha}_k & \bar{b}^k \alpha_k \end{vmatrix} = r. \tag{2.57}$$

Calculation of the two remaining determinants requires a little more effort. We have

$$\begin{vmatrix} a^k \alpha_k & b^k \bar{\alpha}_k \\ \bar{a}^k \alpha_k & \bar{b}^k \bar{\alpha}_k \end{vmatrix} = \begin{vmatrix} a^k \alpha_k & b^k \alpha_k \\ \bar{a}^k \alpha_k & \bar{b}^k \bar{\alpha}_k \end{vmatrix} - g^{ik} \alpha_i \bar{\alpha}_k = p, \tag{2.58}$$

$$\begin{vmatrix} a^k \bar{\alpha}_k & b^k \alpha_k \\ \bar{a}^k \bar{\alpha}_k & \bar{b}^k \alpha_k \end{vmatrix} = \begin{vmatrix} a^k \bar{\alpha}_k & b^k \alpha_k \\ \bar{a}^k \bar{\alpha}_k & \bar{b}^k \alpha_k \end{vmatrix} - g^{ik} \alpha_i \bar{\alpha}_k = -p. \tag{2.59}$$

Thus (2.55) changes to the simple form

$$l \equiv u_\alpha - p u_{\bar\alpha} - r v_{\bar\alpha} + \tilde{f} = 0, \qquad \bar{\bar{l}} \equiv p u_\alpha + u_{\bar\alpha} + r v_\alpha + \bar{\tilde{f}} = 0. \tag{2.60}$$

If we introduce the *normal functions* $U = u$ and $V = pu + rv$, we obtain the *normal form*

$$l \equiv U_\alpha - V_{\bar\alpha} + F = 0, \qquad \bar{l} \equiv U_{\bar\alpha} + V_\alpha + \bar{F} = 0. \tag{2.61}$$

If we suppose u and v in (2.41) to be invariant under transformations of the x^i, then the normal functions U and V turn out to be invariants under (1) coordinate transformations, (2) transformations of the isometric vector field α, $\bar{\alpha}$ (they do not depend on it at all), and (3) conformal transformations.[13] With (2.49), we can write $(ds)^2$ in the form

$$(ds)^2 = g_{ik}\, dx^i\, dx^k = (\omega_1)^2 + (\omega_2)^2. \tag{2.62}$$

We may interpret $\mathfrak{D}$, together with the imposed Riemannian metric, as a nonanalytic surface S and, from the consideration in Section 1.5, we suppose that this surface can be mapped conformally onto the ξ^1, ξ^2-plane by means of the one-to-one transformation $\xi^1 = \varphi^1(x^1,x^2)$, $\xi^2 = \varphi^2(x^1,x^2)$. Then, according to (1.44), we have

$$\begin{aligned}
(ds)^2 &= (\omega_1)^2 + (\omega_2)^2 = \lambda(\xi^1,\xi^2)\{(d\xi^1)^2 + (d\xi^2)^2\}; \\
\omega_1 &= \sqrt{\lambda}\, d\xi^1, \qquad \omega_2 = \sqrt{\lambda}\, d\xi^2.
\end{aligned}
\tag{2.63}$$

[13] By conformal transformations we here understand transformations which change (2.41) to $\eta l = 0$, $\mu l = 0$, where $\eta \neq 0$, $\mu \neq 0$.

Using (2.49) and (2.52), we verify immediately that

$$dS = S_{x^1}\, dx^1 + S_{x^2}\, dx^2 = S_\alpha \omega_1 + S_{\bar\alpha} \omega_2 = S_{\xi^1}\, d\xi^1 + S_{\xi^2}\, d\xi^2, \qquad (2.64)$$

from which $S_\alpha = (1/\sqrt{\lambda})S_{\xi^1}$, $S_{\bar\alpha} = (1/\sqrt{\lambda})S_{\xi^2}$ follows. Hence under hypothesis of the possibility of this conformal mapping, the normal form (2.61) can be brought into the *integrable normal form*

$$U_{\xi^1} - V_{\xi^2} + F^1(\xi^1,\, \xi^2,\, U,\, V) = 0, \qquad U_{\xi^2} + V_{\xi^1} + F^2(\xi^1,\, \xi^2,\, U,\, V) = 0. \qquad (2.65)$$

Here the principal part set equal to zero represents the *Cauchy-Riemann differential equations*. The presentation is similar to the one in W. Haack and G. Hellwig[11]. A somewhat less invariant presentation can also be found in W. A. Hurwitz[12]

3

ON THE NECESSITY OF CLASSIFICATION INTO TYPES

In the theory of partial differential equations use is always made of the classification into types. Therefore some remarks should be made about the delicate question whether this classification into types is necessitated by the nature of the mathematical posing of the problem itself or whether it is caused simply by the fact that we have not yet found sufficiently widely applicable methods in this discipline.

The question is purely mathematical. Therefore it would be more correct for a beginning interpretation to demand of the mathematical problems only the validity of the first two requirements in Section I-2.1, because the third requirement originated from the fact that we connected the description of a phenomenon in nature with the partial differential equation. Further, it will be better to investigate this question in systems of first-order partial differential equations, since general differential equations and systems of higher-order partial differential equations can be reduced to the first-order systems.

3.1 The Existence Theorem of Cauchy—S. Kowalewski

The following considerations do not show any *dependence on dimension*, so that we may restrict ourselves to two independent variables x, y.

THEOREM 1.   Let the special quasi-linear system

$$u_{ix} = \sum_{k=1}^{n} F_{ik}(u_1, u_2, \ldots, u_n)u_{ky} \quad \text{with} \quad i = 1, \ldots, n \tag{3.1}$$

be given. Let $\varphi_i(y)$, $i = 1, 2, \ldots, n$, be analytic functions in a neighborhood of $y = y_0$. We set $\varphi_i(y_0) = u_{i0}$. Let the F_{ik} be analytic functions in a neighborhood of the point $u_0 = (u_{10}, \ldots, u_{n0})$. Then (3.1) has exactly one system of solutions $u_i(x,y)$ which is analytic in a neighborhood of $x = 0$, $y = y_0$ and which satisfies the system (3.1) as well as the initial conditions

$$u_i(0,y) = \varphi_i(y). \tag{3.2}$$

Outline of the Proof. We may assume $y_0 = u_{i0} = 0$, for $y - y_0$ may be introduced as the new variable, and $u_i - u_{i0}$ as the new unknown functions. Then

$$\varphi_i(y) = \sum_{\nu=1}^{\infty} a_{i\nu}y^{\nu}, \qquad F_{ik} = \sum_{\nu_1,\dots,\nu_n=0}^{\infty} f_{ik,\nu_1\nu_2\cdots\nu_n}(u_1)^{\nu_1}(u_2)^{\nu_2}\cdots(u_n)^{\nu_n} \tag{3.3}$$

are convergent power series in a suitable neighborhood $|u| \leqslant r$, $|y| \leqslant \rho$. Let the formal solution of the problem be given by

$$u_i(x,y) = \sum_{j,l=0}^{\infty} c_{ijl}x^j y^l. \tag{3.4}$$

We show that the c_{ijl} can be determined uniquely. First we have

$$j!\,l!\,c_{ijl} = \left.\frac{\partial^{(j+l)}u_i(x,y)}{\partial x^j\,\partial y^l}\right|_{x=y=0}. \tag{3.5}$$

By differentiation with respect to y, we obtain from (3.2) the quantities $\partial^l u_i(0,0)/\partial y^l$. The right-hand side in (3.1) is then known for $x = y = 0$, by substituting $u_i(0,0) = \varphi_i(0)$ and $\partial u_k(0,0)/\partial y = d\varphi_k(0)/dy$. Thus $u_{ix}(0,0)$ can be determined from (3.1). Next we differentiate (3.1) with respect to y. Then $\partial^2 u_i(0,0)/\partial x\,\partial y$ can be determined; on the right-hand side there occur only derivatives of u_i with respect to y which are all known for $x = 0$, $y = 0$. Differentiation of (3.1) with respect to x on the right-hand side gives first derivatives of u_i with respect to x and mixed second derivatives which now are also known for $x = 0$, $y = 0$. Thus $u_{ixx}(0,0)$ can be determined. In this way, an analogous continuation of the procedure determines the c_{ijl} (3.5) uniquely. It merely remains to show now that the power series (3.4) converges in a neighborhood of $x = 0$, $y = 0$. We omit the proof of this (it may be found in R. Courant and D. Hilbert[13]), since the theorem is of no use for later purposes in this book. Quite analogously we prove

THEOREM 2.   Let the second-order partial differential equation

$$u_{xx} = F(x, y, u, u_x, u_y, u_{xy}, u_{yy}) \tag{3.6}$$

be given. Let $\varphi(y)$ and $\psi(y)$ be analytic functions in a neighborhood of $y = y_0$. We set

$$\varphi(y_0) = u_0, \qquad \psi(y_0) = p_0, \qquad \varphi'(y_0) = q_0, \qquad \varphi''(y_0) = t_0, \qquad \psi'(y_0) = s_0,$$

where

$$\varphi' \equiv \frac{d\varphi}{dy}.$$

Let $F(x, y, u, p, q, s, t)$ be an analytic function in a neighborhood of $0, y_0, u_0, p_0, q_0, s_0, t_0$. Then (3.6) has exactly one solution $u(x,y)$ which is analytic in a neighborhood of $x = 0$, $y = y_0$ and which satisfies equation (3.6) as well as the initial conditions

$$u(0,y) = \varphi(y), \qquad u_x(0,y) = \psi(y). \tag{3.7}$$

It is more elegant, however, to reduce this theorem to Theorem 1, as is done in the quoted literature.

Both theorems guarantee the validity of the first and second requirements for the problems we have described, and they are—if we further include the independence of the dimension—quite general. Further, we would like to say that they do not make use of classification into types. The last remark, however, is only conditionally correct. Note that (3.1) and (3.6) have been solved for u_{iz} and u_{zz}, respectively. The outline of the proof shows that this is essential. So the question arises whether this is always possible.

For this purpose we consider an *almost-linear partial differential equation* in the variables x_1, x_2:

$$\sum_{i,k=1}^{2} a_{ik}(x_1,x_2)u_{x_i x_k} + f(x_1, x_2, u, u_{x_1}, u_{x_2}) = 0. \tag{3.8}$$

For $a_{11} \neq 0$ it is possible to solve for $u_{x_1 x_1}$. In the case of $a_{11} = 0$ we try a coordinate transformation $\xi_1 = \varphi(x_1,x_2)$, $\xi_2 = \psi(x_1,x_2)$. We obtain

$$\sum_{i,k=1}^{2} \tilde{a}_{ik}(\xi_1,\xi_2)u_{\xi_i \xi_k} + \tilde{f} = 0 \quad \text{where} \quad \tilde{a}_{11} = \sum_{i,k=1}^{2} a_{ik}\varphi_{x_i}\varphi_{x_k}. \tag{3.9}$$

But according to (1.5), $\tilde{a}_{11} = 0$ just determines the *characteristic manifolds* $\dot{C}$ of (3.8). If (3.8) is of elliptic type, such $\dot{C}$ do not exist, and the form (3.6) can be obtained. For the other types this does not hold in general. This remark shows that not even under restriction to analyticity can we completely free ourselves from the classification into types.

3.2 The Example of O. Perron[14].

THEOREM. Given the problem

$$l^1 \equiv u_x - u_y - v_y = 0, \qquad\qquad u(0,y) = 0,$$
$$\qquad\qquad\qquad\qquad\qquad\qquad\qquad\qquad\qquad a = \text{const.} \tag{3.10}$$
$$l^2 \equiv au_y - v_x + v_y + f(x + y) = 0, \qquad v(0,y) = 0,$$

The system (3.10) is hyperbolic for $a > 0$, parabolic for $a = 0$, and elliptic for $a < 0$. Necessary and sufficient for existence of a pair of solutions u, $v \in C^1$ is $f \in C^0$ for $a > 0$, $f \in C^2$ for $a = 0$, and f analytic in $x + y$ for $a < 0$. This pair of solutions is at the same time the only one so that the first and second requirements are satisfied for the problem (3.10).

Proof. The classification into types is clear from (2.5).

(1) If $a > 0$, the necessity of $f \in C^0$ is trivial because of (3.10). We put (3.10) into the normal form (2.13). We shall not proceed systematically with this; it is left as an exercise for the reader. We form $\bar{l}^1 = \sqrt{a}\, l^1 + l^2$, $\bar{l}^2 = -\sqrt{a}\, l^1 + l^2$ and obtain

$$\bar{l}^1 \equiv \sqrt{a}\, u_x + (a - \sqrt{a})u_y - \frac{1}{\sqrt{a}}[\sqrt{a}\, v_x + (a - \sqrt{a})v_y] + f(x + y) = 0,$$
$$\tag{3.11}$$
$$\bar{l}^2 \equiv -\sqrt{a}\, u_x + (a + \sqrt{a})u_y + \frac{1}{\sqrt{a}}[-\sqrt{a}\, v_x + (a + \sqrt{a})v_y] + f(x + y) = 0.$$

Further, the characteristic directional derivatives are given by

$$\alpha = (\sqrt{a},\, a - \sqrt{a}), \qquad \bar{\alpha} = (-\sqrt{a},\, a + \sqrt{a}). \tag{3.12}$$

Using the normal functions $U = u - \dfrac{1}{\sqrt{a}}\,v$, $V = u + \dfrac{1}{\sqrt{a}}\,v$, from (3.11) we obtain the

normal form (2.13)

$$U_\alpha + f(x + y) = 0,$$
$$V_{\bar\alpha} + f(x + y) = 0, \tag{3.13}$$

with the initial conditions $U(0,y) = 0$, $V(0,y) = 0$.   We look for a solution at the arbitrary point ξ, η and choose the characteristic curves through ξ, η, which of course fit the field (3.12):

$$\alpha: x = \sqrt{a}\,t + \xi, \qquad y = (a - \sqrt{a})t + \eta;$$
$$\bar\alpha: x = - \sqrt{a}\,\tau + \xi, \qquad y = (a + \sqrt{a})\tau + \eta. \tag{3.14}$$

Then to P_1, P_2, P in Figure 12 there belong the parameter values $t = -\xi/\sqrt{a}$, $\tau = \xi/\sqrt{a}$, $t = \tau = 0$.   Furthermore, U_α, $V_{\bar\alpha}$ in (3.13) become U_t, V_τ, and integration along α, $\bar\alpha$

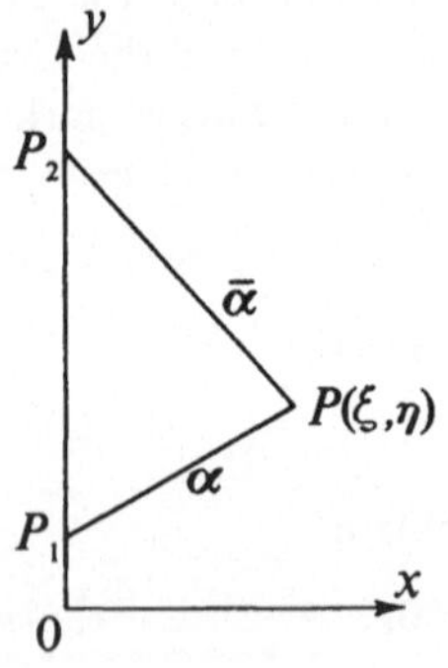

Figure 12

under consideration of the initial conditions gives

$$U(\xi,\eta) = - \int_{-\xi/\sqrt{a}}^{0} f(\sqrt{a}\,t + \xi + (a - \sqrt{a})t + \eta)\,dt = - \frac{1}{a} \int_{\xi+\eta-\sqrt{a}\,\xi}^{\xi+\eta} f(z)\,dz, \tag{3.15}$$

$$V(\xi,\eta) = - \int_{\xi/\sqrt{a}}^{0} f(- \sqrt{a}\,\tau + \xi + (a + \sqrt{a})\tau + \eta)\,d\tau = - \frac{1}{a} \int_{\xi+\eta+\sqrt{a}\,\xi}^{\xi+\eta} f(z)\,dz. \tag{3.16}$$

If, instead of ξ and η, we write x and y, we obtain for u and v:

$$u(x,y) = \frac{1}{2a} \int_{x+y}^{x+y+\sqrt{a}\,x} f(z)\,dz + \frac{1}{2a} \int_{x+y}^{x+y-\sqrt{a}\,x} f(z)\,dz,$$

$$v(x,y) = \frac{1}{2\sqrt{a}} \int_{x+y-\sqrt{a}\,x}^{x+y+\sqrt{a}\,x} f(z)\,dz, \tag{3.17}$$

which completes the proof.

(2) If $a = 0$, we introduce new coordinates $\xi = x$, $\eta = x + y$.   Then (3.10) becomes

$$u_\xi - v_\eta = 0, \qquad v_\xi = f(\eta); \qquad u(0,\eta) = 0, \qquad v(0,\eta) = 0. \tag{3.18}$$

Integration of the last equation gives $v(\xi,\eta) = \xi f(\eta)$.   From the first equation $u(\xi,\eta) =$

$(\xi^2/2)f'(\eta)$ follows and, finally,

$$u(x,y) = \frac{x^2}{2} f'(x+y), \qquad v(x,y) = xf(x+y). \qquad (3.19)$$

From this we can read off the desired result.

(3) The case that $a = -b^2$ with $b > 0$.

(a) The analyticity of $f(x+y)$ is necessary. We introduce new coordinates

$$\xi = bx, \qquad \eta = x+y$$

and new functions

$$U = bu - \frac{1}{b} \int_c^\eta f(t)\, dt, \qquad V = v. \qquad (3.20)$$

After a short elementary calculation we obtain for U and V the problem

$$U_\xi - V_\eta = 0, \qquad U_\eta + V_\xi = 0; \qquad U(0,\eta) = -\frac{1}{b} \int_c^\eta f(t)\, dt, \qquad V(0,\eta) = 0. \qquad (3.21)$$

The system has been transformed into the Cauchy-Riemann differential equations. Thus $F(\zeta) = U + iV$ is an analytic function in $\zeta = \xi + i\eta$ for $\xi \neq 0$ which should tend to

$$-\frac{1}{b} \int_c^\eta f(t)\, dt$$

as $\xi \to 0$. From use of the Schwarz reflection principle it follows that f must necessarily be analytic.

The Schwarz reflection principle reads as follows: If $w = F(\zeta)$ with $\zeta = \xi + i\eta$ is analytic in a domain $\mathfrak{F}$, if $F(\zeta)$ is continuous as ζ approaches an analytic boundary arc $\mathfrak{F}_1$ of $\mathfrak{F}$, and if the image of the arc $\mathfrak{F}_1$, under the mapping $w - F(\zeta)$ is an analytic arc $\mathfrak{F}_1'$ of the w-plane, then $F(\zeta)$ can be continued analytically across $\mathfrak{F}_1$. For the definition of an analytic boundary arc, see Section I-3.1. For $\mathfrak{F}_1$ in the given example we have to use the η-axis. Then $F(\zeta) = U(\xi,\eta) + iV(\xi,\eta)$ turns out to be analytic on the η-axis. Thus $U(0,\eta)$ is analytic in η and this implies the analyticity of $f(t)$ in t. For the exact definition of analyticity in the case of complex functions as well as in the case of real functions, see Section I-3.1.

(b) The analyticity of $f(x+y)$ is sufficient. The unique solution of the system

$$u_{3x} = u_{4y}, \qquad u_{4x} = 0 \quad \text{where} \quad u_3(0,y) = 0, \qquad u_4(0,y) = y \qquad (3.22)$$

is obviously given by $u_3 = x, u_4 = y$. If we put $u = u_1, v = u_2$, then (3.10) can be brought into the form

$$\begin{aligned}
u_{1x} &= u_{1y} + u_{2y}, \qquad u_{2x} = au_{1y} + u_{2y} + f(u_3 + u_4)u_{4y}, \\
u_{3x} &= u_{4y}, \qquad u_{4x} = 0
\end{aligned} \qquad (3.23)$$

where $u_i(0,y) = 0, i = 1, 2, 3, u_4(0,y) = y$. Theorem 1 from Section 3.1 gives the desired result, which of course can also be proved directly. We wanted the reader to become acquainted with the trick (3.22).

This example seems to show that there cannot exist a unified theory unless we are willing to make restrictive assumptions about the coefficients.

Here the following objection can be raised: Although for such systems of hyperbolic and parabolic type the formulation of additional conditions in the form of initial conditions certainly makes sense (this will be shown for hyperbolic systems in Section IV-1), for systems of elliptic and of elliptic-hyperbolic type it will be more reasonable to demand suitable boundary conditions (we shall show this in Section V-4; for equations of mixed type it can be seen from Section III-4). Thus we have to try to find a uniform—but perhaps very abstract—formulation of additional conditions so that these in every case become dependent on the type of the system itself. A very promising start in this direction has been made by K. O. Friedrichs[15].

Problem. Using the trick (3.22), transform the partial differential equation (3.6) with initial conditions (3.7) into the form (3.1) with initial conditions (3.2).

REFERENCES

1. L. Lichtenstein, *Bull. Acad. Cracovie*, 192–217 (1916).

2. F. Tricomi, *Atti Accad. Naz. dei Lincei* [5] **14**, 133 (1923).

3. M. Cibrario, *Atti Accad. Naz. dei Lincei* [6] **15**, 619–625 (1923).

4. R. Courant and K. O. Friedrichs, *Supersonic Flow and Shock Waves* (New York: Interscience Publishers, Inc.), 1948.

5. I. G. Petrovski, *Lectures on Partial Differential Equations* (New York: Interscience Publishers, Inc.), 1954.

6. W. Haack and G. Hellwig, *Math. Z.* **53**, 244–266 (1950).

7. G. Hellwig, *Math. Z.* **68**, 325–337 (1958).

8. W. Blaschke and G. Bol, *Geometrie der Gewebe* (Berlin: Springer Verlag), 1938.

9. W. Haack and G. Hellwig, *Math. Z.* **53**, 340–356 (1950).

10. R. Sauer, *Anfangswertprobleme bei partiellen Differentialgleichungen*, 2nd ed. (Berlin: Springer Verlag), 1958.

11. W. Haack and G. Hellwig, *Math. Nachr.* **4**, 408–418 (1951).

12. W. A. Hurwitz, Dissertation, Göttingen, 1910.

13. R. Courant and D. Hilbert, *Methoden der mathematischen Physik*, Vol. II (Berlin: Springer Verlag), 1937, p. 39 [or R. Courant and D. Hilbert, *Methods of Mathematical Physics*, Vol. II (New York: Interscience Publishers, Inc.), 1962, p. 34—Translator's note].

14. O. Perron, *Math. Z.* **27**, 549–564 (1928).

15. K. O. Friedrichs, *Commun. Pure Appl. Math.* **11**, 333–418 (1958).

Part III. Questions of Uniqueness

It was not without cause that in Section I-2.1 the uniqueness requirement was the first to be made; the problem of uniqueness can be settled without proving existence. This fact is important, since the second requirement is much more profound. Furthermore, the knowledge gained from solving the uniqueness problem often provides the proper point of view for the existence problem.

For settling the first requirement today, in general, we have two methods:

(1) the *maximum-minimum principle* and
(2) the *energy-integral method*.

In principle, both methods are applicable to all types as has been shown in recent researches.

The first method is already familiar to us; the second proceeds as follows: If $Du = 0$ is a partial differential equation with suitable additional conditions $Z_i u = 0$, let $u \equiv 0$ be a solution. To show that there are no other solutions, we start with an expression of the form

$$0 = \int_{\mathfrak{D}} h(x, u, u_{x_1}, \ldots, u_{x_n}, \ldots) \, Du \, dx.$$

By choosing $\mathfrak{D}$ and h suitably and by using—frequently ingenious—integral transformations (for which $Z_i u = 0$ is used) we try to obtain definite expressions. Then it is hoped that the vanishing of these definite expressions will imply $u = 0$. The name "energy-integral method" goes back to the wave equation where the decisive expression is just the total oscillation energy in dx (see Section 3.1).

1

ELLIPTIC AND ELLIPTIC-PARABOLIC TYPE

1.1   The Maximum-Minimum Principle[1]

We consider the linear partial differential equation (II-1.1)

$$Du \equiv Au + au = f \quad \text{where} \quad Au \equiv \sum_{i,k=1}^{n} a_{ik}u_{x_ix_k} + \sum_{i=1}^{n} a_iu_{x_i} \tag{1.1}$$

in the normal domain $\mathfrak{D}$ of R_n and $x = (x_1, x_2, \ldots, x_n)$.   Let

$$a_{ik}(x),\ a_i(x),\ a(x),\ f(x) \ \varepsilon\ C^0 \quad \text{in} \quad \mathfrak{D}.$$

Further let (1.1) be of the elliptic type in $\mathfrak{D}$.   According to Section II-1.1, this fact can be stated as follows: The form

$$\sum_{i,k=1}^{n} a_{ik}(x)y_iy_k \tag{1.2}$$

is positive definite for $x\ \varepsilon\ \mathfrak{D}$.   By solutions $u(x)$ of (1.1) we understand u which satisfy $u\ \varepsilon\ C^0$ in $\mathfrak{D}$, $u\ \varepsilon\ C^2$ in $\mathfrak{D}$, and $Du = f$ in $\mathfrak{D}$.

THEOREM 1.   If $a \leqslant 0$, $f \leqslant 0$ ($f \geqslant 0$) in $\mathfrak{D}$, then every nonconstant solution $u(x)$ attains its negative minimum (its positive maximum)—if it exists—on $\dot{\mathfrak{D}}$ and not in $\mathfrak{D}$.

Note that (1.1) may be of mixed type in $\mathfrak{D}$, namely of elliptic type in $\mathfrak{D}$, and of parabolic or elliptic-parabolic type on $\dot{\mathfrak{D}}$.[2]   For the proof we use E. Hopf's first lemma.

E. HOPF'S FIRST LEMMA.   (1) If $Au \geqslant 0$ in $\mathfrak{D}$ and if x^0 is a point in $\mathfrak{D}$ such that $u(x) \leqslant u(x^0)$ for all $x\ \varepsilon\ \mathfrak{D}$, then $u(x) \equiv u(x^0)$ in $\mathfrak{D}$.   (2) If $Au \leqslant 0$ in $\mathfrak{D}$ and if x^0 is a point in $\mathfrak{D}$ such that $u(x) \geqslant u(x^0)$ for all $x\ \varepsilon\ \mathfrak{D}$, then $u(x) \equiv u(x^0)$ in $\mathfrak{D}$.   We also assume that $u(x)\ \varepsilon\ C^0$ in $\mathfrak{D}$ and $u\ \varepsilon\ C^2$ in $\mathfrak{D}$.

Proof of the Lemma.   It suffices to prove only (1). We put $u(x^0) = M$ and denote by $\mathfrak{M}$ the set of all $x\ \varepsilon\ \mathfrak{D}$ for which $u(x) = M$.   Contrary to the statement we assume that

---

[1] In a form slightly different from the one presented in E. Hopf[1].
[2] These are easy generalizations of the concepts treated in II-1.7.

$u(x) \not\equiv M$ in $\mathfrak{D}$; then because of $u \in C^0$ in $\mathfrak{D}$, $u(x) \not\equiv M$ in $\mathfrak{D}$, and $\mathfrak{M}$ is a proper subset of $\mathfrak{D}$. Then there is an x^* in $\mathfrak{D}$ with $u(x^*) < M$ which has a smaller distance from some points in $\mathfrak{M}$ than from $\dot{\mathfrak{D}}$. This is a consequence of the connectedness of $\mathfrak{D}$. Therefore, because of the continuity of $u(x)$ there is a sphere $\bar{S}$ about x^* which lies entirely in $\mathfrak{D}$ and which contains points of $\mathfrak{M}$ on its boundary $\dot{S}$, and only there. Under a change of notation, if necessary, let one of these points be called x^0.

More precisely, the following reasoning is used: To begin with there is a point $\tilde{x}$ in $\mathfrak{D}$, which does not belong to $\mathfrak{M}$. We join $\tilde{x}$ and x^0 through a continuous curve, the curve being completely contained in $\mathfrak{D}$. This is possible because $\mathfrak{D}$ is a connected set. The points on the curve form a closed bounded set. Hence, there exists $p > 0$ so that every point of the curve has a distance from $\dot{\mathfrak{D}}$ which is $> p$. At the point $\tilde{x}$ we have $u(\tilde{x}) < u(x^0)$ and because of the continuity of $u(x)$ this inequality also holds in a ball with $\tilde{x}$ as center. Its radius is chosen $< \frac{p}{2}$. If we now let the center of this ball move along the curve, the boundary of the ball will contain a point from M at the latest when the point x^0 lies on the boundary of the ball. The center of the first ball, whose boundary contains a point from $\mathfrak{M}$, we call x^*. This is the desired ball.

First Step. We saw that there exists a ball S, for which $\bar{S} \subset \mathfrak{D}$, such that $x^0 \in \dot{S}$ and $u < M$ in S. (For the following, see Figure 13.) Let $S_1 \subset \bar{S}$ be a smaller ball with radius r_1 so that x^0 also lies on $\dot{S}_1$. Then $u < M$ in $\bar{S}_1$, except at the one point x^0 where $u = M$.

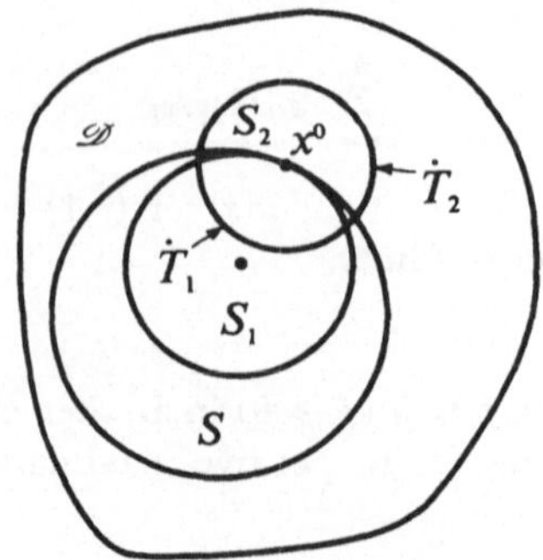

Figure 13

Finally, let $S_2 \subset \mathfrak{D}$ be another ball, with center x^0 and radius $r_2 < r_1$. The boundary $\dot{S}_2$ can be decomposed into $\dot{S}_2 = \dot{T}_1 + \dot{T}_2$ with $\dot{T}_1 = \dot{S}_2 \cap \bar{S}_1$. Thus $\dot{T}_1$ is a closed point set. Then $u < M$ on $\dot{T}_1$ and because $\dot{T}_1$ is closed, $u \leqslant M - \epsilon$ on $\dot{T}_1$ for suitable $\epsilon > 0$, and $u \leqslant M$ on $\dot{T}_2$ by assumption.

Second Step. Place the origin in the center of S_1; a translation does not change the form of Au. We consider the auxiliary function

$$h(x) = e^{-\alpha|x|^2} - e^{-\alpha r_1^2} \quad \text{where} \quad \alpha > 0 \quad \text{and} \quad |x|^2 = (x,x). \tag{1.3}$$

By an elementary calculation we find

$$e^{\alpha|x|^2} Ah = 4\alpha^2 \sum_{i,k=1}^{n} a_{ik}x_i x_k - 2\alpha \sum_{i=1}^{n} (a_{ii} + a_i x_i). \tag{1.4}$$

In any case Au is of the elliptic type in $\bar{S}_2$. Furthermore, since $r_2 < r_1$, the origin is not contained in $\bar{S}_2$. Since $\bar{S}_2$ is closed, we then have

$$\sum_{i,k=1}^{n} a_{ik}x_i x_k \geqslant m > 0 \quad \text{for} \quad x \in \bar{S}_2. \tag{1.5}$$

If we choose $\alpha > 0$ sufficiently large, then from (1.4) it follows that $Ah > 0$ in $\bar{S}_2$.

Third Step. We set $v(x) = u(x) + \delta h(x)$, where $\delta > 0$. If $\delta > 0$ is chosen suitably, then $v < M$ on $\dot{T}_1$ can be achieved according to the first step. But $v < M$ on $\dot{T}_2$, since $h < 0$ by (1.3). Thus $v(x) < M$ is guaranteed on $\dot{S}_2$. But in the center of S_2, $v(x^0) = M$. Therefore $v(x)$ has a maximum in S_2. Suppose it is attained at $x^1 \in S_2$. Then for every vector $(y_1, y_2, \ldots, y_n)$ we have

$$v_{x_i}(x^1) = 0, \qquad i = 1, \ldots, n, \qquad \sum_{i,k=1}^{n} v_{x_i x_k}(x^1)y_i y_k \leqslant 0, \tag{1.6}$$

and because of the ellipticity we have $\sum_{i,k=1}^{n} a_{ik}(x^1)y_i y_k \geqslant 0$. The elementary theorem from the theory of matrices given next therefore yields

$$\sum_{i,k=1}^{n} a_{ik}(x^1)v_{x_i x_k}(x^1) \leqslant 0. \tag{1.7}$$

THEOREM. For every vector $y = (y_1, \ldots, y_n)$, let $\sum_{i,k=1}^{n} \alpha_{ik}y_i y_k \geqslant 0$ where $\alpha_{ik} = \alpha_{ki}$ and $\sum_{i,k=1}^{n} \beta_{ik}y_i y_k \leqslant 0$ where $\beta_{ik} = \beta_{ki}$. Here α_{ik} and β_{ik} are real numbers. Then $\sum_{i,k=1}^{n} \alpha_{ik}\beta_{ik} \leqslant 0$.

Proof. Let A and B be the matrices (α_{ik}) and (β_{ik}). For A there is an orthogonal matrix $C = (\gamma_{ik})$ such that $CAC' = D \equiv (\lambda_i \delta_{ik})$ has diagonal form. The numbers λ_i are the eigenvalues of the matrix, that is, those numbers λ for which there exist vectors $x \neq 0$ satisfying the equation $Ax = \lambda x$. Now

$$\lambda(x,x) = (Ax,x) = \sum_{i,k=1}^{n} \alpha_{ik}x_i x_k \geqslant 0$$

so that $\lambda_i \geqslant 0$. Furthermore, the trace of a product of two matrices is independent of the order of the factors. Therefore

$$\operatorname{tr} AB = \operatorname{tr} CABC' = \operatorname{tr} CAC'CBC' = \operatorname{tr} DCBC' = \sum_{j=1}^{n} \lambda_j \sum_{i,k=1}^{n} \beta_{ik}\gamma_{ji}\gamma_{jk} \leqslant 0$$

where $\lambda_j \geqslant 0$ and $\sum\limits_{i,k=1}^{n} \beta_{ik}\gamma_{ji}\gamma_{jk} \leqslant 0$ for every j.   Since

$$\operatorname{tr} AB = \sum_{i,k=1}^{n} \alpha_{ik}\beta_{ik},$$

everything is proved.

At the point x^1 we then have

$$0 \geqslant Av = Au + \delta Ah. \tag{1.8}$$

From the second step $Ah > 0$ so that $Au < 0$ at x^1.   This is a contradiction to our assumption.   Thus Hopf's lemma is proved.

Proof of Theorem 1.   (See Theorem 1 in I-3.7.)   Contrary to the statement assume that $u(x)$ has a positive maximum at $x^* \, \varepsilon \, \mathfrak{D}$ for which $u(x^*) = M$.   Further we may suppose $u(x) \not\equiv M$ in $\mathfrak{D}$.   We consider the set $\mathfrak{M}$ of all points $x \, \varepsilon \, \mathfrak{D}$ for which $u(x) = M$.

$\mathfrak{M}$ is not empty and it is open relative to $\mathfrak{D}$.   Indeed, if $x^0 \, \varepsilon \, \mathfrak{M}$ then $u(x) \leqslant u(x^0)$ in a suitable ball $\bar{S}$ with center x^0 and $u(x) > 0$ in $\bar{S}$.   In $\bar{S}$ then, from (1.1) we have $Au = -au + f \geqslant 0$.   From E. Hopf's lemma $u(x) \equiv M$ in $\bar{S}$ follows.   But $\mathfrak{M}$ is also closed relative to $\mathfrak{D}$.   This follows as in Section I-3.7.   Thus, because of the connectedness of $\mathfrak{D}$, $\mathfrak{M} \equiv \mathfrak{D}$, and for reasons of continuity $u(x) \equiv M$ in $\bar{\mathfrak{D}}$.   This is a contradiction.

Theorems 2 and 3 are simple corollaries of Theorem 1.

THEOREM 2.   FIRST REQUIREMENT.   The boundary-value problem

$$Du = f \quad \text{in} \quad \mathfrak{D} \quad \text{where} \quad u = \varphi \quad \text{on} \quad \dot{\mathfrak{D}}, \qquad f \, \varepsilon \, C^0 \quad \text{in} \quad \bar{\mathfrak{D}}, \qquad \varphi \, \varepsilon \, C^0 \quad \text{on} \quad \dot{\mathfrak{D}} \tag{1.9}$$

for $a \leqslant 0$ in $\bar{\mathfrak{D}}$ has at most one solution $u(x) \, \varepsilon \, C^0$ in $\bar{\mathfrak{D}}$, $u(x) \, \varepsilon \, C^2$ in $\mathfrak{D}$.

Proof.   If u^1, u^2 are two solutions, then $v = u^1 - u^2$ satisfies

$$Dv = 0 \quad \text{in} \quad \mathfrak{D}, \qquad v = 0 \quad \text{on} \quad \dot{\mathfrak{D}}. \tag{1.10}$$

From Theorem 1, $v \equiv 0$ in $\bar{\mathfrak{D}}$ follows immediately.

Theorems 1 and 2 become false for $a > 0$.

Example 1. $\Delta_2 u + 2u = 0$ with $u = 0$ on the boundary $\dot{\mathfrak{D}}$ of the rectangle $0 \leqslant x_1$, $x_2 \leqslant \pi$ has the solutions $u \equiv 0$ and $u = \sin x_1 \sin x_2$.

Problem.*   Prove that the problem (1.9) has at most one solution $u(x)$ in the case of arbitrary $a(x)$ if there is a solution $w(x) > 0$ of $Dw = 0$ in $\bar{\mathfrak{D}}$.   Notice that here $w(x)$ is subject to no boundary conditions.

THEOREM 3.   THIRD REQUIREMENT.   If $u^i(x)$, $i = 1, 2$ are solutions of

$$Du^i = f \text{ in } \mathfrak{D}, \qquad u^i = \varphi^i(x) \text{ on } \dot{\mathfrak{D}} \text{ where } \max_{x \varepsilon \dot{\mathfrak{D}}} |\varphi^1(x) - \varphi^2(x)| = \epsilon,$$

$$\text{(1.11)}$$

then $|u^1(x) - u^2(x)| \leqslant \epsilon$ in $\mathfrak{D}$.

Proof.   The function $v = u^1 - u^2$ satisfies the problem $Dv = 0$, $v = \varphi^1 - \varphi^2$ on $\dot{\mathfrak{D}}$.   If $v > \epsilon$ in $\mathfrak{D}$ then, because of Theorem 1, the same would be true on $\dot{\mathfrak{D}}$, and so forth.

We formulate E. Hopf's second lemma[2].[3]   We consider equation (1.1) with $a \equiv 0$ and a_{ik}, a_i, $f \varepsilon C^0$ in $\mathfrak{D}$.   Further let (1.1) be either of elliptic type in $\mathfrak{D}$, or elliptic in $\mathfrak{D}$ and parabolic or elliptic-parabolic on $\dot{\mathfrak{D}}$ where those parts of $\dot{\mathfrak{D}}$ on which (1.1) is parabolic must not be characteristic (simple generalization of Section II-1.7).   Finally let $\dot{\mathfrak{D}}$ satisfy the following condition: If $x^0 \varepsilon \dot{\mathfrak{D}}$, then there is a ball S with $S \subset \mathfrak{D}$ and $x^0 \varepsilon \dot{S}$.

LEMMA 2.   Let $u(x) \not\equiv$ const where $u \varepsilon C^1$ in $\mathfrak{D}$, εC^2 in $\mathfrak{D}$, and let u satisfy $Au \geqslant 0$ in $\mathfrak{D}$.   Let $x^0 \varepsilon \dot{\mathfrak{D}}$ be a positive maximum point of $u(x)$.   Then $u_\nu(x^0) > 0$ (u_ν derivative in direction of the outer normal).

The proof is very similar to that of the first lemma.   The lemma becomes false if (1.1) becomes parabolic on parts or all of $\dot{\mathfrak{D}}$ and if at the same time $\dot{\mathfrak{D}}$ is characteristic there.

Example 2. $u_{xx} + (y^2/2)u_{yy} - 0$, $\mathfrak{D}: \pi/4 \leqslant x \leqslant \pi/2$, $0 \leqslant y \leqslant 1$.[4]   Here $y = 0$ is a parabolic curve of type (a); compare with Section II-1.7.   A solution is given by $u = -y^2 \sin x + 1$.   Then $x = 3\pi/8$, $y = 0$ is a positive maximum point but $u_\nu = 0$ there.

The second lemma is needed if we want to prove uniqueness theorems for the second boundary-value problem

$$Au = f \text{ in } \mathfrak{D} \text{ and } u_\nu = \varphi \text{ on } \dot{\mathfrak{D}}. \qquad \text{(1.12)}$$

Things are somewhat different here, since $u \equiv$ const is always a solution of $Au = 0$ where $u_\nu = 0$ on $\dot{\mathfrak{D}}$.

1.2   The Energy-Integral Method

Note that the hypotheses made about (1.1) in Section 1.1 were sufficient for the uniqueness theorem.   Here we shall get acquainted with other sufficient conditions.   In general, necessary and sufficient conditions which could be expressed in a simple way by means of the coefficients are not known.

[3] The above formulation, as well as the other formulations made in Section 1.1 under consideration of differential equations of mixed type, were given by L. Nirenberg in his New York University lecture notes, and by C. Pucci, Rend. Accad. Naz. Lincei, Vol. 23 (1957), Vol. 24 (1958).

[4] Here $\dot{\mathfrak{D}}$, and thus also $\mathfrak{D}$, has to be smoothed by small quarter circles.

THEOREM.   FIRST REQUIREMENT.   Let $a(x)$, $a_{ix_i}(x) \in C^0$ in $\mathfrak{D}$.   Then the boundary-value problem

$$Du \equiv \Delta_n u + \sum_{i=1}^{n} a_i(x) u_{x_i} + a(x) u = f(x) \quad \text{in} \quad \mathfrak{D} \quad \text{where} \quad u = \varphi \quad \text{on} \quad \dot{\mathfrak{D}} \quad (1.13)$$

has at most one solution $u(x) \in C^1$ in $\bar{\mathfrak{D}}$, $\in C^2$ in $\mathfrak{D}$ whenever $2a - \sum_{i=1}^{n} a_{ix_i} \leqslant 0$ in $\mathfrak{D}$.

Proof.   If u^1, u^2 are two solutions, then $v = u^1 - u^2$ satisfies

$$Dv \equiv \Delta_n v + \sum_{i=1}^{n} a_i v_{x_i} + av = 0 \quad \text{with} \quad v = 0 \quad \text{on} \quad \dot{\mathfrak{D}}. \qquad (1.14)$$

Because of the theorem of Gauss—Section I-1.2—and the first Green formula—Section I-1.4—we have

$$\begin{aligned}
\int_{\mathfrak{D}} v\Delta_n v \, dx &= \int_{\dot{\mathfrak{D}}} vv_\nu \, dS - \int_{\mathfrak{D}} |\text{grad } v|^2 \, dx, \\
\int_{\dot{\mathfrak{D}}} v^2 a_i \nu_i \, dS &= 2 \int_{\mathfrak{D}} vv_{x_i} a_i \, dx + \int_{\mathfrak{D}} v^2 a_{ix_i} \, dx,
\end{aligned} \qquad (1.15)$$

and, since $v = 0$ on $\dot{\mathfrak{D}}$, we obtain

$$0 = \int_{\mathfrak{D}} v \, Dv \, dx = \int_{\mathfrak{D}} \left\{ \left(a - \frac{1}{2} \sum_{i=1}^{n} a_{ix_i} \right) v^2 - |\text{grad } v|^2 \right\} dx \leqslant - \int_{\mathfrak{D}} |\text{grad } v|^2 \, dx. \quad (1.16)$$

From this, grad $v \equiv 0$ follows, which is equivalent to $v \equiv$ const in $\mathfrak{D}$.   But $v = 0$ on $\dot{\mathfrak{D}}$ so that $v \equiv 0$.

1.3   Treatment of Existence Problems by Means of the Maximum-Minimum Principle

We outline the proof to show how the maximum-minimum principle can be used to settle the second requirement.   Make the hypotheses of Section 1.1 and consider the *boundary-value problem*

$$Du \equiv Au + au = 0 \quad \text{where} \quad u = \varphi(x) \quad \text{on} \quad \dot{\mathfrak{D}} \quad \text{and} \quad a \leqslant 0 \quad \text{in} \quad \mathfrak{D}. \qquad (1.17)$$

Here it is necessary to assume that Au is of elliptic type in $\mathfrak{D}$.

For the moment let us suppose that the existence problem (1.17) has been settled and that $u(x) \in C^0$ in $\bar{\mathfrak{D}}$, $\in C^2$ in $\mathfrak{D}$ is the solution.   It is possible to construct functions $v^j(x)$ so that $Dv^j \geqslant 0$ in $\mathfrak{D}$ and $v^j \leqslant \varphi$ on $\dot{\mathfrak{D}}$.   We do this in the next section.   Then obviously $D(u - v^j) \leqslant 0$ in $\mathfrak{D}$ and $u - v^j \geqslant 0$ on $\dot{\mathfrak{D}}$.   The function $w = u - v^j$ satisfies the hypotheses for Theorem 1 of Section 1.1 for the negative situation and hence would have its negative minimum on $\dot{\mathfrak{D}}$.   But clearly a negative minimum is impossible, and therefore $w = u - v^j \geqslant 0$ in $\mathfrak{D}$.   Similarly we construct functions $V^k(x)$ for which $DV^k \leqslant 0$ in $\mathfrak{D}$ and $V^k \geqslant \varphi$ on $\dot{\mathfrak{D}}$.   Then $D(u - V^k) \geqslant 0$ and $u - V^k \leqslant 0$ on $\dot{\mathfrak{D}}$.   The function $W = u - V^k$ satisfies the hypotheses of Theorem 1 of Section 1.1 for the positive maximum.   This would have to lie on $\dot{\mathfrak{D}}$, which is impossible.   Hence $u - V^k \leqslant 0$ in $\mathfrak{D}$.   Then we find the estimates

$$v^j(x) \leqslant u(x) \leqslant V^k(x) \quad \text{in} \quad \mathfrak{D}. \qquad (1.18)$$

The functions $v^j(x)$ and $V^k(x)$ are called *lower* and *upper functions*, respectively. As an example we consider the case $a \equiv 0$. Then the constants

$$v = \min_{x \in \mathfrak{D}} \varphi(x), \qquad V = \max_{x \in \mathfrak{D}} \varphi(x) \tag{1.19}$$

are lower and upper functions, respectively. The estimate (1.18) places the solution $u(x)$ between the two bounds. The definition of lower and upper functions was independent of the knowledge of a solution $u(x)$ of the problem. If the lower and upper functions are chosen by ingenious smoothing methods so that $U(x) = \lim_{j \to \infty} v^j(x) = \lim_{k \to \infty} V^k(x)$ in $\mathfrak{D}$, then we may hope that the limit function represents the solution $u(x)$.

In this elegant way O. Perron[3][5] actually solved the existence problem for $\Delta_n u = 0$ in $\mathfrak{D}$ where $u = \varphi$ on $\dot{\mathfrak{D}}$. At the same time, the proof points out what quite natural hypotheses have to be made about $\dot{\mathfrak{D}}$. Later problem (1.17) was settled in this way by G. Tautz[6], and still more general problems by N. Simonoff[7].

1.4   A Priori Estimates

We consider the *boundary-value problem*

$$Du = Au + au = f \quad \text{where} \quad u = \varphi(x) \quad \text{on} \quad \dot{\mathfrak{D}} \quad \text{and} \quad a \leqslant 0 \quad \text{in} \quad \mathfrak{D} \tag{1.20}$$

under the assumptions of Section 1.1. Let K be a bound for the coefficients: $|a_{ik}|$, $|a_i|$, $|a| \leqslant K$. Further let $\varphi \in C^0$ on $\dot{\mathfrak{D}}$, and let Au be *uniformly elliptic* in $\mathfrak{D}$; that is, for all $x \in \mathfrak{D}$

$$\sum_{i,k=1}^{n} a_{ik}(x) y_i y_k \geqslant m \sum_{i=1}^{n} (y_i)^2 \quad \text{with fixed } m > 0. \tag{1.21}$$

We can easily show that, under the assumptions made, the two statements (1) Au is elliptic in $\mathfrak{D}$ and (2) Au is uniformly elliptic in $\mathfrak{D}$ are completely equivalent.

THEOREM.   If $u(x) \in C^0$ in $\dot{\mathfrak{D}}$, $\in C^2$ in $\mathfrak{D}$ is a solution of (1.20) then

$$|u(x)| \leqslant \max_{x \in \dot{\mathfrak{D}}} |\varphi(x)| + M \max_{x \in \dot{\mathfrak{D}}} |f(x)| \tag{1.22}$$

in $\mathfrak{D}$, where $M = M(K,m)$.

Remark. This is the simplest kind of a-priori estimate. Here $u(x)$ is estimated by quantities which are independent of u. Such—though much more precise—estimates today are the key to profound existence proofs. So many mathematicians have done research on these that we shall only mention the book of C. Miranda[8] and the summarizing report of L. Nirenberg[9]. At any rate, (1.22) immediately settles the question of the first and third requirements and the continuous dependence on $f(x)$.

[5] The proof is easily accessible in I. G. Petrovski[4]. See also Remak[5].

Proof. We construct a function $w(x)$ for which $Dw \leqslant - \max |f|$, $w \geqslant \max |\varphi|$ on $\dot{\mathfrak{D}}$. For this we use

$$w(x) = \max |\varphi| + (e^{\alpha\xi} - e^{\alpha x_1}) \max |f|, \tag{1.23}$$

and without loss of generality we may assume that $x_1 \geqslant 0$ in $\mathfrak{D}$. Then we choose $\xi > \max_{x\in\mathfrak{D}} x_1$. The constant $\alpha > 0$ will be chosen later. First we have $w \geqslant \max |\varphi|$ on $\dot{\mathfrak{D}}$ and, using (1.21) and $a \leqslant 0$,

$$-Dw = -a \max |\varphi| + \{-ae^{\alpha\xi} + e^{\alpha x_1}(a_{11}\alpha^2 + a_1\alpha + a)\} \max |f|$$
$$\geqslant \{m\alpha^2 - K(\alpha + 1)\} \max |f| \geqslant \max |f|,$$

since $\{\cdot \cdot \cdot\} \geqslant 1$ for sufficiently large $\alpha = \alpha(K,m)$.

Now $|u| \leqslant w$ in $\mathfrak{D}$. Indeed, for $U = u - w$, obviously $U \leqslant 0$ on $\dot{\mathfrak{D}}$ and $DU = Du - Dw \geqslant f + \max |f| \geqslant 0$, from which $u \leqslant w$ in $\mathfrak{D}$ follows by the maximum-minimum principle. If instead of w we use the function $-w$, then $-w \leqslant u$ in $\mathfrak{D}$ follows. Thus everything is proved.

Problem.* Prove the estimate

$$0 \leqslant g(x,y) \leqslant s(x,y), \qquad\qquad n \geqslant 3,$$
$$\qquad\qquad\qquad\qquad \text{for} \qquad\qquad\qquad\qquad \tag{1.24}$$
$$0 \leqslant g(x,y) \leqslant s(x,y) + \text{const} \qquad n = 2,$$

in $\mathfrak{D}$ for the Green's functions $g(x,y)$ of Section I-3.4 (about whose existence we do not know anything).

1.5 The Analyticity of the Harmonic Functions

According to Section I-3.7, $u(x)$ is called harmonic in $\mathfrak{D}$ if $u(x) \in C^0$ in $\bar{\mathfrak{D}}$, $\in C^2$ in $\mathfrak{D}$ and $\Delta_n u = 0$ in $\mathfrak{D}$.

THEOREM 1. A function harmonic in $\mathfrak{D}$ is analytic in $\mathfrak{D}$.[6]

Proof. It will be carried out in three steps.

First Step. Let $\bar{S} \subset \mathfrak{D}$ be a sphere with center $\bar{x}$ and radius r. By hypothesis $u(x)$ is harmonic in $\mathfrak{D}$ and, according to Section I-3.7, it has the mean-value property (I-3.42a)

$$u(\bar{x}) = \frac{1}{\omega_n r^{n-1}} \int_{|y - \bar{x}| = r} u(y) \, dS. \tag{1.25}$$

From this we have the estimate

$$|u(\bar{x})| \leqslant \max_{x\in\dot{S}} |u(x)|. \tag{1.26}$$

Now, by Poisson's theorem in Section I-3.5, $u(x) \in C^\infty$ in S. Furthermore, $u_{x_i}(x)$ satisfies

[6] That is, in a neighborhood of any point $x \in \mathfrak{D}$ it can be represented by a convergent power series. The precise definition is given in Section I-3.1. The proof above follows that of O. D. Kellogg[10].

the equation $\Delta_n u_{x_i} = 0$. Thus $u_{x_i}(x)$ is harmonic in S and, because of Section I-3.7, has the mean-value property (I-3.42b). So, using (I-1.7), we have

$$u_{x_i}(\bar{x}) = \frac{n}{\omega_n r^n} \int_{|y-\bar{x}| \leqslant r} u_{y_i}(y)\, dy = \frac{n}{\omega_n r^n} \int_{|y-\bar{x}|=r} u(y) \nu_i\, dS, \tag{1.27}$$

$$|u_{x_i}(\bar{x})| \leqslant \frac{n}{\omega_n r^n} \max_{x \in \dot{S}} |u(x)| \omega_n r^{n-1} \leqslant \frac{n}{r} \max_{x \in \dot{S}} |u(x)|, \tag{1.28}$$

because $|\nu_i| \leqslant 1$.

Second Step. Let $\bar{S}_0 \subset \bar{S}$ be a sphere concentric with S, with center $\bar{x}$ and radius $r_0 < r$. If $D^m u(x)$ denotes an arbitrary mth partial derivative of $u(x)$, then we have

$$|D^m u(x)| \leqslant \frac{n^m e^{m-1} m!}{(r - r_0)^m} \max_{y \in \bar{S}} |u(y)| \quad \text{for arbitrary } x \in \bar{S}_0. \tag{1.29}$$

Proof by Induction. First we prove (1.29) for $m = 1$. About the arbitrary point $x \in \bar{S}_0$ we put a sphere with radius $r - r_0$; of course, this sphere is contained in $\bar{S}$. For this sphere, (1.28) gives

$$|D^1 u(x)| \leqslant \frac{n}{r - r_0} \max_{|y-x|=r-r_0} |u(y)| \leqslant \frac{n}{r - r_0} \max_{y \in \bar{S}} |u(y)|. \tag{1.30}$$

Let (1.29) be valid for the index m. About the center $\bar{x}$ we put another sphere $\bar{S}_1$ with radius $r_1 = (1 - \theta)r + \theta r_0$, $0 < \theta < 1$. Then

$$r_1 - r_0 = (1 - \theta)(r - r_0) < r - r_0, \; r_1 > (1 - \theta)r_0 + \theta r_0 = r_0. \tag{1.31}$$

Thus for the three concentric spheres we have $\bar{S}_0 \subset \bar{S}_1 \subset \bar{S}$.

Now we take an arbitrary $x \in \bar{S}_0$ and put a sphere s_1 about x with radius $(1 - \theta)(r - r_0)$. Then, because of (1.31), $\mathfrak{z}_1 \subset \bar{S}_1$. For $\mathfrak{z}_1$ we use formula (1.30). We obtain

$$\begin{aligned}
|D^{m+1} u(x)| &\leqslant \frac{n}{(1 - \theta)(r - r_0)} \max_{y \in \dot{s}} |D^m u(y)| \\
&\leqslant \frac{n}{(1 - \theta)(r - r_0)} \max_{y \in \bar{S}_1} |D^m u(y)|.
\end{aligned} \tag{1.32}$$

Now, for arbitrary $x \in \bar{S}_1$, formula (1.29) gives

$$|D^m u(x)| \leqslant \frac{n^m e^{m-1} m!}{(r - r_1)^m} \max_{y \in \bar{S}} |u(y)| = \frac{n^m e^{m-1} m!}{\theta^m (r - r_0)^m} \max_{y \in \bar{S}} |u(y)|. \tag{1.33}$$

Both formulas together then give for arbitrary $x \in \bar{S}_0$

$$|D^{m+1} u(x)| \leqslant \frac{n^{m+1} e^{m-1} m!}{(1 - \theta)\theta^m (r - r_0)^{m+1}} \max_{y \in \bar{S}} |u(y)|. \tag{1.34}$$

We set $\theta = m/(m + 1)$. Then $1 - \theta = 1/(m + 1)$ and

$$\frac{1}{(1 - \theta)\theta^m} = \frac{(m + 1)(m + 1)^m}{m^m} = (m + 1)\left(1 + \frac{1}{m}\right)^m < (m + 1)e, \tag{1.35}$$

so that finally

$$|D^{m+1}u(x)| \leqslant \frac{n^{m+1}\,e^m(m+1)!}{(r-r_0)^{m+1}}\,\max_{y\in\check{S}}|u(y)| \tag{1.36}$$

follows, which concludes the proof by induction.

Third Step. Taylor's formula at the point $x+h\in\mathfrak{D}$ with $h=(h_1,\ldots,h_n)$, and $x\in\mathfrak{D}$ reads as follows (h is to be chosen so that the straight line segment joining x and $x+h$ lies completely within $\mathfrak{D}$):

$$u(x+h)=u(x)+\sum_{k=1}^{m-1}\frac{1}{k!}\left(h_1\frac{\partial}{\partial x_1}+\cdots+h_n\frac{\partial}{\partial x_n}\right)^k u(x)+R_m, \tag{1.37}$$

$$R_m=\frac{1}{m!}\left(h_1\frac{\partial}{\partial x_1}+\cdots+h_n\frac{\partial}{\partial x_n}\right)^m u(x+\theta h), \tag{1.38}$$

with $0\leqslant\theta\leqslant 1$, where the point $x\in\mathfrak{D}$ is fixed. If for an arbitrarily given $\epsilon>0$ it can be achieved that $|R_m|<\epsilon$ for $m>N(\epsilon)$, then from (1.38) convergence of the Taylor series

$$u(x+h)=u(x)+\sum_{k=1}^{\infty}\frac{1}{k!}\left(h_1\frac{\partial}{\partial x_1}+\cdots+h_n\frac{\partial}{\partial x_n}\right)^k u(x) \tag{1.39}$$

follows and this is indeed a convergent power series in $h_1, h_2, \ldots, h_n$. It suffices to choose $|h|$ sufficiently small since we need convergence of the series only in a neighborhood of the point x.

Let

$$\min_{y\in\mathfrak{D}}|x-y|=A$$

where x is fixed. We choose $d>0$ such that $2d<A$ and then $|h|$ so that $|h|\leqslant d$. About the point x we put two concentric spheres s and s_0, with radii $2d$ and d. From (1.29) it follows that

$$|D^m u(x)|\leqslant\frac{n^m\,e^{m-1}m!}{(2d-d)^m}\,\max_{y\in\check{s}}|u(y)| \quad\text{for arbitrary } x\in\mathfrak{z}_0, \tag{1.40}$$

$$|R_m|\leqslant\frac{|h|^m n^m}{m!}\cdot\frac{n^m\,e^{m-1}m!}{d^m}\,\max_{y\in\check{s}}|u(y)|\leqslant\left(\frac{|h|n^2e}{d}\right)^m\,\max_{y\in\check{s}}|u(y)|<\epsilon \tag{1.41}$$

for $m>N(\epsilon)$ if we choose $|h|<d/n^2e$. Thus everything is proved.

This is the simplest result among a large number of theorems which state for general elliptic partial differential equations that their solutions are analytic in $\mathfrak{D}$ if we only know that the coefficients and solutions are sufficiently often differentiable.

Theorem 2 is an easy generalization of Theorem 1.

THEOREM 2.   If $u(x)\in C^0$ in $\bar{\mathfrak{D}}$, $\in C^2$ in $\mathfrak{D}$ is a solution of

$$\Delta_n u+\lambda u=0 \tag{1.42}$$

with a number $\lambda\geqslant 0$, then $u(x)$ is analytic in $\mathfrak{D}$.

2

PARABOLIC TYPE

2.1   The Maximum-Minimum Principle

We only consider the simplest case in two independent variables x, t and, as in Section II-1.6, we assume the equation in the form (II-1.52)

$$Du \equiv u_{xx} - u_t + a(x,t)u = f(x,t). \tag{2.1}$$

As our domain $\mathfrak{F}$ (not a normal domain) we consider one that goes to infinity for positive t, with the form shown in Figures 14 and 15.

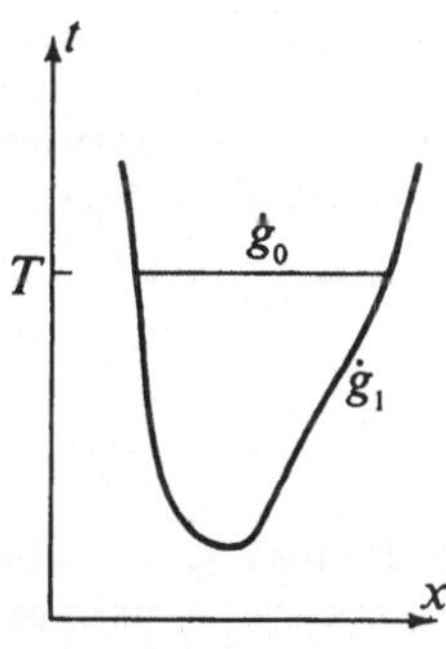

Figure 14

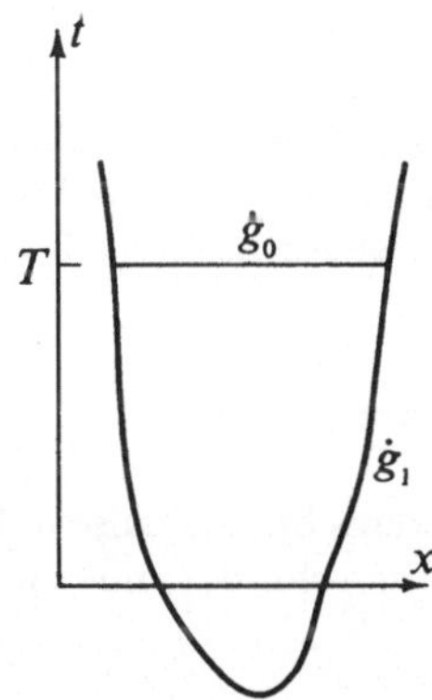

Figure 15

We truncate $\mathfrak{F}$ with a straight line $t = T$. Then $\mathfrak{D}$ with $\dot{\mathfrak{D}} = \mathring{g}_0 + \dot{g}_1$ is a normal domain. Let $u(x,t) \, \varepsilon \, C^0$ in $\bar{\mathfrak{D}}$ with u_{xx}, $u_t \, \varepsilon \, C^0$ in $\mathfrak{D} + \mathring{g}_0$ be a solution of $Du = f$ in $\mathfrak{D} + \mathring{g}_0$ and suppose a, $f \, \varepsilon \, C^0$ in $\bar{\mathfrak{D}}$.

THEOREM 1.   MAXIMUM-MINIMUM PRINCIPLE.   If $a < 0, f \leqslant 0 \ (f \geqslant 0)$ in $\bar{\mathfrak{D}}$, then every nonconstant solution $u(x)$ assumes its negative minimum (its positive maximum)—if one exists—on $\dot{g}_1$. Here we have to take $\dot{g}_1$ as closed.

Proof.   Suppose a negative minimum of u is assumed in x_0, $t_0 \, \varepsilon \, \mathfrak{D}$. Then $u(x_0,t_0) < 0$, $u_x(x_0,t_0) = u_t(x_0,t_0) = 0$, $u_{xx}(x_0,t_0) \geqslant 0$. Equation (2.1), considered at the point (x_0,t_0), gives a contradiction.   Now we suppose the negative minimum is assumed at (x_0,t_0) ε $\dot{g}_0$ ($\dot{g}_0$ is open).   Then $u(x_0,t_0) < 0$, $u_x(x_0,t_0) = 0$, $u_t(x_0,t_0) \leqslant 0$, $u_{xx}(x_0,t_0) \geqslant 0$ also contradicts (2.1) at the point (x_0,t_0).   Hence the negative minimum must lie on $\dot{g}_1$.   Analogous arguments have to be made for the positive maximum.

THEOREM 2.   FIRST REQUIREMENT.   The boundary-value problem

$$Du = f \quad \text{in} \quad \mathfrak{D} + \dot{g}_0 \quad \text{where} \quad u = \varphi(x,t) \quad \text{on} \quad \dot{g}_1, \qquad \varphi \,\varepsilon\, C^0 \quad \text{on} \quad \dot{g}_1 \quad (2.2)$$

for arbitrary $a(x,t)$ has at most one solution $u(x,t)$ for which $u \,\varepsilon\, C^0$ in $\mathfrak{D}$ and u_{xx}, $u_t \,\varepsilon\, C^0$ in $\mathfrak{D} + \dot{g}_0$.

Proof.   If u^1, u^2 are two solutions, then $v = u^1 - u^2$ satisfies

$$Dv = 0 \quad \text{in} \quad \mathfrak{D} + \dot{g}_0, \qquad v = 0 \quad \text{on} \quad \dot{g}_1. \tag{2.3}$$

For $a < 0$, $v \equiv 0$ follows from Theorem 1.   If $a < 0$ is not satisfied, we choose a constant α so large that $\alpha > a(x,t)$ in $\mathfrak{D}$ and put $v = w\, e^{\alpha t}$.   Then w satisfies the problem

$$w_{xx} - w_t + b(x,t)w = 0 \quad \text{where} \quad b = a - \alpha < 0 \quad \text{and} \quad w = 0 \quad \text{on} \quad \dot{g}_1. \quad (2.4)$$

Hence Theorem 1 can be applied and gives $w \equiv 0$, and this means that $v \equiv 0$.

Since T was arbitrary, Theorem 2 holds for $\mathfrak{F}$.   Analogously to Section 1.1 we show that the third requirement is satisfied.   Exactly as in Section 1.3 the maximum-minimum principle can be used to settle the second requirement by means of O. Perron's method of upper and lower functions[11,12].

2.2   Counterexample

These theorems become false if the domain $\mathfrak{F}$ reaches infinity for $t < 0$.   This means that we consider the solutions, not for the future, but for the past.   This of course was to be

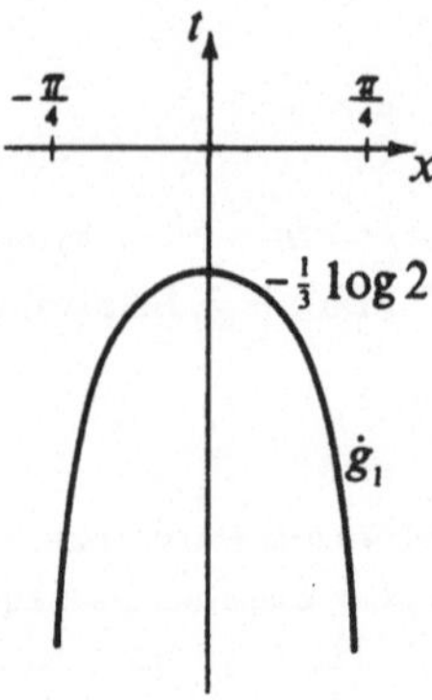

Figure 16

expected, since the solution operator for the heat equation has only the semigroup property [see Section I-4.4].

Example[13].   The equation $u_{xx} - u_t = 0$ has the solution

$$u(x,t) = e^{-t} \cos x - \tfrac{1}{2} e^{-4t} \cos 2x. \tag{2.5}$$

We have $u = 0$ on the curve

$$t = \frac{1}{3} \log \frac{1}{2} \frac{\cos 2x}{\cos x}.$$

For $-\pi/4 < x < \pi/4$ this curve is shown in Figure 16.   Thus the boundary-value problem $u_{xx} - u_t = 0$, where $u = 0$ on $\dot{g}_1$, has both $u \equiv 0$ and (2.5) for solutions.

3

HYPERBOLIC TYPE

3.1   The Energy Integral Method for the Wave Equation

In Section I-2.5 we showed that the wave equation satisfies the first requirement for the initial-value problem; however, the proof was closely linked to the construction of a solution of the initial-value problem.   Since such a method is not appropriate if we only want to treat the first requirement, we proceed differently in the following.   It suffices to consider the wave equation in R_2, since there is no dependence on dimension in the first requirement.

THEOREM.   FIRST REQUIREMENT.   If $u(x,t) \in C^2$ in $-\infty < x_1,\ x_2,\ t < \infty$ is a solution of

$$Du \equiv u_{x_1 x_1} + u_{x_2 x_2} - u_{tt} = 0 \quad \text{where} \quad u(x,0) = 0, \quad u_t(x,0) = 0, \qquad (3.1)$$

then $u \equiv 0$.   According to Section I-2.8, a better formulation is given by: If $u(x,0) = 0$, $u_t(x,0) = 0$ in

$$\bar{S}: |x - a| \leqslant r,$$

then $u \equiv 0$ in $\bar{\mathfrak{B}}(\bar{S}): |x - a| \leqslant r - |t|$.

Proof.   We may and shall assume $t \geqslant 0$.   First we have

$$0 = 2 \int_{\mathfrak{B}} u_t Du\, dx\, dt = \int_{\mathfrak{B}} \{2(u_t u_{x_1})_{x_1} + 2(u_t u_{x_2})_{x_2} - [(u_{x_1})^2 + (u_{x_2})^2 + (u_t)^2]_t\}\, dx\, dt, \quad (3.2)$$

where $\mathfrak{B}$ is the cone in Figure 6 with $t \geqslant 0$.   However, in (3.2) we integrate only over the truncated cone $\underset{\sim}{\mathfrak{B}}$ cut off by the plane $t = r_1 < r$.   Using (I-1.7) we obtain

$$0 = \int_{\dot{\mathfrak{B}}_1} \{2u_t u_{x_1} \nu_1 + 2u_t u_{x_2} \nu_2 - [(u_{x_1})^2 + (u_{x_2})^2 + (u_t)^2]\nu_3\}\, dS$$

$$- \int_{\dot{\mathfrak{B}}_2} [(u_{x_1})^2 + (u_{x_2})^2 + (u_t)^2]\, dS, \quad (3.3)$$

where $\dot{\mathfrak{B}}_1$ is the lateral surface of the truncated cone with the normal vector $\nu = (\nu_1,\ \nu_2,\ \nu_3)$ and $\dot{\mathfrak{B}}_2$ is the top surface of the truncated cone with $\nu = (0,\ 0,\ 1)$.   An integral over the base surface $|x - a| \leqslant r,\ t = 0$ does not occur, since the first derivatives of $u(x,t)$ vanish there.   An easy calculation gives

$$\int_{\dot{\mathcal{B}}_1} \{\cdot \cdot \cdot\} \, dS$$

$$= - \int_{\dot{\mathcal{B}}_1} \frac{1}{\nu_3} \{(u_{x_1}\nu_3 - u_t\nu_1)^2 + (u_{x_2}\nu_3 - u_t\nu_2)^2 + (u_t)^2[(\nu_3)^2 - (\nu_1)^2 - (\nu_2)^2]\} \, dS. \quad (3.4)$$

But now on $\dot{\mathcal{B}}_1$ we have $\nu_3 = \frac{1}{2}\sqrt{2}$ and $(\nu_1)^2 + (\nu_2)^2 = 1 - (\nu_3)^2 = \frac{1}{2}$.   Thus the right-hand side in (3.3) is

$$\leqslant - \int_{\dot{\mathcal{B}}_2} [(u_{x_1})^2 + (u_{x_2})^2 + (u_t)^2] \, dS$$

so that from (3.3)

$$\int_{\dot{\mathcal{B}}_2} [(u_{x_1})^2 + (u_{x_2})^2 + (u_t)^2] \, dS = 0, \quad (3.5)$$

and thus $(u_{x_1})^2 + (u_{x_2})^2 + (u_t)^2 = 0$ on $\dot{\mathcal{B}}_2$.   But now $r_1 < r$ was chosen arbitrarily; hence $(u_{x_1})^2 + (u_{x_2})^2 + (u_t)^2 = 0$ in $\mathcal{B}$, or $u \equiv$ const in $\overline{\mathcal{B}}$; because of the initial condition, the constant can only be zero, and so $u \equiv 0$ in $\overline{\mathcal{B}}(\overline{S})$.

3.2   The Energy-Integral Method for General Systems

We consider the most general *linear system* of $n + 1$ partial differential equations of the first order in $n + 1$ independent variables $x = (x_1, x_2, \ldots, x_{n+1})$ and $n + 1$ unknown functions $u(x) = (u_1(x), u_2(x), \ldots, u_{n+1}(x))$.   According to Section II-2.4 it is of the form

$$l^i = \sum_{j,k=1}^{n+1} a_{jk}^i(x)u_{jx_k} + \sum_{j=1}^{n+1} b_j^i(x)u_j + c^i(x) = 0; \qquad i = 1, 2, \ldots, n + 1. \quad (3.6)$$

We use matrix notation and set

$$\mathcal{A}_k = (a_{jk}^i), \qquad \mathcal{B} = (b_j^i), \qquad c = (c^1, \ldots, c^{n+1}), \qquad l = (l^1, \ldots, l^{n+1}). \quad (3.7)$$

Then by matrix multiplication, (3.6) appears in the form[7]

$$l = \sum_{k=1}^{n+1} \mathcal{A}_k u_{x_k} + \mathcal{B}u + c = 0. \quad (3.8)$$

We only pursue the special case $\mathcal{A}_{n+1} = -I$ (I = identity matrix) and distinguish x_{n+1} in this way.   This is useful for the applications since x_{n+1} will then appear as time t.   For convenience we assume $x_{n+1} = t \geqslant 0$ and introduce the vector $\tilde{x} = (x_1, x_2, \ldots, x_n)$ in R_n.   Now we finally consider the *symmetric system*

$$u_t = \sum_{k=1}^{n} \mathcal{A}_k(x)u_{x_k} + \mathcal{B}(x)u + c(x) \quad (3.9)$$

with the symmetric matrices $\mathcal{A}_k$, $\mathcal{B}$ ($a_{jk}^i = a_{ik}^j$, $b_j^i = b_i^j$).

---

[7] $\mathcal{A}_k u_{x_k}$ denotes matrix multiplication of $\mathcal{A}_k$ with the matrix (or rather the vector) u_{x_k}.

THEOREM 1.   Let $\mathcal{A}_k(x) \in C^1$; $\mathcal{B}(x)$, $c(x) \in C^0$ in R_{n+1}. Let $u(x) \in C^1$ for $\tilde{x} \in R_n$ and $t \geqslant 0$ be a solution of (3.9). For this vector solution we define a norm by

$$N(t) = \int_{R_n} s(x)(u,u)\, d\tilde{x} \quad \text{where} \quad s(x) = \max\ \{0,\, 1 - \mu t - |\tilde{x}|\}, \qquad (3.10)$$

where[8] $\mu > 0$ is a number yet to be determined. Then for sufficiently large μ we have the *a priori* estimate

$$\dot{N}(t) \leqslant \int_{R_n} s(x)(c,c)\, d\tilde{x} \quad \text{where} \quad \cdot \equiv \frac{d}{dt}. \qquad (3.11)$$

Later we interpret this theorem geometrically. First we mention a well-known lemma from matrix theory.

LEMMA. Suppose the symmetric matrix $\mathcal{A} = (a_j^i)$ with $i, j = 1, \ldots, n + 1$ has exactly the numbers $\lambda_1, \lambda_2, \ldots, \lambda_{n+1}$ for its eigenvalues.[9] Let

$$|\lambda_1| \geqslant |\lambda_2| \geqslant \cdots \geqslant |\lambda_{n+1}|. \qquad (3.12)$$

Then for each vector $x \in R_{n+1}$ the following estimate holds:

$$|(x,\mathcal{A}x)| \leqslant |\lambda_1|(x,x). \qquad (3.13)$$

Proof. As is well known, the eigenvectors $x^1, x^2, \ldots, x^{n+1}$ belonging to $\mathcal{A}$ can be chosen so that they form an orthonormal basis in R_{n+1}; in particular,

$$(x^i,x^k) = \delta_{ik} = \begin{cases} 1 & \text{for} \quad i = k, \\ 0 & \text{for} \quad i \neq k. \end{cases}$$

Furthermore, for arbitrary $x \in R_{n+1}$ we have

$$x = \sum_{i=1}^{n+1} \alpha_i x^i, \qquad \alpha_i = (x,x^i), \qquad \mathcal{A}x = \sum_{i=1}^{n+1} \alpha_i \mathcal{A}x^i = \sum_{i=1}^{n+1} \alpha_i \lambda_i x^i, \qquad (3.14)$$

$$(x,x) = \sum_{i=1}^{n+1} \alpha_i^2, \qquad (x,\mathcal{A}x) = \sum_{i=1}^{n+1} \alpha_i^2 \lambda_i, \qquad (3.15)$$

and thus we obtain

$$|(x,\mathcal{A}x)| = \Big|\sum_{i=1}^{n+1} \alpha_i^2 \lambda_i\Big| \leqslant |\lambda_1| \sum_{i=1}^{n+1} \alpha_i^2 = |\lambda_1|(x,x). \qquad (3.16)$$

Proof of the Theorem. We have $s(x) \geqslant 0$ and $s > 0$ exactly, where

$$|\tilde{x}| < 1 - \mu t \quad \text{or} \quad |\tilde{x}| < \mu(t_0 - t) \quad \text{with} \quad t_0 = \frac{1}{\mu}. \qquad (3.17)$$

This set of points is the interior of a cone with vertex at $\tilde{x} = 0$, $t = t_0$ (see Figure 17 in R_3).

---

[8] $(u,v) = (v,u)$ is the real inner product of two vectors. See Section I-1.2.

[9] Vectors $x \neq 0$ that satisfy $\mathcal{A}x = \lambda x$ are called eigenvectors. Such are possible only for certain values of λ. These values are called eigenvalues.

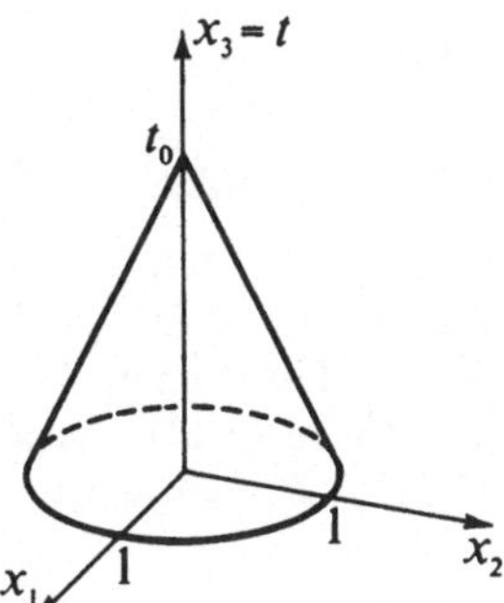

Figure 17

From (3.10), by consideration of (3.9), it follows that

$$\dot{N}(t) = \int_{R_n} \{ s_t(u,u) + 2s(u_t,u) \}\, d\tilde{x}$$

$$= \int_{R_n} \{ s_t(u,u) + 2s(\sum_{k=1}^{n} \mathcal{C}_k u_{x_k} + \mathcal{B}u + c,\, u) \}\, d\tilde{x}. \tag{3.18}$$

Now, for fixed t, we apply the theorem of Gauss (Section I-1.7) to an integral over the sphere $|\tilde{x}| \leqslant 1 - \mu t$. We have

$$\int_{|\tilde{x}| = 1 - \mu t} s(\mathcal{C}_k u, u) \nu_k\, dS = \int_{|\tilde{x}| \leqslant 1 - \mu t} \{ (s_{x_k} \mathcal{C}_k u + s\mathcal{C}_{k x_k} u + s\mathcal{C}_k u_{x_k},\, u) + (s\mathcal{C}_k u, u_{x_k}) \}\, d\tilde{x}. \tag{3.19}$$

The left-hand side in (3.19) becomes zero because of (3.17), and the integration on the right-hand side may be taken over all of R_n.

Because of symmetry we have $(\mathcal{C}_k u, u_{x_k}) = (u, \mathcal{C}_k u_{x_k})$. From (3.19) we now compute the term with $\mathcal{C}_k u_{x_k}$ and substitute its value in (3.18). Then for (3.18) we obtain

$$\dot{N}(t) = \int_{R_n} \{ s_t(u,u) + 2s(c,u) - \sum_{k=1}^{n} s_{x_k}(\mathcal{C}_k u, u) + s(2\mathcal{B}u - \sum_{k=1}^{n} \mathcal{C}_{k x_k} u,\, u) \}\, d\tilde{x}. \tag{3.20}$$

We estimate (3.20). To begin with we need carry out the integration only over the region where $s > 0$. But because of (3.10), we have $|s| \leqslant 1$, $s_t = -\mu$, and $s_{x_k} = -(x_k/|\tilde{x}|)$; thus $|s_{x_k}| \leqslant 1$, $k = 1, \ldots, n$. Furthermore we have

$$0 \leqslant (c - u,\, c - u) = (c,c) + (u,u) - 2(c,u),$$

so that

$$2(c,u) \leqslant (c,c) + (u,u). \tag{3.21}$$

Finally, because of the lemma, we have

$$\left| \sum_{k=1}^{n} (s_{x_k} \mathcal{C}_k u, u) \right| \leqslant |\lambda_1|(u,u), \tag{3.22}$$

where $|\lambda_1|$ is the largest of the moduli of the eigenvalues of the matrix $\mathcal{C} = \sum_{k=1}^{n} s_{x_k} \mathcal{C}_k$. But since its elements a_j^i depend on x, it is also true that $\lambda_1 = \lambda_1(x)$. Setting

$$\mu_1 = \max_{|\tilde{x}| \leqslant 1 - t} |\lambda_1(x)|$$

from (3.22) for $|\tilde{x}| \leqslant 1 - t$ we obtain

$$\left| \sum_{k=1}^{n} (s_{z_k}\mathcal{Q}_k u, u) \right| \leqslant \mu_1(u,u), \qquad \left| (2\mathcal{B}u - \sum_{k=1}^{n} \mathcal{Q}_{kz_k}u,\, u) \right| \leqslant \mu_2(u,u), \qquad (3.23)$$

the last formula for analogous reasons.

Now we fix μ by $\mu \geqslant 1 + \mu_1 + \mu_2$. The domain (3.17) belonging to this μ is contained in $|\tilde{x}| < 1 - t$. Therefore (3.23) holds in (3.17) and also in the domain of integration of (3.20) because $s \equiv 0$ outside of (3.17). Consequently we have

$$\dot{N}(t) \leqslant \int_{R_n} \{-\mu + 1 + \mu_1 + \mu_2\}\, (u,u)\, d\tilde{x} + \int_{R_n} s(c,c)\, d\tilde{x} \leqslant \int_{R_n} s(c,c)\, d\tilde{x}, \qquad (3.24)$$

and everything is proved.

From the theorem the first and third requirements for the initial-value problem of the symmetric system (3.9) follow immediately.

THEOREM 2.   FIRST REQUIREMENT.   Given the system

$$u_t = \sum_{k=1}^{n} \mathcal{Q}_k(x)u_{z_k} + \mathcal{B}(x)u \qquad (3.25)$$

with symmetric matrices $\mathcal{Q}_k$, $\mathcal{B}$ and $\mathcal{Q}_k \,\varepsilon\, C^1$, $\mathcal{B} \,\varepsilon\, C^0$ in R_{n+1}. If $u(x) \,\varepsilon\, C^1$ in $\tilde{x} \,\varepsilon\, R_n$ and $t \geqslant 0$ is a solution of (3.25) where $u(\tilde{x},0) = 0$ for $|\tilde{x}| \leqslant 1$, then there is a $\mu > 0$ such that $u \equiv 0$ in the cone $|\tilde{x}| < 1 - \mu t$.

Proof.   According to (3.11), the estimate $\dot{N}(t) \leqslant 0$ is true since $c = 0$. From the initial conditions we obtain $N(0) = 0$. Therefore

$$N(t) = \int_0^t \dot{N}(\tau)\, d\tau \leqslant 0.$$

But according to (3.10), $N(t) \geqslant 0$ so that $N(t) \equiv 0$. But for all x for which $s(x) > 0$ this implies $u \equiv 0$.

Problem.   Investigate the third requirement.

The cone we have mentioned has to be considered as part of the *domain of determinateness* $\mathcal{B}(|\tilde{x}| \leqslant 1)$ of the system. This part can be made larger by better choice of $s(x)$, but this method is not suitable for finding the entire domain of determinateness; that problem remains a delicate question even when other methods are used. It is remarkable that for (3.25) we did not assume hyperbolicity; but according to Theorem 2 it will turn out that the special form of (3.25) guarantees hyperbolicity, though perhaps in a sense still more general than that in Section II-2.4.

These considerations are the starting point for the method of K. O. Friedrichs[14] with which he could also settle the second requirement for the initial-value problem (3.9).[10]

---

[10] The estimate (3.11) is not as general as that of K. O. Friedrichs but it is easier to obtain.   P. Lax used it in his lectures at New York University in 1951.

3.3 The Radiation Problem

Let $\mathfrak{D}$ in R_n be a normal domain. We consider the exterior of $\mathfrak{D}$, that is, $R_n - \mathfrak{D}$, and denote this point set by $\mathfrak{F}$ (not a normal domain). In $\mathfrak{F}$ we look for a solution of

$$\Delta_n u + \lambda u = f(x) \quad \text{where} \quad u = \varphi(x) \quad \text{on} \quad \dot{\mathfrak{D}}. \tag{3.26}$$

Here f, φ, and the number λ are prescribed and $x = (x_1, x_2, \ldots, x_n)$. This problem does not satisfy the first requirement (uniqueness).

Example. $n = 3$, $\lambda = k^2$, $k > 0$; $\mathfrak{D}$: $|x| = \pi/k$, $\mathfrak{F}$: $|x| > \pi/k$.

The problem

$$\Delta_3 v + \lambda v = 0 \quad \text{where} \quad v = 0 \quad \text{on} \quad \dot{\mathfrak{D}} \tag{3.27}$$

has the solutions $v(x) = c[(\sin k|x|)/|x|]$ with an arbitrary constant c. Thus for this case (3.26) has, together with u, infinitely many solutions $u + v$. A. Sommerfeld[15] found two additional conditions which together guarantee the first requirement. These are the *radiation condition*:

$$\lim_{r \to \infty} r^{(n-1)/2}|u_r - i\sqrt{\lambda}\,u| = 0 \quad \text{where} \quad r = |x|, \tag{3.28}$$

which should be valid uniformly in all directions, and the *finiteness condition*:

$$|r^{(n-1)/2}u(x)| \leqslant \text{const}$$

in $\mathfrak{F}$.

If in R_3 we consider the solutions $f = e^{ikr}/r$, $g = e^{-ikr}/r$ of $\Delta_3 v + k^2 v = 0$, $k > 0$ and set $\sqrt{\lambda} = k$ in (3.28), then f satisfies the radiation condition but g does not. This is so because we have

$$r|f_r - ikf| = \frac{1}{r}, \qquad r|g_r - ikg| = \left| -2ik\,e^{-ikr} - \frac{e^{-ikr}}{r} \right|. \tag{3.29}$$

However, in order to understand condition (3.28) properly we must pass to the wave equation.[11] According to Section I-2.11,

$$u^1 = \frac{e^{ik(r-t)}}{r}, \qquad u^2 = \frac{e^{-ik(r+t)}}{r} \tag{3.30}$$

are solutions[12] of the wave equation in R_3. The function u^1 represents a *spherical wave*, *going out* from the origin, and u^2 one *coming in* to the origin. The outgoing wave satisfies the radiation condition, but the incoming one does not. Since with such a wave a transport of energy takes place, (3.28) means, in physical terms, that no energy must be radiated in from infinity.

[11] This is the reason we put the above considerations in the section on hyperbolic equations.

[12] For physical purposes, the real or imaginary parts have to be taken. The solutions (3.30) are found as follows: If we substitute the product solution $u = e^{-ikt}v(x)$ into the wave equation $\Delta_3 u - u_{tt} = 0$, then for v we obtain the equation $\Delta_3 v + k^2 v = 0$. If we look for solutions of the form $v = v(r)$ with $r = |x|$, then, from (I-2.59) the equation $(rv)_{rr} + k^2(rv) = 0$ results. If finally we put $z = rv$, then we obtain $z_{rr} + k^2 z = 0$ with the solutions $z = e^{\pm ikr}$.

Outline of Proof for the First Requirement.   Let $u^1(x)$, $u^2(x)$ be two solutions of (3.26) which in addition satisfy (3.28).   Then $v = u^1 - u^2$ is a solution of

$$\Delta_n v + \lambda v = 0 \quad \text{in} \quad \mathfrak{F}, \qquad v = 0 \quad \text{on} \quad \dot{\mathfrak{D}}, \qquad \lim_{r \to \infty} r^{(n-1)/2}|v_r - i\sqrt{\lambda}\,v| = 0.$$

Further let S be the ball $|x| < r$, where r is chosen so that $S \supset \mathfrak{D}$.   The open point set $S - \mathfrak{D}$ is denoted by B_r, and $\dot{B}_r = \dot{S} + \dot{\mathfrak{D}}$ (Figure 18).   In B_r we apply the first Green

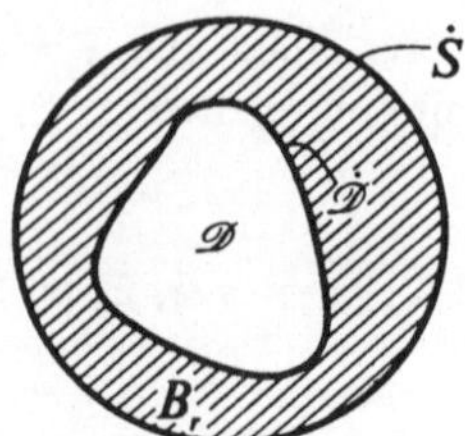

Figure 18

formula (I-1.16) and note that $v = 0$ on $\dot{\mathfrak{D}}$.   Since we consider complex-valued functions v, by $\bar{v}(x)$ we always understand the conjugate function to v.   Then (I-1.16) gives

$$\int_{B_r} \bar{v}\Delta_n v \, dx = \int_{\dot{S}} \bar{v}v_r \, dS - \int_{B_r} |\text{grad } v|^2 \, dx. \tag{3.31}$$

In the following theorem we replace $v = 0$ on $\dot{\mathfrak{D}}$ by the condition that v satisfies equation (3.31).   The advantages are that instead of one domain $\mathfrak{D}$ we can take arbitrarily many $\mathfrak{D}_i$, and also that $\mathfrak{F}$ may be the entire space.   Finally we replace (3.28) by the more appropriate version of the *radiation condition* given by F. Rellich:

$$\lim_{r \to \infty} \int_{|x| = r} |u_r - i\sqrt{\lambda}\,u|^2 \, dS = 0; \tag{3.32}$$

the finiteness condition is completely superfluous.

THEOREM.[13]   FIRST REQUIREMENT.   (1) Let $\mathfrak{F}$ be an open connected point set in R_n which contains the surfaces of all spheres $\dot{S}_r$: $|x| = r$ with $r \geqslant r_0$.   Let B_r be the points of $\mathfrak{F}$ with $|x| < r$.   (2) Let $v(x) \in C^2$ be a solution of $\Delta_n v + \lambda v = 0$ in $\mathfrak{F}$ where $\lambda \neq 0$ is a complex number.   (3) For all $r \geqslant r_0$ suppose

$$\int_{B_r} \bar{v}\Delta_n v \, dx = \int_{|x| = r} \bar{v}v_r \, dS - \int_{B_r} |\text{grad } v|^2 \, dx. \tag{3.33}$$

(4) Assume the radiation condition

$$\lim_{r \to \infty} R(r) = 0 \quad \text{where} \quad R(r) = \int_{|x| = r} |v_r - ikv|^2 \, dS \tag{3.34}$$

is satisfied, where k is the complex number uniquely determined by $k^2 = \lambda$, Im $k \geqslant 0$ and Re $k > 0$ if Im $k = 0$.
    Then $v \equiv 0$ in $\mathfrak{F}$.

[13] See F. Rellich[16] (here only for $\lambda > 0$).   For complex λ in the above presentation, see F. Rellich[17].

Proof. We put $k = a + ib, b \geqslant 0$. Then $\lambda = a^2 - b^2 + 2abi$, and from (3.34) we have

$$R(r) = \int_{|x|=r} (\bar{v}_r + ik\bar{v})(v_r - ikv)\, dS = \int_{|x|=r} (\bar{v}_r + b\bar{v} + ia\bar{v})(v_r + bv - iav)\, dS. \quad (3.35)$$

An elementary calculation (always remember that $|v|^2 = v\bar{v}$) gives

$$R(r) = \int_{|x|=r} |v_r + bv|^2\, dS + a^2 \int_{|x|=r} |v|^2\, dS + ia \int_{|x|=r} (\bar{v}v_r - v\bar{v}_r)\, dS. \quad (3.36)$$

By using the partial differential equation, from (3.33) it follows that

$$\int_{|x|=r} \bar{v}v_r\, dS - \int_{B_r} |\operatorname{grad} v|^2\, dx = -\lambda \int_{B_r} v\bar{v}\, dx. \quad (3.37)$$

From (3.37) we obtain—by taking the conjugate of the whole equation (3.37) and subtracting the new formula from (3.37)—

$$\int_{|x|=r} (\bar{v}v_r - v\bar{v}_r)\, dS = -4abi \int_{B_r} v\bar{v}\, dx. \quad (3.38)$$

Using (3.36) we then have

$$R(r) = \int_{|x|=r} |v_r + bv|^2\, dS + a^2 \int_{|x|=r} |v|^2\, dS + 4a^2 b \int_{B_r} |v|^2\, dx. \quad (3.39)$$

(a) Let $b = 0$, $a \neq 0$. From $\lim\limits_{r \to \infty} R(r) = 0$ and because of (3.39),

$$\lim_{r \to \infty} \int_{|x|=r} |v|^2\, dS = 0$$

follows. We use the first lemma of F. Rellich for the remainder of the proof.

FIRST LEMMA OF F. RELLICH. Let $v(x) \in C^2$ for $|x| > r_0$ $(r_0 \geqslant 0)$ be a (complex-valued) solution of $\Delta_n v + \lambda v = 0$ with positive λ. Let

$$\lim_{r \to \infty} \int_{|x|=r} |v|^2\, dS = 0.$$

Then v vanishes identically in $|x| > r_0$.

Since $\lambda = a^2$ is positive, from this lemma $v \equiv 0$ in $|x| > r_0$. But $\mathfrak{F}$ is connected and, according to Theorem 2 in Section 1.5, $v(x)$ is analytic in x. Hence $v(x)$ vanishes in all of $\mathfrak{F}$.[14]

(b) Let $b > 0$, $a \neq 0$. From (3.39), $R(r) \geqslant 4a^2 b \int_{B_r} |v|^2\, dx$. But $\lim\limits_{r \to \infty} R(r) = 0$, and from this $v \equiv 0$ in $\mathfrak{F}$ follows.

(c) Let $b > 0$, $a = 0$. In this case from (3.37) we find—by adding the two formulas leading to (3.38)—that

$$\int_{|x|=r} (v\bar{v})_r\, dS = 2 \int_{B_r} |\operatorname{grad} v|^2\, dx + 2b^2 \int_{B_r} |v|^2\, dx. \quad (3.40)$$

---

[14] It would be desirable to avoid using Theorem 2 of Section 1.5. Here it is applied to the exterior of a normal domain.

The computation of (3.39) becomes

$$R(r) = \int_{|x|=r} (|v_\nu|^2 + b^2|v|^2)\, dS + b \int_{|x|=r} (v\bar{v})_\nu\, dS \tag{3.41}$$

and finally, using (3.40), we obtain

$$R(r) \geqslant 2b \int_{B_r} |\operatorname{grad} v|^2\, dx + 2b^3 \int_{B_r} |v|^2\, dx, \tag{3.42}$$

from which $v \equiv 0$ in $\mathfrak{F}$ follows immediately.

Thus everything is proved and only in the case $\lambda > 0$ did we use the fact that $\mathfrak{F}$ is connected.

The theorem is false for $\lambda = 0$.

Example 1. $n = 2$, $v = \log |x|$, $\mathfrak{F}: |x| > 1$. We have $\Delta_2 v = 0$ in $\mathfrak{F}$. Relation (3.33) is satisfied with $r_0 = 1$ since $v = 0$ on $|x| = 1$. Relation (3.34) is satisfied, since

$$\int_{|x|=r} |v_\nu|^2\, dS = \frac{2\pi}{r}. \tag{3.43}$$

Example 2. $n > 2$, $v = 1 - |x|^{2-n}$, $\mathfrak{F}: |x| > 1$. We have $\Delta_n v = 0$ in $\mathfrak{F}$. Relation (3.33) is satisfied with $r_0 = 1$, since $v = 0$ on $|x| = 1$. Relation (3.34) is satisfied, since

$$\int_{|x|=r} |v_\nu|^2\, dS = (n - 2)^2 \omega_n r^{1-n}. \tag{3.44}$$

3.4   Proof of F. Rellich's First Lemma

The proof is carried out using the following lemma:

SECOND LEMMA OF F. RELLICH.   In $|x| > \rho_0$, let $v(x) \in C^2$, $v(x) \not\equiv 0$ be a (not necessarily real) solution of $\Delta_n v + \lambda v = 0$ with $\lambda > 0$. Then there is a positive number p such that for all sufficiently large ρ (and an arbitrary but fixed $\rho_1 > \rho_0$) we have

$$\int_{\rho_1 \leqslant |x| \leqslant \rho} |v|^2\, dx \geqslant p\rho. \tag{3.45}$$

Proof of the First Lemma.   Contrary to the statement of the first lemma we assume that $v(x)$ does not vanish identically in $|x| > r_0$. If we set

$$\int_{|x|=r} |v(x)|^2\, dS = f(r),$$

then according to the second lemma there would be a $p > 0$ such that

$$\int_{\rho_1}^{\rho} f(r)\, dr \geqslant p\rho$$

for sufficiently large ρ and fixed $\rho_1 > r_0$. But by hypothesis we ought to have $\lim_{\rho \to \infty} f(\rho) =$

0; hence[15]

$$\lim_{\rho \to \infty} \frac{1}{\rho} \int_{\rho_1}^{\rho} f(r)\, dr = 0, \tag{3.46}$$

which contradicts $p > 0$.

We remark that the second lemma is stronger than the first. For questions of existence in the radiation problem we refer to Cl. Müller.[16]

Proof of Rellich's Second Lemma. First we give the proof for $n = 2$. Obviously it suffices to restrict ourselves to real solutions $v(x)$. We consider $v(x)$ on the circle $|x| = r > \rho_0$. On it we have $x_1 = r \cos \varphi$, $x_2 = r \sin \varphi$ with $0 \leqslant \varphi < 2\pi$. By hypothesis there is an $r_0 > \rho_0$ such that $v(x)$ does not vanish identically on the circle with radius r_0. We expand $v(x)$ on $|x| = r$ in a Fourier series and obtain

$$v(r \cos \varphi, r \sin \varphi) = \tfrac{1}{2} a_0(r) + \sum_{l=1}^{\infty} (a_l(r) \cos l\varphi + b_l(r) \sin l\varphi) \tag{3.47}$$

with the Fourier coefficients

$$a_l(r) = \frac{1}{\pi} \int_0^{2\pi} v(r \cos \varphi, r \sin \varphi) \cos l\varphi\, d\varphi, \qquad l = 0, 1, 2, \ldots ,$$
$$b_l(r) = \frac{1}{\pi} \int_0^{2\pi} v(r \cos \varphi, r \sin \varphi) \sin l\varphi\, d\varphi, \qquad l = 1, 2, \ldots . \tag{3.48}$$

For our Fourier series, the Parseval equation is of the form

$$\int_0^{2\pi} v^2(r \cos \varphi, r \sin \varphi)\, d\varphi = \pi \left\{ \frac{a_0^2(r)}{2} + \sum_{l=1}^{\infty} (a_l^2(r) + b_l^2(r)) \right\}. \tag{3.49}$$

Since, in particular, $v \not\equiv 0$ on the circle with radius r_0, not all $a_l(r_0)$, $b_l(r_0)$ can vanish. Without loss of generality let us suppose $a_j(r_0) \neq 0$ for a fixed $j > 0$. We consider the function

$$w(\rho) = \int_0^{2\pi} v(\rho \cos \varphi, \rho \sin \varphi) \cos j\varphi\, d\varphi, \qquad \rho > \rho_0. \tag{3.50}$$

Because of (3.48), $w(r_0) \neq 0$. Using Schwarz's inequality we obtain the estimate

$$w^2(\rho) \leqslant \pi \int_0^{2\pi} v^2(\rho \cos \varphi, \rho \sin \varphi)\, d\varphi. \tag{3.51}$$

Now, for arbitrary but fixed $\rho_1 > \rho_0$,

$$\int_{\rho_1 \leqslant |x| \leqslant \rho} v^2(x)\, dx = \int_{\rho_1}^{\rho} \left(\int_0^{2\pi} v^2(r \cos \varphi, r \sin \varphi)\, d\varphi \right) r\, dr \geqslant \frac{1}{\pi} \int_{\rho_1}^{\rho} w^2(r) r\, dr. \tag{3.52}$$

---

[15] For sufficiently large ρ, where $\sqrt{\rho} > \rho_1$, we have

$$\left| \frac{1}{\rho} \int_{\rho_1}^{\rho} f(r)\, dr \right| = \left| \frac{1}{\rho} \int_{\rho_1}^{\sqrt{\rho}} f(r)\, dr + \frac{1}{\rho} \int_{\sqrt{\rho}}^{\rho} f(r)\, dr \right|$$

$$\leqslant \left| \frac{\text{const}}{\rho} (\sqrt{\rho} - \rho_1) + \frac{\rho - \sqrt{\rho}}{\rho} f(\sqrt{\rho} + \theta(\rho - \sqrt{\rho})) \right|;$$

where the last term results from the mean-value theorem of the calculus of integration with $0 \leqslant \theta \leqslant 1$. For $\rho \to \infty$ we obtain (3.46).

[16] See Section I-1.2.

Thus the lemma will be proved if only $\int_{\rho_1}^{\rho} w^2(r)r\,dr \geq \tilde{p}\rho$ has been shown for some $\tilde{p} > 0$ where ρ is sufficiently large. This we do by deducing an ordinary differential equation for $w(\rho)$ and solving it. Since $v(x)$ is a solution of $\Delta_2 v + \lambda v = 0$, writing $w' = dw/d\rho$ we have

$$
\begin{aligned}
0 &= \int_0^{2\pi} \cos j\varphi \{\Delta_2 v + \lambda v\}_{\substack{x_1 = \rho\cos\varphi \\ x_2 = \rho\sin\varphi}} d\varphi \\
&= \int_0^{2\pi} \cos j\varphi \left\{ v_{\rho\rho} + \frac{1}{\rho} v_\rho + \frac{1}{\rho^2} v_{\varphi\varphi} + \lambda v \right\} d\varphi \\
&= w'' + \frac{1}{\rho} w' + \lambda w + \int_0^{2\pi} \cos j\varphi \frac{1}{\rho^2} v_{\varphi\varphi} \, d\varphi.
\end{aligned}
\tag{3.53}
$$

If in the last term we integrate by parts twice, then using (3.50) we obtain the expression $-(j^2/\rho^2)w(\rho)$ for it, so that $w(\rho)$ must satisfy the equation

$$
w'' + \frac{1}{\rho} w' + \left(\lambda - \frac{j^2}{\rho^2} \right) w = 0 \quad \text{in} \quad \rho_0 < \rho < \infty.
\tag{3.54}
$$

This is a Bessel differential equation. If we let $\lambda = k^2$ with $k > 0$, $\rho k = s$, $w(\rho) = z(s)$, then for $z(s)$ we obtain

$$
\frac{d^2 z}{ds^2} + \frac{1}{s} \frac{dz}{ds} + \left(1 - \frac{j^2}{s^2} \right) z = 0.
\tag{3.55}
$$

As is well known, this equation has the general solution

$$
C_1 J_j(s) + C_2 N_j(s),
\tag{3.56}
$$

where J_j (N_j) is the jth Bessel (Neumann) function. To remind the reader we notice that j is a fixed index. We still need a very simple statement about the behavior of these functions for large values of the argument. For $s \geq s_0 > 0$ we have the representations

$$
\begin{aligned}
J_j(s) &= \frac{\alpha}{\sqrt{s}} \sin (s + \beta_j) + \Phi_j(s) \quad \text{with} \quad |\Phi_j(s)| \leq c s^{-3/2}, \\
N_j(s) &= \frac{\bar{\alpha}}{\sqrt{s}} \sin (s + \tilde{\beta}_j) + \tilde{\Phi}_j(s) \quad \text{with} \quad |\tilde{\Phi}_j(s)| \leq \tilde{c} s^{-3/2},
\end{aligned}
\tag{3.57}
$$

where $\alpha,\ \bar{\alpha},\ \beta_j,\ \tilde{\beta}_j\ c,\ \tilde{c}$ are suitable numbers with $\alpha,\ \bar{\alpha} \neq 0$. Now our function $z(s)$ can be represented as a linear combination (3.56). Since it does not vanish identically—because we have $z(r_0 k) \neq 0$—from (3.57) we find

$$
z(s) = \frac{\gamma}{\sqrt{s}} \sin (s + \delta) + \Psi(s) \quad \text{where} \quad |\Psi(s)| \leq C s^{-3/2}
\tag{3.58}
$$

and $\gamma \neq 0$ in $\rho_1 k \leq s < \infty$. Now, using the decisive expression $\int_{\rho_1}^{\rho} w^2(r)r\,dr$ from (3.52), we obtain

$$
A(\rho) \equiv \int_{\rho_1}^{\rho} w^2(r)r\,dr = k^{-2} \int_{\sigma_1}^{\sigma} z^2(s)s\,ds
\tag{3.59}
$$

Here we put $\sigma_1 = \rho_1 k$, $\sigma = \rho k$. Thus

$$A(\rho) = k^{-2} \int_{\sigma_1}^{\sigma} s \left\{ \frac{\gamma^2}{s} \sin^2 (s + \delta) + \frac{2\gamma}{\sqrt{s}} \sin (s + \delta)\Psi(s) + \Psi^2(s) \right\} ds,$$

$$k^2 A(\rho) \geqslant \gamma^2 \int_{\sigma_1}^{\sigma} \sin^2 (s + \delta)\, ds - \text{const} \int_{\sigma_1}^{\sigma} \frac{1}{s}\, ds, \tag{3.60}$$

$$k^2 A(\rho) \geqslant \gamma^2 \left\{ \frac{\sigma}{2} - \frac{\sin (2\sigma + 2\delta)}{4} \right\} - \text{const} \log \sigma + \text{const}$$

follows, where the constants do not depend on σ.   But now it is obvious that, for sufficiently large ρ, $A(\rho) \geqslant \tilde{p}\rho$ if for instance we set $\tilde{p} = \gamma^2/4k$.   Thus for $n = 2$ everything is proved with $p = \tilde{p}/\pi$.

The proof is almost exactly the same for $n > 2$.   Then instead of a Fourier expansion on the circle we must use an expansion in spherical functions on the sphere, as is done in F. Rellich[16].

3.5   The Radiation Problem for the Whole Space

For the special case of the entire space we give a new proof of the central theorem of Section 3.3 which is surprisingly elementary and yet contains at the same time a proof of Rellich's Second Lemma in this case.   Here it will suffice to consider the case $n = 3$.

THEOREM.   Let $v(x) \in C^2$ in R_3 be a complex-valued solution of $\Delta_3 v + \lambda v = 0$ with a complex number $\lambda \neq 0$.   For the complex number k, uniquely defined by $k^2 = \lambda$, $\text{Im } k \geqslant 0$, and $\text{Re } k > 0$ if $\text{Im } k = 0$, we assume that the radiation condition

$$\lim_{r \to \infty} R(r) = 0 \quad \text{where} \quad R(r) = \int_{|x| = r} |v_r - ikv|^2\, dS. \tag{3.61}$$

is satisfied.   Then $v \equiv 0$.

Proof.   After setting $k = a + ib$ we need only consider the case (a) $(b = 0, a \neq 0)$ from Section 3.3.   In all other cases the proof was already completely elementary in Section 3.3   We apply the Green formula $(I-1.16)$ to the functions $\bar{v}$ and v in the domain $|x| < r$ and obtain

$$\int_{|x| \leqslant r} \bar{v}\, \Delta_3 v\, dx = \int_{|x| = r} \bar{v} v_r\, dS - \int_{|x| \leqslant r} |\text{grad } v|^2\, dx. \tag{3.62}$$

This is formula (3.33) for our case.   Following the calculations of Section 3.3, we then obtain (3.39) with $b = 0$ and finally

$$\lim_{r \to \infty} \int_{|x| = r} |v|^2\, dS = 0.$$

If we set $f(r) = \int_{|x| = r} |v|^2\, dS$, then from (3.46) and the mean-value theorem of integration

$$\lim_{r \to \infty} \frac{1}{r} \int_0^r f(\tau)\, d\tau = 0$$

results, which in detail reads

$$\lim_{r \to \infty} \frac{1}{r} \int_0^r \left(\int_{|x|=\tau} |v|^2 \, dS \right) d\tau = \lim_{r \to \infty} \frac{1}{r} \int_{|x| \leqslant r} |v|^2 \, dx = 0. \tag{3.63}$$

Contrary to the conclusion we suppose $v(x) \not\equiv 0$. Then there is at least one point x^0 at which $v(x^0) \not= 0$. About this point we put a sphere $|x - x^0| < \sigma$. Since v is a solution of $\Delta_3 v + k^2 v = 0$, Theorem 3 of Section I-3.10 gives

$$v(x^0) \frac{\sin k\sigma}{k\sigma} = \frac{1}{4\pi\sigma^2} \int_{|x-x^0|=\sigma} v(x) \, dS \tag{3.64}$$

or

$$\left| \int_{|x-x^0|=\sigma} v(x) \, dS \right|^2 = \frac{16\pi^2\sigma^2}{k^2} |v(x^0)|^2 \sin^2 k\sigma. \tag{3.65}$$

From Schwarz's inequality we have

$$\left| \int_{|x-x^0|=\sigma} v(x) \, dS \right|^2 \leqslant \int_{|x-x^0|=\sigma} dS \int_{|x-x^0|=\sigma} |v(x)|^2 \, dS = 4\pi\sigma^2 \int_{|x-x^0|=\sigma} |v(x)|^2 \, dS. \tag{3.66}$$

Thus from (3.65) we obtain the estimate

$$\int_{|x-x^0|=\sigma} |v|^2 \, dS \geqslant \frac{1}{4\pi\sigma^2} \left| \int_{|x-x^0|=\sigma} v \, dS \right|^2 = \frac{4\pi |v(x^0)|^2}{k^2} \sin^2 k\sigma, \tag{3.67}$$

and furthermore

$$\int_{|x-x^0| \leqslant \rho} |v|^2 \, dx = \int_0^\rho \left(\int_{|x-x^0|=\sigma} |v|^2 \, dS \right) d\sigma \geqslant \frac{4\pi |v(x^0)|^2}{k^2} \int_0^\rho \sin^2 k\sigma \, d\sigma, \tag{3.68}$$

which is valid for every $\rho \geqslant 0$. After computing the last integral we finally obtain

$$\int_{|x-x^0| \leqslant \rho} |v|^2 \, dx \geqslant \frac{2\pi |v(x^0)|^2}{k^2} \left\{ \rho - \frac{\sin 2k\rho}{2k} \right\}. \tag{3.69}$$

We choose a sequence of spheres s_j: $|x - x^0| < \rho_j$, where $0 < \rho_1 < \rho_2 < \rho_3 < \cdots$ and $\lim_{j \to \infty} \rho_j = \infty$ and determine a corresponding sequence of spheres S_j: $|x| < r_j$, where $r_j = |x^0| + \rho_j$. Then $S_j \supset s_j$. From (3.69) we obtain

$$\frac{1}{r_j} \int_{|x| \leqslant r_j} |v|^2 \, dx \geqslant \frac{1}{r_j} \int_{|x-x^0| \leqslant \rho_j} |v|^2 \, dx \geqslant \frac{2\pi |v(x^0)|^2}{k^2} \left\{ \frac{\rho_j}{|x^0| + \rho_j} - \frac{\sin 2k\rho_j}{2k(|x^0| + \rho_j)} \right\} \tag{3.70}$$

and thus

$$\lim_{j \to \infty} \frac{1}{r_j} \int_{|x| \leqslant r_j} |v|^2 \, dx \geqslant \frac{2\pi |v(x^0)|^2}{k^2} \not= 0, \tag{3.71}$$

which contradicts (3.63). Thus everything is proved.

4

MIXED TYPE

4.1   The Energy-Integral Method for Equations of Elliptic-Parabolic-Hyperbolic Type

Our starting position here is much less favorable than in the previous cases, since we do not know yet which posing of the problem will be reasonable. However, a uniqueness theorem, which will work with as few assumptions as possible, already makes an essential contribution toward clearing up these questions.

We consider a very simple equation in the variables x, y:

$$Du \equiv yu_{xx} + u_{yy} + f(x,y) = 0, \tag{4.1}$$

which, according to Section II-1.3, is elliptic for $y > 0$, parabolic for $y = 0$, and hyperbolic for $y < 0$. The characteristic $\dot{C}$ are given as solutions of

$$\dot{C}: \quad (a)\ \frac{dx}{dy} = -\sqrt{-y}, \quad (b)\ \frac{dx}{dy} = \sqrt{-y} \quad \text{for}\ \ y \leqslant 0. \tag{4.2}$$

We consider the domain $\mathfrak{D}$ in Figure 19. Here c_1, c_3 are solutions of (4.2a) and c_2, c_4 are

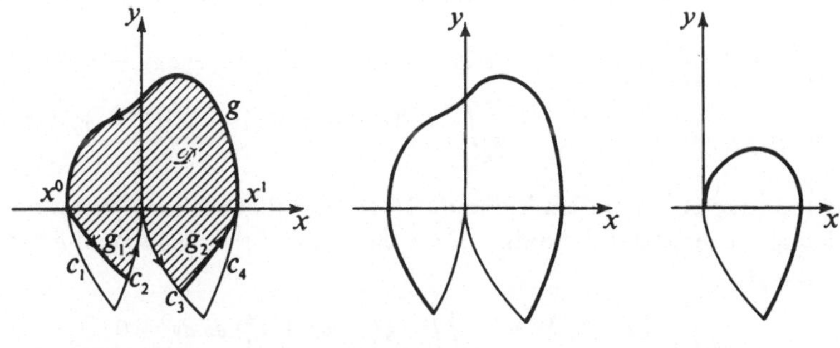

Figure 19   Figure 20   Figure 21

solutions of (4.2b). Let g_1, g_2 be monotone curves which lie in the wedges bounded by c_1, c_2 or c_3, c_4, respectively:

$$g_1: \frac{dx}{dy} \leqslant -\sqrt{-y}, \qquad g_2: \frac{dx}{dy} \geqslant \sqrt{-y}. \tag{4.3}$$

Let g be an arbitrary curve through the points x^0, x^1 where $x^0 < x^1$, which lies in $y \geqslant 0$ and is star-shaped: $x\,dy - y\,dx \geqslant 0$ for g. For (4.1) we consider the boundary-value problem (Frankl's problem[18]), that is, $u = \varphi$ is given on $g + g_1 + g_2$ and we suppose there are two solutions u^1, $u^2 \in C^1$ in $\bar{\mathfrak{D}}$, $\in C^2$ in $\mathfrak{D}$ where $\bar{\mathfrak{D}}$ is bounded by g, g_1, c_2, c_3, g_2. Then the difference $u = u^1 - u^2$ satisfies (4.1) with $f \equiv 0$ and $\varphi = 0$ on $g + g_1 + g_2$.

THEOREM.[17]   FIRST REQUIREMENT.   If $u(x,y) \in C^1$ in $\bar{\mathfrak{D}}$, $\in C^2$ in $\mathfrak{D}$ is a solution of

$$yu_{xx} + u_{yy} = 0 \quad \text{where} \quad u = 0 \quad \text{on} \quad g + g_1 + g_2, \tag{4.4}$$

then $u \equiv 0$ in $\bar{\mathfrak{D}}$.

Proof.   It results from a skillful application of the energy-integral method.   The reader need only verify the validity of the formulas; but there is no motivation for this procedure.

We take $\alpha = y$ for $y \geqslant 0$ and $\alpha = 0$ for $y \leqslant 0$, and from the Gauss theorem (I-1.2) in the form

$$\int_{\dot{\mathfrak{D}}} (v(x,y)\,dx + w(x,y)\,dy) = \int\!\!\int_{\mathfrak{D}} (w_x(x,y) - v_y(x,y))\,dx\,dy \tag{4.5}$$

we obtain[18] with the notations $(u_x)^2 \equiv u_x^2$, $(u_y)^2 \equiv u_y^2$

$$0 = \int\!\!\int_{\mathfrak{D}} (xu_x + \alpha u_y)(yu_{xx} + u_{yy})\,dx\,dy = \int\!\!\int_{\mathfrak{D}} \{\tfrac{1}{2}y(xu_x^2)_x + x(u_xu_y)_y$$
$$- \tfrac{1}{2}(xu_y^2)_x + \alpha y(u_xu_y)_x - \tfrac{1}{2}(\alpha yu_x^2)_y + \tfrac{1}{2}(\alpha u_y^2)_y$$
$$+ \tfrac{1}{2}u_x^2(-y + (\alpha y)_y) + \tfrac{1}{2}u_y^2(1 - \alpha_y)\}\,dx\,dy \tag{4.6}$$
$$= \int_{\dot{\mathfrak{D}}} \{(\tfrac{1}{2}yxu_x^2 - \tfrac{1}{2}xu_y^2 + \alpha yu_xu_y)\,dy + (-xu_xu_y + \tfrac{1}{2}\alpha yu_x^2 - \tfrac{1}{2}\alpha u_y^2)\,dx\}$$
$$+ \int\!\!\int_{\mathfrak{D}} \{\tfrac{1}{2}u_x^2(-y + (\alpha y)_y) + \tfrac{1}{2}u_y^2(1 - \alpha_y)\}\,dx\,dy.$$

On $g + g_1 + g_2$ we have $du = u_x\,dx + u_y\,dy = 0$; on c_2, $dx = \sqrt{-y}\,dy$; on c_3, $dx = -\sqrt{-y}\,dy$.   If we use this in (4.6), we obtain

$$0 = \int_{g+g_1+g_2} \frac{1}{2}\left[y + \left(\frac{dx}{dy}\right)^2\right] u_x^2(x\,dy - \alpha\,dx) + \int_{c_2+c_3} \frac{1}{2}\left(\frac{du}{dy}\right)^2 (-x\,dy - \alpha\,dx)$$
$$+ \int\!\!\int_{\mathfrak{D}} \{\tfrac{1}{2}u_x^2[-y + (\alpha y)_y] + \tfrac{1}{2}u_y^2(1 - \alpha_y)\}\,dx\,dy. \tag{4.7}$$

Considering the sense of orientation in Figure 19 we find that each integral is nonnegative. Hence the integral over $\mathfrak{D}$ must vanish.   If we take note of the meaning of α we can write this fact as follows:

$$\int\!\!\int_{D \cap \{y \geqslant 0\}} \tfrac{1}{2}yu_x^2\,dx\,dy + \int\!\!\int_{\mathfrak{D} \cap \{y \leqslant 0\}} \tfrac{1}{2}(-yu_x^2 + u_y^2)\,dx\,dy = 0. \tag{4.8}$$

From this we see that $u_x \equiv 0$ in $\bar{\mathfrak{D}}$.   If (x,y) is an arbitrary point in $\mathfrak{D}$, then there corresponds to it a point (x^*,y) which lies on $g + g_1 + g_2$ so that the entire segment between the

---

[17] See C. S. Morawetz[19].   A similar proof for the Tricomi problem is found in M. H. Protter[20].

[18] It is not true that $\alpha \in C^1$ in $\mathfrak{D}$.   Hence in the following calculations we first have to decompose the integrals over $\mathfrak{D}$ into such over $\mathfrak{D}_1 = \mathfrak{D} \cap \{y > 0\}$ and over $\mathfrak{D}_2 = \mathfrak{D} \cap \{y < 0\}$.   But the resulting boundary integrals over $\dot{\mathfrak{D}}_1$ and $\dot{\mathfrak{D}}_2$, taken together, give the integral over $\dot{\mathfrak{D}}$ which is given in (4.6).

two is contained in $\mathfrak{D}$. Since $u = 0$ on $g + g_1 + g_2$ was given, we then have $u(x,y) = \pm \int_{x*}^{x} u_x(t,y)\, dt$; thus $u(x,y) = 0$, which implies $u \equiv 0$ in $\mathfrak{D}$.

We have to remark that in the present form the theorem has little meaning, for it happens that the solutions to the few existence problems which have so far been solved do not have the nice properties which we assumed. The arguments carried out here, however, can be maintained using limit considerations if $u(x,y)$ satisfies weaker hypotheses. This is done in C. S. Morawetz[19].

If in Figure 19 we let g_1 coincide with c_1, and g_2 with c_4, then the problem of Gellerstedt arises (Figure 20). If we put x^0 at the origin we finally get Tricomi's problem (Figure 21), with which the whole development started.[19] In the figures the heavy curves are the carriers for the prescriptions of u.

Heuristically it is easily seen that the prescription for u must not be made on the closed curve in Figure 21. Otherwise, for $y \leqslant 0$ we would have a characteristic problem as was described in Section I-2.2 for the wave equation. If we assume these considerations to be valid for $y \leqslant 0$, then on the x-axis for $0 \leqslant x \leqslant x^1$ not only u but also its first derivatives would be known. For $y \geqslant 0$ we then would have a boundary-value problem for an equation which is elliptic for $y > 0$ and parabolic for $y = 0$. If the considerations of Section I-3 are valid—which indeed is the case—then, in particular, on the x-axis only the values of u may be prescribed. But there we should also be permitted to prescribe the first derivatives, which yields a senseless problem.

From Section II-1.3 we know that the equations of mixed type govern the flow of gas. But the mathematical results secured until now do not satisfy the technical requirements. Even the very simple question whether for the equation $yu_{xx} + u_{yy} + \lambda u = 0$ with the number $\lambda < 0$ the uniqueness theorem will remain correct for any one of the problems characterized by the figures is still an open question. It is correct for $\lambda \geqslant 0$ [23].

4.2 The Maximum-Minimum Principle for Equations of Elliptic-Parabolic-Hyperbolic Type

We consider the equation

$$Du \equiv a(x_1,x_2)u_{x_1x_1} + u_{x_2x_2} = f(x_1,x_2) \tag{4.9}$$

and again set $x = (x_1,x_2)$. We suppose $a \, \varepsilon \, C^1$, $f \, \varepsilon \, C^0$ in $-\infty < x_1, x_2 < \infty$, and $a > 0$ for $x_2 > 0$, $a = 0$ for $x_2 = 0$, and $a < 0$ for $x_2 < 0$. Hence (4.9) is elliptic for $x_2 > 0$, parabolic for $x_2 = 0$ and hyperbolic for $x_2 < 0$. Furthermore, $k: x_2 = 0$ is a parabolic curve and noncharacteristic (Section II-1.7). The characteristic curves for (4.9) are given by

$$c_1: \frac{dx_2}{dx_1} = -\frac{1}{\sqrt{-a}}, \qquad c_2: \frac{dx_2}{dx_1} = \frac{1}{\sqrt{-a}} \quad \text{for} \quad x_2 \leqslant 0, \tag{4.10}$$

and the characteristic directional derivatives by

$$u_\alpha, u_\beta \quad \text{where} \quad \alpha = (-\sqrt{-a},\, 1), \quad \beta = (\sqrt{-a},\, 1). \tag{4.11}$$

[19] See F. Tricomi[21]. We also find existence considerations in F. Tricomi[22].

For (4.9) we want to consider the Tricomi problem.   Through the points p_1, p_2 on the x_1-axis (Figure 22) we choose the characteristic curves c_1 and c_2, respectively, up to their intersection.   For $x_2 \geqq 0$ we join p_1 and p_2 by a curve g such that the domain $\mathfrak{D}$ with the

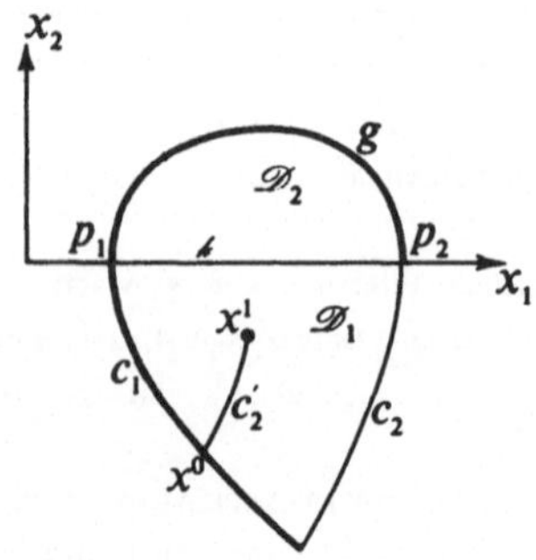

Figure 22

boundary $\dot{\mathfrak{D}} = c_1 + c_2 + g$ is a normal domain.   By $\mathfrak{D}_1$ we denote the domain which has the boundary $\dot{\mathfrak{D}}_1 = c_1 + c_2 + k$.   Then we have the maximum-minimum principle.

THEOREM 1.   MAXIMUM-MINIMUM PRINCIPLE.   In $\mathfrak{D}_1 - k$ let $a \in C^2$, $a_\beta > 0$, $((-a)^{-\frac{1}{4}})_{\beta\beta} \geqq 0$.   Furthermore let $u \in C^0$ in $\mathfrak{D}_1$, $u \in C^1$ in $\mathfrak{D}_1 - p_1 - p_2$,[20] $u \in C^2$ in $\mathfrak{D}_1$ be a solution of $Du = f$ where $f \geqq 0$ $(f \leqq 0)$.   Suppose that on c_1, u is a monotonically increasing (decreasing) function[21] with respect to x_2.   Then the maximum (minimum) of u in $\mathfrak{D}_1$ lies on k where k has to be considered closed by adding p_1 and p_2.

Proof.   If we put $(-a)^{-\frac{1}{4}} = b$, then, as is verified by an elementary calculation, we obtain the identity

$$(bu_\alpha)_\beta = b\, Du + b_\beta u_\beta \quad \text{in} \quad \mathfrak{D}_1. \tag{4.12}$$

Contrary to the conclusion, suppose the maximum of u in $\mathfrak{D}_1$ does not lie on k.   If it were on c_1 (Figure 22) then, since u is monotone increasing, it must be at p_1.   But p_1 belongs to k; hence the maximum of $u(x)$ would be assumed at $x^1 \in \mathfrak{D}_1 + c_2 - p_2$.   But through x^1 we choose the characteristic c_2' which meets c_1 at, say, x^0 (Figure 22).   The curve c_2' may be described by

$$\dot{x}_1 = \sqrt{-a}, \qquad \dot{x}_2 = 1 \quad \text{where} \quad \cdot \equiv \frac{d}{dt}.$$

Then $u_\beta = \dot{u}$ on c_2'.   Let the parameter values t_1, t_0 belong to x^1 and x^0; then $t_0 < t_1$. Integration of (4.12) along c_2', using $f \geqq 0$ and an integration by parts, then yields

$$bu_\alpha \Big|_{t_0}^{t_1} = \int_{t_0}^{t_1} bf\, dt + \int_{t_0}^{t_1} b\dot{u}\, dt \geqq \int_{t_0}^{t_1} b\dot{u}\, dt, \tag{4.13}$$

$$bu_\alpha \Big|_{t_0}^{t_1} \geqq bu \Big|_{t_0}^{t_1} - \int_{t_0}^{t_1} u\dot{b}\, dt + u(t_1)\int_{t_0}^{t_1} \ddot{b}\, dt - u(t_1)\dot{b}\Big|_{t_0}^{t_1}, \tag{4.14}$$

---

[20] The points p_1, p_2 have been removed from $\mathfrak{D}_1$.

[21] This means monotone in the wider sense so that the function may remain constant.

where the last two terms (since their sum is zero) have been added.   Furthermore

$$ bu_\alpha \Big|_{t_0}^{t_1} \geqslant \int_{t_0}^{t_1} \{u(t_1) - u(t)\}\ddot{b}(t)\, dt + \{u(t_1) - u(t_0)\}\dot{b}(t_0). \tag{4.15} $$

Since the maximum of u is at t_1, we have $u(t_1) - u(t) \geqslant 0$ and $u(t_1) - u(t_0) \geqslant 0$.   The equal sign in the last inequality may be excluded, since otherwise the maximum would also be assumed at t_0, and this means that, as was shown already, it would have to be at p_1.   By hypothesis, $\ddot{b}(t) \geqslant 0$ and $\dot{b}(t_0) > 0$ so that $bu_\alpha \Big|_{t_0}^{t_1} > 0$.   Thus we have

$$ u_\alpha(t_1) > \frac{b(t_0)}{b(t_1)} u_\alpha(t_0) \geqslant 0 \tag{4.16} $$

where the inequality $\geqslant 0$ follows from $b > 0$ and $u_\alpha \geqslant 0$ (monotonicity of u on c_1).   The inequality $u_\alpha(t_1) > 0$ is the desired contradiction, because u increases in the direction α and the maximum is not at t_1..   With respect to the minimum we only have to point out that $-u(x)$ satisfies the hypotheses for the statement about the maximum.

Now we shall prove a maximum-minimum principle for the Tricomi problem.   In accord with Figure 22 let $\mathfrak{D}$ be the open point set with boundary $\dot{\mathfrak{D}} = g + c_1 + c_2$.   The curve g is to be considered closed with addition of the points p_1, p_2.

THEOREM 2.   Under the hypotheses of Theorem 1 let $u(x) \in C^0$ in $\bar{\mathfrak{D}}$, where $u(x) \in C^1$ in $\mathfrak{D} - g$ and $u(x) \in C^2$ in $\mathfrak{D}$, be a solution of $Du = f$ for $f \geqslant 0$ $(f \leqslant 0)$ which on c_1 is monotone increasing (decreasing) relative to x_2.   Then the solution $u(x)$ assumes its positive maximum—if one exists—(its negative minimum—if one exists—) on g.

Proof.   Let $\mathfrak{D}_2$ be the domain which is bounded by $g + k$.   The expression Du is elliptic in $\mathfrak{D}_2$; however, on k it is parabolic, but k as a parabolic curve is not characteristic.

Contrary to the conclusion we assume that a positive maximum of u lies in $\mathfrak{D} - g$.   By Theorem 1 it cannot lie in $\mathfrak{D}_1 - k$.   Hence it can only be in $\mathfrak{D}_2 + (k - p_1 - p_2)$.   But Du is elliptic in $\mathfrak{D}_2$.   Application of Theorem 1 from Section 1.1 shows that it must lie on the boundary of $\mathfrak{D}_2$: $\dot{\mathfrak{D}}_2 = g + (k - p_1 - p_2)$.   Since g is excluded by our assumption, the positive maximum finally would have to be at a point x^0 on $k - p_1 - p_2$.   There Du is parabolic and $k - p_1 - p_2$ is noncharacteristic.   Hence the second lemma of Section 1.1 gives $u_\nu(x^0) > 0$ where ν is the outer normal relative to $\mathfrak{D}_2$; thus $\nu = (0, -1)$.   So we find $u_{x_2}(x^0) < 0$.   But indeed this must be false, because at a maximum point $x^0 \in \mathfrak{D}$ we necessarily have $u_{x_2}(x^0) = 0$.   Thus u assumes the positive maximum on g.   The statement about the negative minimum is obtained by applying what was just proved to $-u(x)$.

THEOREM 3.   FIRST REQUIREMENT FOR THE TRICOMI PROBLEM.   The boundary-value problem

$$ Du = f \quad \text{where} \quad u = \varphi \in C^0 \quad \text{on} \quad g + c_1 \tag{4.17} $$

has at most one solution $u \in C^0$ in $\mathfrak{D}$, $\in C^1$ in $\mathfrak{D} - g$, $\in C^2$ in $\mathfrak{D}$. We summarize the hypotheses: $a(x) \in C^1$ in $\mathfrak{D}$, $\in C^2$ in $\mathfrak{D}_1 - k$; $a > 0$ for $x_2 > 0$, $a = 0$ for $x_2 = 0$, $a < 0$ for $x_2 < 0$; $a_\beta > 0$, $((-a)^{-\frac{1}{4}})_{\beta\beta} \geqslant 0$ in $\mathfrak{D}_1 - k$; $f(x) \in C^0$ in $\mathfrak{D}$.

Proof. If u^1, u^2 are two solutions, then $v = u^1 - u^2$ satisfies the problem

$$Dv = 0 \quad \text{where} \quad v = 0 \quad \text{on} \quad g + c_1. \tag{4.18}$$

At any rate v is monotone on c_1. By Theorem 2 a positive maximum of v in $\mathfrak{D}$ would have to be assumed on g, thus $v \leqslant 0$ in $\mathfrak{D}$. The eventual negative minimum implies $v \geqslant 0$ in $\mathfrak{D}$, thus $v \equiv 0$ in $\mathfrak{D}$.

4.3   Remarks

(1) All hypotheses are satisfied for $a(x_1,x_2) = (\operatorname{sgn} x_2)|x_2|^r$ with $r \geqslant 1$ where $\operatorname{sgn} x_2 = 1, 0, -1$ for $x_2 > 0$, $x_2 = 0$, $x_2 < 0$. (2) The derivatives of the solution may become discontinuous in p_1, p_2. Here is the great advantage of this method of proof, compared with that in Section 4.1, since in the treatment of the second requirement for (4.17) only such solutions occur that have discontinuous derivatives at p_1, p_2. On the other hand, the method in Section 4.1 gives a uniqueness proof for the more general Frankl-Problem.

For $a(x_1,x_2) = x_2$ the maximum-minimum principle was first proved by P. Germain and R. Bader[24]. Our presentation followed S. Agmon, L. Nirenberg, and M. H. Protter[23], where more general equations are considered.

REFERENCES

1. E. Hopf, *Sitz. Preuss. Akad. Wiss. Berlin* **19**, 147–152 (1927).

2. E. Hopf, *Proc. Am. Math. Soc.* **3**, 791–793 (1952).

3. O. Perron, *Math. Z.* **18**, 42-54 (1923).

4. I. G. Petrovsky, *Lectures on Partial Differential Equations* (New York: Interscience Publishers, Inc.), 1954.

5. R. Remak, *Math. Z.* **20**, 126-130 (1924).

6. G. Tautz, *Math. Ann.* **118**, 733–770 (1943).

7. N. Simonoff, *Bull. Math. Univ. Moscow* **2**, No. 1 (1933).

8. C. Miranda, *Equazioni alle derivate parziali di tipo ellittico* (Berlin: Springer Verlag), 1955.

9. L. Nirenberg, *Transactions of the Symposium on Partial Differential Equations* (Berkeley, California, 1955) (New York: Interscience Publishers, Inc.), 1956, pp. 211–232.

10. O. D. Kellogg, *Trans. Am. Math. Soc.* **33**, 486–510 (1931).

11. W. Sternberg, *Math. Ann.* **101**, 394–398 (1929).

12. E. Kamke, *Jahresber. Deut. Math. Ver.* **62**, 1–33. (1959).

13. L. Bieberbach, *Lehrbuch der Differentialgleichungen* (Berlin: Springer Verlag), 1930.

14. K. O. Friedrichs, *Commun. Pure Appl. Math.* **7**, 345–392 (1954).

15. A. Sommerfeld, *Jahresber. Deut. Math. Ver.* **21**, 309–353 (1912).

16. F. Rellich, *Jahresber. Deut. Math. Ver.* **53**, 57–65 (1943).

17. F. Rellich, *Eigenwerttheorie partieller Differentialgleichungen*, Math. Inst. Univ. Göttingen, 1952–53.

18. F. I. Frankl, *Učn. Zap. Mosk. Gos. Univ.* **152**, 99–116 (1951).

19. C. S. Morawetz, *Commun. Pure Appl. Math.* **7**, 697–703 (1954).

20. M. H. Protter, *J. Rat. Mech. Anal.* **2**, 107–114 (1953).

21. F. Tricomi, *Atti Accad. Naz. dei Lincei* [5] **14**, 133 (1923).

22. F. Tricomi, *Equazioni a derivate parziali* (Rome: Edizioni Cremonese), 1957.

23. S. Agmon, L. Nirenberg, and M. H. Protter, *Commun. Pure Appl. Math.* **6**, 455–470 (1953).

24. P. Germain and R. Bader, *Office Nationale d'Etudes Recherches Aéronautiques No. 54*, 1952.

Part IV.   Questions of Existence

Questions of existence are quite difficult if they are to be treated with the necessary generality. In many cases existence proofs require deep theorems from other fields of mathematics, in particular from functional analysis, the theory of Hilbert spaces, the theory of integral equations, and topology. However, some existence questions accessible to us can be treated by modest tools.

1

EQUATIONS OF HYPERBOLIC TYPE IN TWO INDEPENDENT VARIABLES

1.1   The Initial-Value Problem for Linear Systems in Two Unknown Functions

In the simply connected domain $\mathfrak{D}$ of R_2 we consider the system

$$l^i \equiv \sum_{j,k=1}^{2} a^i_{jk} u_{j x_k} + \sum_{j=1}^{2} b^i_j u_j + c^i = 0, \qquad i = 1, 2, \tag{1.1}$$

where the coefficients a^i_{jk}, b^i_j, c^i are functions of $x = (x_1, x_2)$ and the $u_j(x)$ are the unknown solutions. We assume the hypotheses

$$H_0: \quad a^i_{jk} \, \varepsilon \, C^1; \qquad b^i_j, \, c^i \, \varepsilon \, C^0 \quad \text{in} \quad \mathfrak{D}. \tag{1.2}$$

Furthermore we always assume that the system (1.1) is of *hyperbolic type* in $\mathfrak{D}$. The simply connected domains occurring in Section 1 need not necessarily be normal domains. In particular, by $\mathfrak{D}$ or $\bar{\mathfrak{D}}$ we may always understand the whole of R_2.

After the considerations in Section II-2.1 we may assume (1.1) to have the normal form

$$\underset{\alpha}{U^\tau_\tau} + \sum_{j=1}^{2} A^\tau_j U^j + C^\tau = 0 \quad \text{where} \quad U^\tau = \sum_{j=1}^{2} g^\tau_j u_j, \qquad \tau = 1, 2, \tag{1.3}$$

and the determinant $|g^\tau_j| \neq 0$ in $\mathfrak{D}$. The *characteristic 2-web* is given by

$$\dot{x}_1 = \overset{\tau}{\alpha}_1(x), \qquad \dot{x}_2 = \overset{\tau}{\alpha}_2(x); \tag{1.4}$$

the $\overset{\tau}{\alpha}_k$ are determined from the a^i_{jk} by elementary calculation as was done in Section II-2.1. We also recall that we set

$$\underset{\tau}{U^\tau_\tau} = \overset{\tau}{\alpha}_1 U^\tau_{x_1} + \overset{\tau}{\alpha}_2 U^\tau_{x_2}. \tag{1.5}$$

Under hypotheses H_0 we find for the terms of the normal form (1.3) that A^τ_j, $C^\tau \, \varepsilon \, C^0$ in $\mathfrak{D}$, $\overset{\tau}{\alpha}_k \, \varepsilon \, C^1$ in $\mathfrak{D}$, and the determinant $|\overset{\tau}{\alpha}_k| \neq 0$.

In (1.1) we may use a conformal transformation which carries the system (1.1) into $\sigma l^1 = 0$, $\mu l^2 = 0$ where $\sigma > 0$, $\mu > 0$. Therefore we may always assume that

$$(\overset{\tau}{\alpha}_1)^2 + (\overset{\tau}{\alpha}_2)^2 = 1, \qquad \tau = 1, 2. \tag{1.6}$$

But then the differentiation parameter in (1.4) is exactly the arc length $\underset{\tau}{s}$.

We now consider the *Cauchy initial-value problem* for (1.1) or (1.3).   For this, there is given a smooth connected curve $k \subset \mathfrak{D}$ whose direction is nowhere characteristic.   We may imagine k to be continued as $\bar{k}$ to the boundary of $\mathfrak{D}$, and $\bar{k}$ embedded in a suitable non-characteristic vector field β which is given by

$$\dot{x}_1 = \beta_1(x), \qquad \dot{x}_2 = \beta_2(x), \qquad \beta_i(x) \, \varepsilon \, C^1 \quad \text{in} \quad \mathfrak{D}. \tag{1.7}$$

Then the $\overset{\tau}{\alpha}$, β ($\tau = 1, 2$) form a 3-web in $\mathfrak{D}$.   In particular, through every point $x \, \varepsilon \, \mathfrak{D}$ there pass exactly two characteristic curves, and these are defined to be solutions of (1.4). By K we denote the set of all points $x \, \varepsilon \, \mathfrak{D}$ which can be connected to k by two distinct characteristic curves lying entirely in $\mathfrak{D}$.   We denote these connecting curves by $c_1 = c_1(x)$ and $c_2 = c_2(x)$; their end points which lie on k shall be designated by $x^1 = x^1(x)$ and $x^2 = x^2(x)$; see Figure 23.   We note immediately that K is an open set.

For each $x \, \varepsilon \, K$ we denote the closed set bounded by $c_1(x)$, k, and $c_2(x)$ by $\bar{\mathfrak{B}}(x)$ (see Figure 23).   Furthermore, we assume that the arc length $\underset{\tau}{s}(x)$ of $c_\tau(x)$, $\tau = 1, 2$, satisfies $\underset{\tau}{s}(y) \leqslant \underset{\tau}{s}(x)$ for all $y \, \varepsilon \, \bar{\mathfrak{B}}(x)$.   This additional hypothesis about the characteristic 2-web is denoted by H_1.

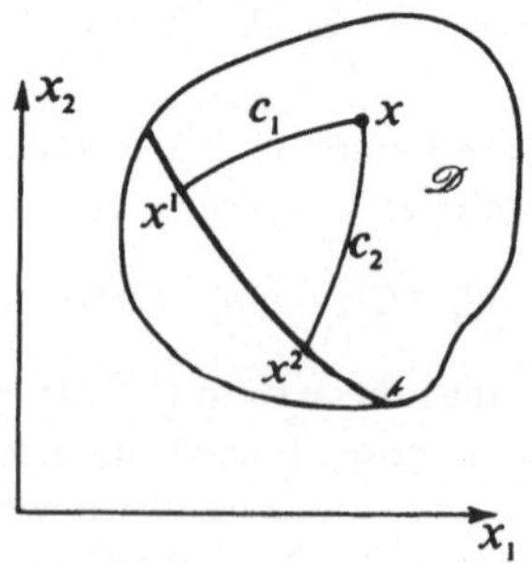

Figure 23

THEOREM 1.   If Ψ^1 and Ψ^2 are two continuous and bounded functions on k, then under hypotheses H_0 and H_1 (above) there is in K exactly one system of solutions $U^\tau(x)$ of the system (1.3) for which $U^\tau(x)$, $\underset{\alpha}{U^\tau_\tau}(x) \, \varepsilon \, C^0$ in K and which satisfies the initial conditions $U^\tau = \Psi^\tau$ on k.   At the point $x \, \varepsilon \, K$ the system of solutions $U^\tau(x)$ depends only on the values of the functions Ψ^1, Ψ^2 that are given on the part of the initial curve k which is cut out by the two characteristic curves c_1 and c_2 passing through x.

Proof. By $\underset{\tau}{s} = \underset{\tau}{s}(x)$ we denote the arc length of the characteristic joining curves $c_\tau = c_\tau(x)$. Along c_1 we have $\underset{\alpha}{U^1_1} = dU^1/d\underset{1}{s}$, and $\underset{\alpha}{U^2_2} = dU^2/d\underset{2}{s}$ along c_2. We denote by $\int_{c_\tau} \{ \cdot \cdot \cdot \} \, d\underset{\tau}{s}$ the integral over c_τ with respect to the parameter $\underset{\tau}{s}$. Then if we integrate

(1.3) along c_τ and note that

$$\int_{c_\tau} U^\tau_{\underset{\alpha}{s}} \, ds = U^\tau(x) - U^\tau(x^\tau) = U^\tau(x) - \Psi^\tau(x^\tau)$$

it becomes obvious that

$$U^\tau(x) \, \varepsilon \, C^0 \quad \text{in} \quad K,$$
$$U^\tau(x) = - \int_{c_\tau} \{ \sum_{j=1}^{2} A^\tau_j U^j + C^\tau \} \, ds + \Psi^\tau(x^\tau) \quad \text{in} \quad K, \qquad \tau = 1, 2, \tag{1.8}$$

is equivalent to the initial-value problem under consideration because if we differentiate (1.8) with respect to s we obtain system (1.3). Furthermore a solution of (1.8) automatically satisfies the initial conditions.

The problem (1.8) is solved by an iteration process, initially, however, not for all of K but only for those $x \, \varepsilon \, K$ which are sufficiently close to k. That is, if M denotes an upper bound for $|A^\tau_j|$ in $\mathfrak{D}$, then (1.8) is solved for those $x \, \varepsilon \, K$ which satisfy the condition $4Ms(x) < 1$. This set is denoted by $\tilde{K}$.

We define the functions $\overset{\nu}{U}^\tau(x)$, $\nu = 0, 1, 2, \ldots$ by

$$\overset{0}{U}^\tau(x) = - \int_{c_\tau} C^\tau \, ds + \Psi^\tau(x^\tau),$$
$$\overset{\nu}{U}^\tau(x) = - \int_{c_\tau} \{ \sum_{j=1}^{2} A^\tau_j \overset{\nu-1}{U^j} + C^\tau \} \, ds + \Psi^\tau(x^\tau), \qquad \nu = 1, 2, \ldots \tag{1.9}$$

We first have to show that all $\overset{\nu}{U}^\tau(x)$ defined in this manner are continuous functions of $x \, \varepsilon \, \tilde{K}$. This is nearly obvious for $\overset{0}{U}^\tau(x)$, but then it also follows for $\overset{1}{U}^\tau(x)$ by means of the second formula in (1.9), since it has already been proved that the braces contain a known continuous function, and so forth. Thus in $\tilde{K}$ we generate a sequence of continuous functions $\overset{0}{U}^\tau(x), \overset{1}{U}^\tau(x), \ldots$. In particular, $\overset{0}{U}^\tau(x)$ and $\overset{1}{U}^\tau(x)$ are bounded functions in $\tilde{K}$. If we put

$$|\overset{1}{U}^\tau(x) - \overset{0}{U}^\tau(x)| \leqslant T \quad \text{for} \quad x \, \varepsilon \, \tilde{K}, \tag{1.10}$$

then

$$|\overset{\nu+1}{U}^\tau(x) - \overset{\nu}{U}^\tau(x)| \leqslant T(\tfrac{1}{2})^\nu \quad \text{for} \quad x \, \varepsilon \, \tilde{K}, \tag{1.11}$$

as is shown immediately by complete induction, using the equation

$$\overset{\nu+1}{U}^\tau(x) - \overset{\nu}{U}^\tau(x) = - \int_{c_\tau} \sum_{j=1}^{2} A^\tau_j (\overset{\nu}{U^j} - \overset{\nu-1}{U^j}) \, ds \tag{1.12}$$

for $x \, \varepsilon \, \tilde{K}$. From the validity of the estimate (1.11) for the index $\nu = l$ and from (1.12) it follows that

$$|\overset{l+2}{U}^\tau(x) - \overset{l+1}{U}^\tau(x)| \leqslant 2MT(\tfrac{1}{2})^l s(x) \leqslant T(\tfrac{1}{2})^{l+1}$$

for $\nu = l + 1$, which proves (1.11).

The uniform convergence of the infinite series

$$\overset{0}{U}^\tau(x) + \sum_{\nu=1}^{\infty} (\overset{\nu}{U}^\tau(x) - \overset{\nu-1}{U}^\tau(x)) \tag{1.13}$$

to the continuous limit function $U^\tau(x)$ follows from (1.11). Now (1.13) is the limit of the functions $\overset{\nu}{U}{}^\tau(x)$ as $\nu \to \infty$ so that $\lim_{\nu \to \infty} \overset{\nu}{U}{}^\tau(x) = U^\tau(x)$ uniformly for $x \in \tilde{K}$. Hence in the second equation of (1.9) we may pass to the limit $\nu \to \infty$ under the integral sign, and so the limit functions $U^\tau(x)$ form a solution of (1.8) for all $x \in \tilde{K}$.

Uniqueness is proved as follows. Suppose $U_1^\tau(x)$ and $U_2^\tau(x)$ are two systems of solutions of the initial-value problem in a closed subdomain $\hat{K}$ of $\tilde{K}$; then from (1.8) we have

$$U_1^\tau(x) - U_2^\tau(x) = - \int_{c_\tau} \sum_{j=1}^{2} A_j^\tau (U_1^j - U_2^j) \, ds_\tau. \tag{1.14}$$

Let $\mu^\tau = \sup_{x \in \hat{K}} |U_1^\tau(x) - U_2^\tau(x)|$. Then (1.14) gives

$$\mu^1 \leqslant M(\mu^1 + \mu^2) \underset{1}{s}(x), \qquad \mu^2 \leqslant M(\mu^1 + \mu^2) \underset{2}{s}(x) \tag{1.15}$$

or

$$\mu^1 + \mu^2 \leqslant \tfrac{1}{2}(\mu^1 + \mu^2), \tag{1.16}$$

which implies $\mu^1 = \mu^2 = 0$.

This essentially concludes the proof; however, construction of the solution and the proof of uniqueness have been established only in $\tilde{K}$. But from the noncharacteristic family (1.7) it is possible to select a curve $\tilde{\beta}$ whose points $\bar{x}$, as far as they belong to K, satisfy the condition $\tfrac{1}{2} \leqslant 4M \underset{\tau}{s}(\bar{x}) < 1$; in particular, they lie in $\tilde{K}$; and if $x \in K$, then $\tilde{\beta}$ intersects $c_\tau(x)$ at those points which belong to K. For if $x \in K$, then c_1 and c_2 belong to K, since $\mathfrak{D}$ is simply connected.

If we now take $\beta = \tilde{\beta} \cap \tilde{K}$—"$\cap$" indicates the intersection of the two point sets—as our new initial curve and the continuous and bounded $U^\tau(x)$, which are already uniquely determined on β, as initial values, then the preceding arguments give the existence and uniqueness of a solution of the old problem for all $x \in K$ for which $4M \underset{\tau}{s}(x) < \tfrac{3}{2}$. Proceeding in this way we finally obtain the conclusion of Theorem 1 in a finite number of steps.

Remark. This also gives an existence theorem for the original system (1.1), because, if in (1.1) we prescribe the initial values ψ_τ, $\tau = 1, 2$ on k, as continuous and bounded, then, using $\Psi^\tau = \sum_{j=1}^{2} g_j^\tau \psi_j$ on k, we obtain an initial-value problem for (1.3) which satisfies the hypotheses of Theorem 1. The functions

$$u_\tau(x) = \sum_{j=1}^{2} h_{\tau j}(x) U^j(x) \quad \text{with} \quad \sum_{j=1}^{2} g_j^\tau h_{jl} = \delta_l^\tau = \begin{cases} 1 & \text{for} \quad \tau = l, \\ 0 & \text{for} \quad \tau \neq l, \end{cases} \tag{1.17}$$

defined by the solution $U^1(x)$, $U^2(x)$, however, only form a *weak solution* of (1.1); that is, they have the properties $u_\tau(x)$, $\left(\sum_{j=1}^{2} g_j^\tau(x) u_j(x)\right)_\alpha \in C^0$, and therefore they cannot be substituted immediately in (1.1). In order to arrive at solutions $u_\tau(x) \in C^1$ of (1.1) we must know that $U^\tau(x) \in C^1$.

THEOREM 2.   If ψ_1 and ψ_2 are two functions in C^1 which are defined and bounded on k, then, under hypotheses H_1 and

$$H_2: a^i_{jk} \in C^2; \qquad b^i_j, c^i \in C^1 \quad \text{in } \mathfrak{D} \tag{1.18}$$

on the system (1.1) which is hyperbolic in $\mathfrak{D}$, there is in K exactly one system of solutions $u_j(x) \in C^1$ which satisfies the initial conditions $u_j = \psi_j$ on k. The system of solutions $u_j(x)$, $j = 1, 2$ at the point $x \in K$ depends only on the values of the functions ψ_1, ψ_2 that are prescribed on that segment of the initial curve k which is cut out by the two characteristic curves c_1, c_2 passing through x.

Proof.   Passing to the normal form we obtain A^τ_j, $C^\tau \in C^1$ in $\mathfrak{D}$ and $\Psi^\tau \in C^1$ on k. From $g^\tau_j \in C^1$ and (1.17) we have $h_{\tau j} \in C^1$ in $\mathfrak{D}$, and it is sufficient to prove $U^\tau \in C^1$. By Theorem 1 we have $U^\tau_\tau \in C^0$, so that the proof would be complete if, for instance, we could show $U^1_{\underset{\alpha}{2}}(x)$, $U^2_{\underset{\alpha}{1}}(x) \in C^0$ for $x \in K$, for from two linearly independent directional derivatives we can obtain any other directional derivative.

For this purpose we introduce the characteristic 2-web as new coordinates ξ_1, ξ_2. Here we must note that the ds_τ are not complete differentials. Therefore we use the much more appropriate linear forms $\underset{\tau}{\omega}$ from Section II-2.2. Then, according to (II-2.10a) we have on c_τ

$$\underset{\tau}{\omega} = \sum_{k=1}^{\mathfrak{D}} \underset{\tau}{\alpha_k}\, dx_k = \underset{\tau}{ds}. \tag{1.19}$$

Now the forms $\underset{\tau}{\omega}$ can be made into complete differentials by suitable multipliers $M_\tau(x)$. We consider the coordinate transformation $\xi_\tau = \xi_\tau(x)$ and require $d\xi_\tau = M_\tau\underset{\tau}{\omega}$. On the other hand,

$$d\xi_\tau = \sum_{k=1}^{2} \xi_{\tau x_k}\, dx_k$$

so that we finally obtain the condition

$$\xi_{\tau x_k} = M_\tau\underset{\tau}{\alpha_k} \quad \text{or} \quad (M_\tau\underset{\tau}{\alpha_1})_{x_2} - (M_\tau\underset{\tau}{\alpha_2})_{x_1} = 0, \qquad \tau = 1, 2. \tag{1.20}$$

For every M_τ this is a first-order partial differential equation of a very simple form. If $\underset{\tau}{\alpha_k} \in C^1$, then it is well known that there exist solutions[1] $M_\tau(x) \in C^1$ which do not vanish in a neighborhood of k. We even have $\underset{\tau}{\alpha_k} \in C^2$ at our disposal, which is not needed at this moment.

Thus $M_\tau\underset{\tau}{\omega} = d\xi_\tau$ has been achieved and by the considerations in Section II-2.2 the characteristic 2-web can now be described by

$$\xi_\tau(x) = \text{const,}$$

as is shown in Figure 24. In these coordinates, the initial curve k admits the two representations $\xi_1 = f(\xi_2)$ and $\xi_2 = g(\xi_1)$. For the convenience of the reader we rewrite the

---

[1] See, for instance, E. Kamke[1].

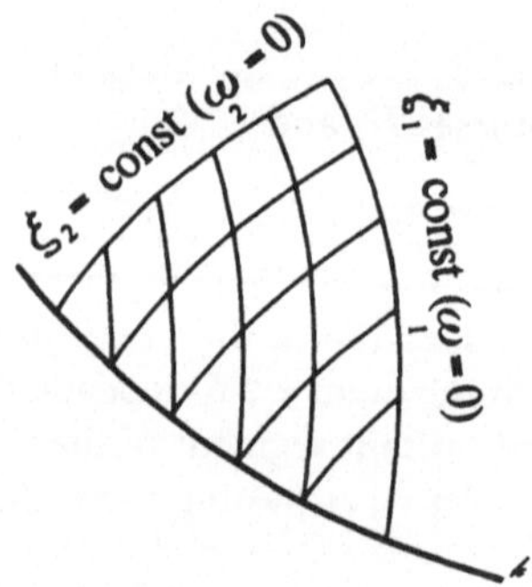

Figure 24

first equation in (1.8) in the new coordinates.   If to an arbitrary point x in Figure 23 there correspond the coordinate values ξ_1, ξ_2, then we have

$$U^1(\xi_1,\xi_2) = - \int_{f(\xi_2)}^{\xi_1} \{ \sum_{j=1}^{2} A_j^1(t,\xi_2)U^j(t,\xi_2) + C^1(t,\xi_2) \} \frac{dt}{M_1(t,\xi_2)} + \Psi^1(\xi_2). \qquad (1.21)$$

If we further note that $U_\tau^1 = M_\tau U_{\xi_r}^1$, then it is clear that everything will be proved for U^1 if we can prove $U_{\xi_2}^1 \in C^0$.   To this end we write the first equation in (1.3) in the form

$$U_{\xi_1}^1 + \frac{A_1^1}{M_1} U^1 = - \frac{1}{M_1} (C^1 + A_2^1 U^2). \qquad (1.22)$$

This is a linear ordinary differential equation in ξ_1.   Under consideration of the initial values its integration gives

$$U^1(\xi_1,\xi_2) = \left[\exp \left\{ - \int_{f(\xi_2)}^{\xi_1} \frac{A_1^1(t,\xi_2)}{M_1(t,\xi_2)} dt \right\} \right]$$
$$\cdot \left[\Psi^1(\xi_2) - \int_{f(\xi_2)}^{\xi_1} \frac{C^1(t,\xi_2) + A_2^1(t,\xi_2)U^2(t,\xi_2)}{M_1(t,\xi_2)} \exp \left\{ \int_{f(\xi_2)}^{t} \frac{A_1^1(\tau,\xi_2)}{M_1(\tau,\xi_2)} d\tau \right\} dt \right]. \qquad (1.23)$$

Now $U_{\xi_2}^1$ can be formed forthwith, and because $U_{\xi_2}^2 \in C^0$ (by Theorem 1), $U_{\xi_2}^1 \in C^0$ follows. For $U_{\xi_1}^2 \in C^0$ we proceed quite analogously, and the theorem is proved at least in a neighborhood of k.   But, as in Theorem 1, the procedure may be repeated in suitable strips.

Of course this result could have been obtained by a new iteration process.   But it is of principal importance that, starting with weak solutions, we can arrive at strong solutions without going through another proof of existence.

Existence questions in the Cauchy initial-value problem for a linear system in n unknown functions may be treated quite analogously.[2]

1.2   Supplements

The preceding existence theorems also contain an existence theorem for the Cauchy initial-value problem of the most general linear hyperbolic differential equation of the

<hr>

[2] See References [4–7] of Part II.

second order in two independent variables; according to (II-1.28) this problem may be written in the form

$$u_{x_1 x_2} + \sum_{k=1}^{2} a_k(x)u_{x_k} + a(x)u + f(x) = 0, \qquad (1.24)$$

where the characteristic manifolds $\acute{C}$ are given by the two families of curves $x_1 = \text{const}$, $x_2 = \text{const}$. Hence we require that the initial curve k can be represented in the form $x_2 = g(x_1)$ as well as in the form $x_1 = h(x_2)$. If we describe k by the parametric representation $x_i = x_i(\lambda)$, then the formulation of the initial-value problem for (1.24) reads as follows: u and its first derivatives are prescribed on k so that the *strip relation*

$$\dot{u} = u_{x_1}\dot{x}_1 + u_{x_2}\dot{x}_2, \qquad \cdot \equiv \frac{d}{d\lambda}, \qquad (1.25)$$

is satisfied. If, for instance, on k we prescribe $u = \psi_1$, $u_{x_1} = \psi_2$, where $\psi_1, \psi_2 \in C^1$ and are bounded on k, then u_{x_2} can be calculated on k from the strip relation (1.25). We find $u_{x_2} = 1/\dot{x}_2(\dot{\psi}_1 - \psi_2\dot{x}_1)$. Setting $u = u_1$, $u_{x_1} = u_2$, we obtain the system

$$\begin{aligned} u_{1x_1} - u_2 &= 0, \\ a_2 u_{1x_2} + u_{2x_2} + au_1 + a_1 u_2 + f &= 0, \end{aligned} \qquad (1.26)$$

with the initial values

$$u_1 = \psi_1, \qquad u_2 = \psi_2 \quad \text{on} \quad k; \qquad (1.27)$$

this is an initial-value problem of the kind treated in Section 1.1. Of course, other substitutions (for example, $u = u_1$, $u_{x_2} = u_2$) are possible.

Another question is when a solution of the initial-value problem can be obtained by quadratures only. We assume a linear system to be in the normal form

$$\underset{\alpha}{U^\tau_\tau} + \sum_{j=1}^{n} A^\tau_j U^j + C^\tau = 0, \qquad \tau = 1, 2, \ldots, n, \qquad (1.28)$$

and suppose that the characteristic n-web is known explicitly. The values of $U^\tau = \Psi^\tau$ are prescribed on the initial curve k whose direction is nowhere characteristic. If the matrix (A^τ_j) in (1.28) is triangular,

$$A^\tau_j = 0 \quad \text{for} \quad \tau < j, \qquad (1.29)$$

then the first equation in (1.28) reads

$$\underset{\alpha}{U^1_1} + A^1_1 U^1 + C^1 = 0. \qquad (1.30)$$

For the first family of the web of characteristic curves this is an ordinary linear differential equation with the variable $\underset{1}{s}$ (see Section 1.1); its unique solution at an arbitrary point x can be found by quadratures since the initial values $U^1 = \Psi^1$ on k are given. The second equation in (1.28) then appears in the form

$$\underset{\alpha}{U^2_2} + A^2_2 U^2 + \tilde{C}^2 = 0 \quad \text{with} \quad \tilde{C}^2 = C^2 + A^2_1 U^1, \qquad (1.31)$$

where $\tilde{C}^2$ has to be considered as already known. Hence (1.31), as the equation for the second family of the web of characteristic curves, is of the same form as (1.30) and consequently an ordinary linear differential equation. The procedure is continued in an analogous way. The quantities A^τ_j, $\tau < j$, are called *generalized Laplace invariants*.

Problem. Determine the Laplace invariants of (1.24) by writing (1.24) in two different ways as a system. (Solution: $a - a_1a_2 - a_{1x_1}$ and $a - a_1a_2 - a_{2x_2}$.)

Of course the Laplace invariants of (1.24) can be found directly without passing to systems. For this purpose we write (1.24) in the form

$$(u_{x_2} + a_1 u)_{x_1} + a_2(u_{x_2} + a_1 u) - hu + f = 0, \tag{1.31a}$$

where we set $h = a_{1x_1} + a_1a_2 - a$. Using the abbreviation $v = u_{x_2} + a_1 u$ we obtain

$$v_{x_1} + a_2 v - hu + f = 0.$$

If $h = 0$, this is an ordinary linear differential equation for v, all of whose solutions can be found by quadratures. Here x_2 plays the role of a parameter. But if we determine $v(x_1,x_2)$, then $u(x_1,x_2)$ in turn is obtained from an ordinary linear differential equation. Thus h is a Laplace invariant. The other Laplace invariant mentioned in the problem will be easily found by the reader, by rewriting (1.31a).

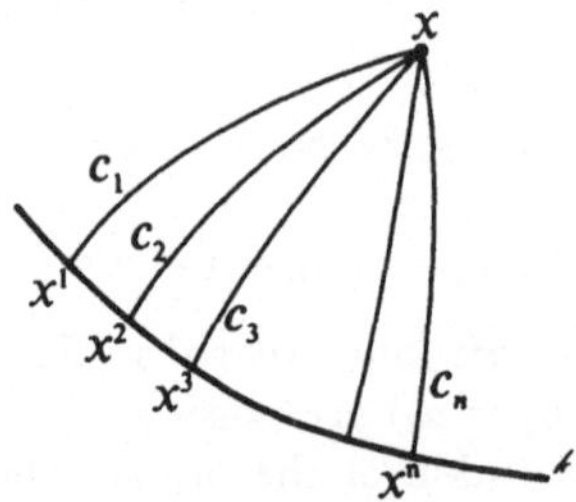

Figure 25

If furthermore (A_j^r) is a diagonal matrix: $A_j^r = 0$ for $r \neq j$, then the rth equation, $r = 1, 2, \ldots, n$ in (1.28), is an ordinary linear differential equation with the variable s on c_r. Therefore the solutions U^r of the initial-value problem at an arbitrary point x depend only on the initial values Ψ^r at the point x^r (Figure 25). In other words, U^r depends only on those initial values Ψ^r which lie in the intersection $k \cap c_r$. By Section I-2.9, in an analogous interpretation, this means that the system is of *Huygen's type*. This remains true for the original system; however, then, at the point x, u_j depends on the initial values at the points x^1, x^2, $\ldots$, x^n (Figure 25) which can be expressed in terms of the domain of dependence $\bar{\mathfrak{a}}(x)$ as follows:

$$\bar{\mathfrak{a}}(x) = (c_1 + c_2 + \cdots + c_n) \cap k = \dot{C} \cap k,$$

which brings out the Huygens-like character. With this we also show that the conditions for the Huygens character of the system can be formulated in a proper way only by means of the normal form.

The question of such necessary and sufficient conditions for the Huygens character of a general linear partial differential equation or of a general linear system of partial differential equations in more than two independent variables is one of the central problems still open in the theory of partial differential equations. It is difficult because, in the case of more than two independent variables, no such suitable normal forms can be constructed (see also Section II-2.4).

1.3 The Characteristic Initial-Value Problem

We consider the linear hyperbolic system

$$l^i \equiv \sum_{j=1}^{n} \sum_{k=1}^{2} a_{jk}^i u_{jx_k} + \sum_{j=1}^{n} b_j^i u_j + c^i = 0, \qquad i = 1, \ldots, n, \tag{1.32}$$

in the domain $\mathfrak{D}$ under hypotheses H_0 (1.2). To this system we assign the normal form

$$U_\tau^\tau + \sum_{j=1}^{n} A_j^\tau U^j + C^\tau = 0 \quad \text{where} \quad U^\tau = \sum_{j=1}^{n} g_j^\tau u_j, \qquad \tau = 1, \ldots, n. \tag{1.33}$$

The Cauchy initial-value problem treated in Section 1.1 (carried out there only for $n = 2$) permits only the one formulation for the normal form (1.33): U^τ are prescribed on the initial curve k as arbitrary functions Ψ^τ. For the original system (1.32) the formulation of the Cauchy initial-value problem is quite analogous: $u_j, j = 1, \ldots, n$, are prescribed on k as arbitrary functions ψ_j.

We investigate the possibility of making prescriptions on the characteristic curves of (1.32) or (1.33), and we retain hypotheses H_0 from Section 1.1. For this purpose we choose the n characteristic curves $c_1, c_2, \ldots, c_n$ through the interior point $x^0 \in \mathfrak{D}$ which are given as solutions of the ordinary differential equations

$$\dot{x}_1 = \overset{\tau}{\alpha}_1(x), \qquad \dot{x}_2 = \overset{\tau}{\alpha}_2(x).$$

Furthermore we consider these n curves only for $x_2 \leqslant x_2^0$ and assume that the line $x_2 = x_2^0$ is not a characteristic curve itself (this, if necessary, can be avoided by using a coordinate transformation). Qualitatively we then obtain the picture of Figure 25 if we identify x with x^0 and omit the curve k. Then c_1 and c_n are called the *outer* and $c_2, \ldots, c_{n-1}$ the *inner characteristics*. Then we have

THEOREM 1.   On the characteristic curves $c_1, c_2, \ldots, c_n$ through the point $x^0 \in \mathfrak{D}$ (Figure 25) the following continuous prescriptions are made:

(1) $U^1 = \Psi^1$ is prescribed arbitrarily on exactly one of the characteristic curves $c_2, \ldots, c_n$, and $U^n = \Psi^n$ on exactly one of the characteristic curves $c_1, \ldots, c_{n-1}$.

(2) $U^\sigma = \Psi^\sigma$, for $\sigma = 2, \ldots, n - 1$, is prescribed arbitrarily on exactly one of the characteristic curves $c_1, \ldots, c_{\sigma-1}$.

(3) $U^\sigma = \widetilde{\Psi}^\sigma$, for $\sigma = 2, \ldots, n - 1$, is prescribed arbitrarily on exactly one of the characteristic curves $c_{\sigma+1}, \ldots, c_n$ but so that $\widetilde{\Psi}^\sigma$ coincides at x^0 with the prescription Ψ^σ from (2).

Then, under hypothesis H_0 from Section 1.1, for all points $x \in \mathfrak{D}$ which lie in the angular space enclosed between the outer characteristics c_1 and c_n (Figure 25) and in a sufficiently small neighborhood of x^0, there is exactly one system of solutions $U^\tau(x)$ of (1.33) for which $U^\tau(x), U_\tau^\tau(x) \in C^0$ and which satisfies the prescriptions made on the characteristic curves.

THEOREM 2.   If in addition the prescriptions Ψ, $\tilde{\Psi}$ are in C^1 on the characteristic curves $c_1, \ldots, c_n$ and if we assume hypotheses H_2 from Section 1.1, then $U^\tau(x) \in C^1$ for all admissible points $x \in \mathfrak{D}$.

We draw some simple *conclusions:*

(a) From (2) and (3) it follows that U^2 must be prescribed on c_1, and U^{n-1} on c_n.

(b) U^σ must not be prescribed on c_σ, $\sigma = 1, \ldots, n$.

(c) For n characteristic curves, exactly $2n - 2$ prescriptions are needed.

For a better understanding of Theorem 1 we give a few examples in the case in which four characteristic curves $c_1, \ldots, c_4$ pass through the point x^0.

Examples for $n = 4$.   Figures 26 through 29 each show admissible prescriptions on the characteristic curves.

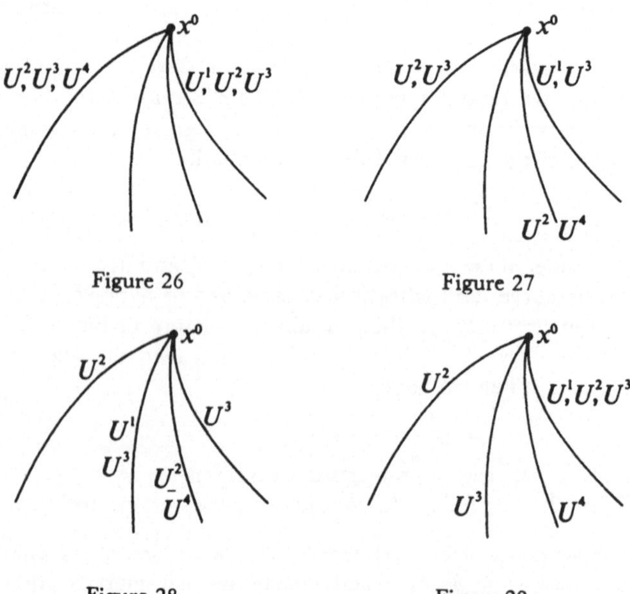

<table>
<tr><td align="center">Figure 26</td><td align="center">Figure 27</td></tr>
<tr><td align="center">Figure 28</td><td align="center">Figure 29</td></tr>
</table>

Outline of the Proof of Theorem 1.   The proof can be given by an iteration process.   First we consider the simple system

$$U^\tau_{\underset{\alpha}{\tau}} + C^\tau = 0, \qquad \tau = 1, \ldots, n. \tag{1.33a}$$

From the existence theorems for ordinary differential equations we may suppose the characteristic curves $c_1, \ldots, c_n$ through the point x^0 to be known.   But then this simple system can be solved by quadratures as we saw in Section 1.2.   To (1.33a) we assign such a characteristic initial-value problem.   To make the technique of integration to be used quite clear, we suppose that $n = 4$ and that the characteristic initial-value problem is given as in Figure 29.   At the arbitrary point x of the closed angular space c_1 and c_2 (Figure 30) we want to determine the system of solutions

$$U^\tau(x), \qquad \tau = 1, \ldots, 4.$$

We draw the four characteristic curves c_1', c_2', c_3', c_4' through x.   If x is sufficiently close to x^0, then the position of the curves c_1', $\ldots$, c_4' is changed only a little relative to the position of the curves c_1, $\ldots$, c_4.   This follows from the well-known continuity theorem for solutions of ordinary differential equations.   Now we integrate (1.33a) on c_τ'.   In order to determine $U^1(x)$ we arrive at an ordinary linear differential equation on c_1' (Figure 30).

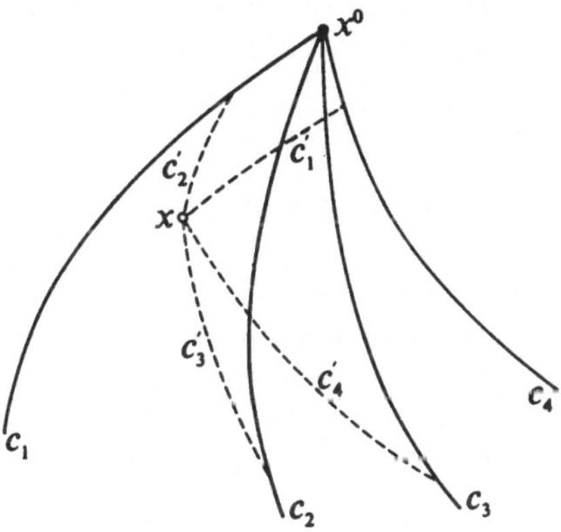

Figure 30

Since the value of U^1 is known on c_4, the value of $U^1(x)$ at one point of c_1' is known, namely at the intersection of c_1' and c_4.   Thus $U^1(x)$ can be computed.   In order to obtain $U^2(x)$ we integrate the second equation in (1.33a) on c_2'.   Here the value of U^2 at the intersection of c_2' and c_1 is known.   Continuing in this way we obtain all U^τ at the point x.   The procedure is repeated in an analogous way if x lies in the angular sector bounded by c_2, c_3 or in that bounded by c_3, c_4.

In order to solve the characteristic initial-value problem (1.33) we imagine a sequence of functions $\overset{\nu}{U}{}^\tau(x)$ defined by

$$\overset{\nu}{\underset{\alpha}{U}}{}^\tau_\tau + \sum_{j=1}^{n} A^\tau_j \overset{\nu-1}{U}{}^j + C^\tau = 0, \qquad \overset{0}{\underset{\alpha}{U}}{}^\tau_\tau + C^\tau = 0, \qquad \nu = 1, 2, 3, \ldots ,$$

all of which satisfy the same prescriptions on the characteristics.   The characteristic initial-value problem for $\overset{0}{U}(x)$ can be solved by quadratures as we have shown.   But using the iteration process, this state of affairs repeats itself for every $\overset{\nu}{U}(x)$ since the expression $\sum_{j=1}^{n} A^\tau_j \overset{\nu-1}{U}{}^j + C^\tau = \tilde{C}^\tau$ may be considered as a known function.   Therefore it is plausible that the solution will be given by $U^\tau(x) = \lim_{\nu \to \infty} \overset{\nu}{U}{}^\tau(x)$.

By Theorems 1 and 2, characteristic initial-value problems are found for the original system (1.32) by prescribing the corresponding linear combinations $U^\tau = \sum\limits_{j=1}^{n} g_j^\tau u_j$.

Those characteristic initial-value problems for (1.32) for whose formulation the g_j^τ are not needed are called *distinguished*. Thus the u_j are to be prescribed separately on suitable characteristic curves. Obviously only such problems for the normal form, for which prescriptions were made merely on the outer characteristics, lead to such distinguished characteristic initial-value problems. (For an example, see Figure 26.)

The formulation then could be made as follows: u_2, u_3, . . . , u_n are prescribed on c_1 and u_1, u_2, . . . , u_{n-1} on c_n. Other formulations are also possible, for instance, the following one: u_2, . . . , u_n are prescribed on c_1 and u_2, . . . , u_n on c_n, and u_1 is prescribed at some point on $c_1 + c_n$. For then U^2, . . . , U^n can be determined on c_1, and U^1, . . . , U^{n-1} on c_n.

1.4   The Initial-Value Problem for Quasi-Linear Systems

We consider the most general *quasi-linear system* in two unknown functions in a domain $\mathfrak{D}$ of the x_1, x_2-plane. Here it may be useful not to make too much use of the summation notation. We write the system as in Section II-2.5:

$$l \equiv \sum_{k=1}^{2} \{a_k u_{x_k} + b_k v_{x_k}\} + f = 0,$$

$$\bar{l} \equiv \sum_{k=1}^{2} \{\bar{a}_k u_{x_k} + \bar{b}_k v_{x_k}\} + \bar{f} = 0,$$

$$(1.34)$$

where a_k, . . . , $\bar{f}$ are functions of x, u, and v; and where a_k, . . . , $\bar{f} \in C^2$ for $x \in \bar{\mathfrak{D}}$ and $-\infty < u, v < \infty$. We assume that $\mathfrak{D}$ contains the curve

$$k: x_i = \Phi_i(\lambda), \qquad 0 < \lambda < 1, \quad \text{where} \quad \dot{\Phi}_1^2 + \dot{\Phi}_2^2 > 0 \quad \text{and} \quad \Phi_i(\lambda) \in C^1 \quad \text{and} \quad \cdot \equiv \frac{d}{d\lambda},$$

$$(1.35)$$

on which we pose a *Cauchy initial-value problem*:

$$u = \varphi(\lambda), \qquad v = \psi(\lambda) \quad \text{where} \quad \varphi(\lambda), \psi(\lambda) \in C^1 \quad \text{on} \quad k. \qquad (1.36)$$

The reader may note the great difficulty that without knowledge of a system of solutions nothing can be said about the type of (1.34). We use the notation of Section II-2.5.

THEOREM.   Under the foregoing hypotheses, let the prescribed φ and ψ be such that the system, which arises from (1.34) by substituting the prescribed φ, ψ for u, v in the coefficients a_k, . . . , $\bar{f}$, is of hyperbolic type on k whose direction is nowhere characteristic. According to Section II-2.5, this means

$$d = d(\lambda) = d(x_1, x_2, \varphi, \psi) \equiv A(x_1, x_2, \varphi, \psi)C(x_1, x_2, \varphi, \psi) - \tfrac{1}{4}B^2(x_1, x_2, \varphi, \psi) < 0$$

and

$$A(\dot{x}_2)^2 - B\dot{x}_1\dot{x}_2 + C(\dot{x}_1)^2 \neq 0, \qquad (1.37)$$

both on k. Then the initial-value problem has a system of solutions $u(x)$, $v(x) \in C^1$ in a neighborhood of an arbitrary segment[3] of k. At the point x, the system of solutions depends only on the values of the functions φ and ψ that were prescribed on that part of the initial curve k which is cut out by the two characteristics c_1 and c_2 passing through x. However, c_1 and c_2 are not known *a priori* because they contain the unknown system of solutions $u(x)$, $v(x)$.

Proof. *First Step.* Under the initial conditions made on k we may interpret the quantities d, A, B, C from Section II-2.5 as known functions of λ on k; this guarantees that we may speak about hyperbolic type on k. We may assume $A \neq 0$ and $C \neq 0$ on some piece of k, for if $A = 0$ or $C = 0$ or both at a point of k, then $B \neq 0$ there. By a rotation of the coordinate system it is then possible to achieve $A \neq 0$ and $C \neq 0$. Using (II-2.44), we determine the characteristic directions on k. We find

$$\frac{dx_2}{dx_1} = \frac{1}{A}\left\{\frac{B}{2} \pm \sqrt{-d}\right\} \quad \text{or} \quad \frac{dx_1}{dx_2} = \frac{1}{C}\left\{\frac{B}{2} \pm \sqrt{-d}\right\}, \qquad \text{respectively.}$$

Because $A \neq 0$ and $C \neq 0$ on k these do not coincide with the directions of the x_1- and x_2-axes.

Second Step. We assume that the problem has been solved and that u, $v \in C^1$ is a solution in a neighborhood of a segment[3] of k. For reasons of continuity it is clear that $d = d(x_1, x_2, u, v) < 0$ in a neighborhood and that the system is hyperbolic there. Section II-2.1 guarantees that (1.34) can be put into the invariant form (II-2.4). If we denote the characteristic directional derivatives by u_α, $u_{\bar{\alpha}}$, then, by (II-2.4), the invariant form has qualitatively the form

$$\rho u_\alpha + \sigma v_\alpha + g = 0, \qquad \bar{\rho} u_{\bar{\alpha}} + \bar{\sigma} v_{\bar{\alpha}} + \bar{g} = 0 \quad \text{where} \quad \begin{vmatrix} \rho & \sigma \\ \bar{\rho} & \bar{\sigma} \end{vmatrix} \neq 0, \qquad (1.38)$$

and where the coefficients $\rho, \ldots, \bar{g}$ belong to C^2 in x_1, x_2, u, v. By assumption they may be interpreted as functions of x_1 and x_2 only; and then they belong to C^1.

In this neighborhood the characteristic 2-web may be described according to (II-2.8), in the parametric form

$$\dot{x}_i = \alpha_i, \qquad \dot{x}_i = \bar{\alpha}_i \quad \text{where} \quad \begin{vmatrix} \alpha_1 & \alpha_2 \\ \bar{\alpha}_1 & \bar{\alpha}_2 \end{vmatrix} \neq 0, \qquad \alpha_i, \bar{\alpha}_i \neq 0. \qquad (1.39)$$

By assumption the right-hand sides may be interpreted as functions of x_1 and x_2 only. By the existence theorems for ordinary differential equations the solutions of (1.39) can be represented in the form

$$\xi_1 = \xi_1(x_1, x_2) = \text{const}, \qquad \xi_2 = \xi_2(x_1, x_2) = \text{const.} \qquad (1.40)$$

In a neighborhood of a point of k we can introduce ξ_1, ξ_2 as new coordinates because the system is hyperbolic there and the characteristic curves form a 2-web. Furthermore, in (1.39) we use ξ_1 as a new parameter in one family of characteristic curves and ξ_2 in the other family, so that for fixed ξ_1 and variable ξ_2, $x_1(\xi_1, \xi_2)$ and $x_2(\xi_1, \xi_2)$ describe a curve

---

[3] By a segment of k we understand any closed connected subset of k which consists of more than one point.

of the first family, and for fixed ξ_2 and variable ξ_1 a curve of the second family. Then, in a somewhat heuristic way, we find from (1.39) that

$$\frac{dx_2}{dx_1} = \frac{\dot{x}_2}{\dot{x}_1} = \frac{x_{2\xi_1}}{x_{1\xi_1}} = \frac{\alpha_2}{\alpha_1} \quad \text{or} \quad \alpha_2 x_{1\xi_1} - \alpha_1 x_{2\xi_1} = 0,$$
$$\frac{dx_2}{dx_1} = \frac{\dot{x}_2}{\dot{x}_1} = \frac{x_{2\xi_2}}{x_{1\xi_2}} = \frac{\bar{\alpha}_2}{\bar{\alpha}_1} \quad \text{or} \quad \bar{\alpha}_2 x_{1\xi_2} - \bar{\alpha}_1 x_{2\xi_2} = 0. \tag{1.41}$$

Now we represent u_α, $u_{\bar{\alpha}}$ in the new coordinates. Using (1.41) we find

$$u_{\xi_1} = u_{x_1} x_{1\xi_1} + u_{x_2} x_{2\xi_1} = u_{x_1} x_{1\xi_1} + u_{x_2} \frac{\alpha_2}{\alpha_1} x_{1\xi_1}$$
$$= \frac{x_{1\xi_1}}{\alpha_1} (u_{x_1}\alpha_1 + u_{x_2}\alpha_2) = \frac{x_{1\xi_1}}{\alpha_1} u_\alpha, \tag{1.42}$$
$$u_{\xi_2} = u_{x_1} x_{1\xi_2} + u_{x_2} x_{2\xi_2} = \frac{x_{1\xi_2}}{\bar{\alpha}_1} (u_{x_1}\bar{\alpha}_1 + u_{x_2}\bar{\alpha}_2) = \frac{x_{1\xi_2}}{\bar{\alpha}_1} u_{\bar{\alpha}}.$$

Corresponding formulas are valid for v_α and $v_{\bar{\alpha}}$. If we use these relations in (1.38), then in the new coordinates we obtain a system of four partial differential equations

$$\alpha_2 x_{1\xi_1} - \alpha_1 x_{2\xi_1} = 0,$$
$$\alpha_1(\rho u_{\xi_1} + \sigma v_{\xi_1}) + g x_{1\xi_1} = 0,$$
$$\bar{\alpha}_2 x_{1\xi_2} - \bar{\alpha}_1 x_{2\xi_2} = 0,$$
$$\bar{\alpha}_1(\bar{\rho} u_{\xi_2} + \bar{\sigma} v_{\xi_2}) + \bar{g} x_{1\xi_2} = 0. \tag{1.43}$$

Third Step. The assumption that our initial-value problem had been solved will be dropped, and we consider (1.43) as a system of four partial differential equations in the four unknown functions

$$x_1 = x_1(\xi_1,\xi_2), \qquad x_2 = x_2(\xi_1,\xi_2), \qquad u = u(\xi_1,\xi_2), \qquad v = v(\xi_1,\xi_2),$$

where the coefficients α_1, α_2, . . . , $\bar{g}$ are the original functions from (1.38) and (1.39). They are functions of x_1, x_2, u, v and belong to C^2 so that (1.43) is a quasi-linear system. We shall pose an initial-value problem for (1.43). We look for solutions of (1.43) which satisfy the following initial conditions: The open segment $\xi_1 = \xi_1(\lambda) \equiv \lambda$, $\xi_2 = \xi_2(\lambda) \equiv \lambda$, $0 < \lambda < 1$, is chosen as the initial curve $\tilde{k}$ in the ξ_1, ξ_2-plane. Then x_1 and x_2 are to assume values so that they describe the curve k, that is,

$$x_1(\lambda,\lambda) = \Phi_1(\lambda), \qquad x_2(\lambda,\lambda) = \Phi_2(\lambda), \qquad 0 < \lambda < 1. \tag{1.44}$$

Furthermore u and v are to assume the corresponding φ- and ψ-values on $\tilde{k}$, that is,

$$u(\lambda,\lambda) = \varphi(\lambda), \qquad v(\lambda,\lambda) = \psi(\lambda), \qquad 0 < \lambda < 1. \tag{1.45}$$

The solvability of this initial-value problem is guaranteed by the following

LEMMA. The quasi-linear system (1.43) is considered in a neighborhood of an initial curve[4] $\tilde{k}$ that can be represented in the form $\xi_2 = \xi_2(\xi_1)$ as well as in the form $\xi_1 = \xi_1(\xi_2)$. The values of x_1, x_2, u, v are prescribed by (1.44) and (1.45) on $\tilde{k}$ and belong to C^1. This initial-value problem can be solved uniquely in a suitable

[4] This is a point set which contains a neighborhood of every point on $\tilde{k}$.

neighborhood of $\tilde{k}$. The system of solutions $x_1 = x_1(\xi_1,\xi_2)$, $x_2 = x_2(\xi_1,\xi_2)$, $u = u(\xi_1,\xi_2)$, $v = v(\xi_1, \xi_2)$ belongs to C^1.

Using this strong lemma, we can bring the proof of our theorem to a quick end. Although the new system (1.43) seems to be more complicated than the original system (1.34), it is of a very special structure which will be used in an essential way later in the proof. First we convince ourselves that the functional determinant $x_{1\xi_1}x_{2\xi_2} - x_{1\xi_2}x_{2\xi_1}$ determined by the solution of (1.43), (1.44), and (1.45) is different from zero in a neighborhood of $\tilde{k}$. By (1.44), we have the strip relation

$$\dot{\Phi}_1 = x_{1\xi_1} + x_{1\xi_2}, \qquad \dot{\Phi}_2 = x_{2\xi_1} + x_{2\xi_2} \quad \text{on} \quad \tilde{k};$$

from this and from (1.43) we obtain

$$x_{1\xi_1}x_{2\xi_2} - x_{1\xi_2}x_{2\xi_1} = - \frac{x_{2\xi_2}}{\bar{\alpha}_2}(\bar{\alpha}_1\dot{\Phi}_2 - \bar{\alpha}_2\dot{\Phi}_1) = \frac{x_{1\xi_1}}{\alpha_1}(\alpha_1\dot{\Phi}_2 - \alpha_2\dot{\Phi}_1)$$

on $\tilde{k}$. Both expressions in parentheses are different from zero, otherwise k would be partly characteristic. Furthermore, $x_{2\xi_2}$ does not vanish, because otherwise $x_{1\xi_1}$ would vanish, and then from (1.43) $x_{1\xi_2}$ and $x_{2\xi_1}$ would vanish; hence according to (1.35) this violates the strip relation. Since the functional determinant is different from zero on $\tilde{k}$ continuity ensures that it is nonzero in a neighborhood of $\tilde{k}$. The segment of k mentioned in the theorem is represented by (1.35) if we restrict λ to $\lambda_1 \leqslant \lambda \leqslant \lambda_2$. Then, from the non-vanishing of the functional determinant, and since, by (1.44) and (1.35), $\tilde{k}$ is mapped one-to-one onto k, there is a neighborhood of the part of $\tilde{k}$ belonging to this λ-interval which by $x_1(\xi_1,\xi_2)$ and $x_2(\xi_1,\xi_2)$ is mapped one-to-one and continuously differentiable in both directions onto a neighborhood of the part of k. Hence, by passing to the inverse mapping $\xi_1 = \xi_1(x_1,x_2)$ and $\xi_2 = \xi_2(x_1,x_2)$, we may interpret the quantities u, v as functions of x_1, x_2 in C^1. The relations (1.42) which follow from (1.43) only, and the second and fourth equations in (1.43), give us the existence of the system (1.38). But after the considerations in Section II-2.1, this is equivalent to the original system (1.34). That the initial conditions are satisfied follows from (1.44) and (1.45), since we choose for the initial prescriptions of u and v on $\tilde{k}$ exactly these φ and ψ.

The proof given here is not a constructive one. The idea of the proof follows that of H. Lewy[5] for a differential equation of the second order.

1.5 Proof of the Lemma

We begin with the existence theorem for the initial-value problem of a hyperbolic differential equation of the second order in the form

$$u_{x_1x_2} = f(x_1, x_2, u, p, q) \quad \text{where} \quad p = u_{x_1}, \qquad q = u_{x_2}, \tag{1.46}$$

in two independent variables x_1, x_2 and where the characteristic families of curves $\dot{C}$ of (1.46) are $x_i = \text{const}$. As initial curve k we choose

$$k: x_i = x_i(\lambda) \quad \text{where} \quad x_i(\lambda) \in C^1 \quad \text{in} \quad \lambda_1 < \lambda < \lambda_2, \tag{1.47}$$

<hr>

[5] H. Lewy[2], K. Friedrichs and H. Lewy[3], and R. Courant and D. Hilbert[4].

and assume that it coincides nowhere with a characteristic curve. This means that k can be represented in the form $x_1 = x_1(x_2)$ as well as in the form $x_2 = x_2(x_1)$. This in turn is equivalent to the fact that the $x_i(\lambda)$ are strictly monotone functions of λ.

The values of $u \in C^1$, $p \in C^0$, $q \in C^0$ are prescribed arbitrarily on k so that the *strip relation*

$$\dot{u} = p\dot{x}_1 + q\dot{x}_2 \quad \left(\text{where} \quad \cdot = \frac{d}{d\lambda}\right) \quad \text{on} \quad k \tag{1.48}$$

is satisfied. Hence exactly two functions may be prescribed arbitrarily. A different, though equivalent formulation, proceeds as follows: Because of the strict monotonicity, u and therefore also p may be represented on k as continuous functions of x_1, and we require that $p = \kappa'(x_1)$, where κ' is the derivative of a function $\kappa(x_1)$. Correspondingly q can be represented on k in the form $q = \tau'(x_2)$. Then the strip relation becomes

$$\dot{u} = \kappa'(x_1)\dot{x}_1 + \tau'(x_2)\dot{x}_2 = \frac{d}{d\lambda}\{\kappa(x_1(\lambda)) + \tau(x_2(\lambda))\}, \tag{1.49}$$

or, if the constant of integration is included in κ or τ,

$$u(\lambda) = \kappa(x_1(\lambda)) + \tau(x_2(\lambda)) \quad \text{on} \quad k. \tag{1.50}$$

THEOREM 1.[6]   Let the initial curve k (1.47) be given.   Let the smallest rectangle with sides parallel to the axes which contain k be

$$R: \alpha < x_1 < \beta, \quad \gamma < x_2 < \delta \quad (\alpha, \gamma = -\infty; \quad \beta, \delta = +\infty \text{ are admissible}).$$
$$\tag{1.51}$$

Furthermore let $f(x_1, x_2, u, s, t) \in C^0$ for $x_1, x_2 \in R$ and $-\infty < u, s, t < \infty$ and suppose f is bounded on every closed subrectangle $\bar{r}$ of R and satisfies a Lipschitz condition with respect to u, s, and t in $\bar{r}$:

$$|f(x_1, x_2, u^1, s^1, t^1) - f(x_1, x_2, u^2, s^2, t^2)| \leqslant M(|u^1 - u^2| + |s^1 - s^2| + |t^1 - t^2|). \tag{1.52}$$

If $\kappa(x_1)$, $\tau(x_2) \in C^1$ for $\alpha < x_1 < \beta$ and $\gamma < x_2 < \delta$, then (1.46) has exactly one solution in R for which $u \in C^1$, $u_{x_1 x_2} \in C^0$ and which satisfies the prescription (1.50) on k.

Remark.   For the first part of the conclusion (existence) a Lipschitz condition in s and t alone will suffice:

$$|f(x_1, x_2, u, s^1, t^1) - f(x_1, x_2, u, s^2, t^2)| \leqslant M(|s^1 - s^2| + |t^1 - t^2|) \tag{1.52a}$$

as was shown by J. Conlan[6].   For the second part (uniqueness), (1.52a) does not suffice.

Example[6]. $u_{x_1 x_2} = -|u|^{\frac{1}{2}}$, $k: x_1 = \lambda$, $x_2 = \lambda$ where $-\infty < \lambda < \infty$, $R: -\infty < x_1 < \infty$, $-\infty < x_2 < \infty$; $\kappa(x_1) = 0$, $\tau(x_2) = 0$.

This initial-value problem has the solutions $u \equiv 0$ and $u = \frac{1}{144}(x_2 - x_1)^4$. The function $f = |u|^{\frac{1}{2}}$ satisfies (1.52a) but not (1.52).

---

[6] The presentation follows exactly the one in E. Kamke[5].

Proof of Theorem 1.   It will be given by an iteration process.   For $\overset{v}{u}(x)$ we use

$$\overset{0}{u}(x) = \kappa(x_1) + \tau(x_2), \tag{1.53}$$

$$\overset{v}{u}(x) = \overset{0}{u}(x) \pm \int_{\mathfrak{B}(x)} f(x, \overset{v-1}{u}, \overset{v-1}{u_{x_1}}, \overset{v-1}{u_{x_2}})\, dx, \qquad v = 1, 2, \ldots, \tag{1.54}$$

where the integration has to be taken over the domain $\mathfrak{B}(x)$ of Figure 31.   The upper or lower sign is to be used depending on whether k is decreasing or increasing.   If $\overset{v-1}{u}(x) \in C^1$ in R, then clearly so is $\overset{v}{u}(x)$.   We find (see Figure 31)

$$\overset{v}{u_{x_1}}(x) = \overset{0}{u_{x_1}}(x) + \int_{\bar{x}_2}^{x_2} f(x, \overset{v-1}{u}, \overset{v-1}{u_{x_1}}, \overset{v-1}{u_{x_2}})\, dx_2,$$
$$\overset{v}{u_{x_2}}(x) = \overset{0}{u_{x_2}}(x) + \int_{\bar{x}_1}^{x_1} f(x, \overset{v-1}{u}, \overset{v-1}{u_{x_1}}, \overset{v-1}{u_{x_2}})\, dx_1. \tag{1.55}$$

Let $\xi = (\xi_1, \xi_2) \in R$ be an arbitrary but fixed point.   By $\mathfrak{B}(\xi)$ we denote the domain which is bounded by k and the parallels to the coordinate axes through the point (ξ_1, ξ_2).   We now show that the sequence $\overset{v}{u}(x)$, together with its first derivatives, converges uniformly for all

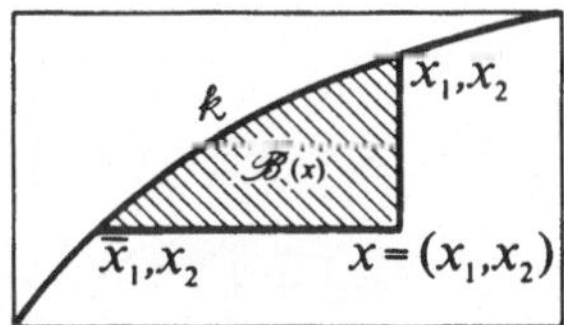

Figure 31

$x \in \overline{\mathfrak{B}}(\xi)$.   Let the parallels to the x_1- and x_2-axes passing through $\xi = (\xi_1, \xi_2)$ intersect k at the points (α_0, ξ_2) and (ξ_1, β_0).   Then for every $x \in \overline{\mathfrak{B}}(\xi)$ the area $|\overline{\mathfrak{B}}(x)|$ of $\overline{\mathfrak{B}}(x)$ can be estimated by

$$|\overline{\mathfrak{B}}(x)| \leqslant |x_1 - \alpha_0|\,|x_2 - \beta_0|. \tag{1.56}$$

Furthermore we have

$$|x_1 - \bar{x}_1| \leqslant |x_1 - \alpha_0|, \qquad |x_2 - \bar{x}_2| \leqslant |x_2 - \beta_0|. \tag{1.57}$$

Now f is bounded on $\overline{\mathfrak{B}}(\xi)$: $|f| \leqslant A$.   If we set

$$2K = \max\,(2, |\xi_1 - \alpha_0|, |\xi_2 - \beta_0|), \tag{1.58}$$

we have

$$\frac{1}{|x_1 - \alpha_0|} + \frac{1}{|x_2 - \beta_0|} \geqslant \frac{1}{|\xi_1 - \alpha_0|} + \frac{1}{|\xi_2 - \beta_0|} \geqslant \frac{1}{2K} + \frac{1}{2K} = \frac{1}{K}, \tag{1.59}$$

$$|\overline{\mathfrak{B}}(x)| \leqslant |x_1 - \alpha_0|\,|x_2 - \beta_0| \leqslant K(|x_1 - \alpha_0| + |x_2 - \beta_0|). \tag{1.60}$$

From (1.54) and (1.55) it follows that

$$|\overset{1}{u}(x) - \overset{0}{u}(x)| \leqslant A|\overline{\mathfrak{B}}(x)| \leqslant AK(|x_1 - \alpha_0| + |x_2 - \beta_0|),$$
$$|\overset{1}{u_{x_i}}(x) - \overset{0}{u_{x_i}}(x)| \leqslant A(|x_1 - \alpha_0| + |x_2 - \beta_0|) \leqslant AK(|x_1 - \alpha_0| + |x_2 - \beta_0|). \tag{1.61}$$

In general for every $x \in \overline{\mathcal{B}}(\xi)$ we have

$$\left.\begin{array}{l} |\overset{\nu}{u}(x) - \overset{\nu-1}{u}(x)| \\[2mm] |\overset{\nu}{u}_{x_i}(x) - \overset{\nu-1}{u}_{x_i}(x)| \end{array}\right\} \leqslant \frac{A}{3M} \frac{(3KM)^\nu}{\nu!} (|x_1 - \alpha_0| + |x_2 - \beta_0|)^\nu, \tag{1.62}$$

as will be shown by complete induction. The inequality (1.62) is true for $\nu = 1$. Using (1.52) and (1.62) we obtain from (1.54)

$$|\overset{\nu+1}{u}(x) - \overset{\nu}{u}(x)| \leqslant \int_{\mathcal{B}(x)} |f(x, \overset{\nu}{u}, \overset{\nu}{u}_{x_1}, \overset{\nu}{u}_{x_2}) - f(x, \overset{\nu-1}{u}, \overset{\nu-1}{u}_{x_1}, \overset{\nu-1}{u}_{x_2})| \, dx$$

$$\leqslant M \int_{\mathcal{B}(x)} \{|\overset{\nu}{u} - \overset{\nu-1}{u}| + |\overset{\nu}{u}_{x_1} - \overset{\nu-1}{u}_{x_1}| + |\overset{\nu}{u}_{x_2} - \overset{\nu-1}{u}_{x_2}|\} \, dx$$

$$\leqslant \frac{A}{3M} \frac{(3KM)^\nu}{\nu!} 3M \left| \int_{\alpha_0}^{x_1} \int_{\beta_0}^{x_2} (|x_1 - \alpha_0| + |x_2 - \beta_0|)^\nu \, dx_1 \, dx_2 \right|. \tag{1.63}$$

For estimating the integral we use the elementary inequality

$$(r_1 + r_2)^{\nu+2} - r_1^{\nu+2} - r_2^{\nu+2} \leqslant (\nu + 2) r_1 r_2 (r_1 + r_2)^\nu \quad \text{with} \quad r_1, r_2 \text{ real} \geqslant 0, \tag{1.64}$$

which follows immediately by induction. Then we have

$$\left| \int_{\alpha_0}^{x_1} \int_{\beta_0}^{x_2} \cdots \, dx_1 x_2 \right|$$

$$= \frac{1}{(\nu + 1)(\nu + 2)} \{(|x_1 - \alpha_0| + |x_2 - \beta_0|)^{\nu+2} - |x_1 - \alpha_0|^{\nu+2} - |x_2 - \beta_0|^{\nu+2}\}$$

$$\leqslant \frac{1}{\nu + 1} |x_1 - \alpha_0| \, |x_2 - \beta_0| (|x_1 - \alpha_0| + |x_2 - \beta_0|)^\nu$$

$$\leqslant \frac{K}{\nu + 1} (|x_1 - \alpha_0| + |x_2 - \beta_0|)^{\nu+1}; \tag{1.65}$$

the last inequality follows from (1.60). If we use this estimate in (1.63) the induction proof is finished. Quite analogously we take care of the validity of the second estimate in (1.62) by noting that $K \geqslant 1$.

The uniform convergence of the infinite series

$$u(x) = \overset{0}{u}(x) + \sum_{\nu=1}^{\infty} [\overset{\nu}{u}(x) - \overset{\nu-1}{u}(x)], \tag{1.66}$$

together with the first derivatives to a limit function $u(x) \in C^1$ for every $x \in \overline{\mathcal{B}}(\xi)$ follows. But now we see that $\lim_{\nu \to \infty} \overset{\nu}{u}(x) = u(x)$ uniformly, including the first derivatives. Hence in (1.54) we may pass to the limit under the integral sign and obtain

$$u(x) = \overset{0}{u}(x) \pm \int_{\mathcal{B}(x)} f(x, u, u_{x_1}, u_{x_2}) \, dx. \tag{1.67}$$

The complete conclusion about existence in the theorem now follows immediately from (1.67) by differentiation after the integral has been written out in detail. If $\pmb{\ell}$ is represented in the form $x_1 = g(x_2)$ and $x_2 = h(x_1)$, then, in the case of Figure 31, we obtain

$$u(x_1, x_2) = \overset{0}{u}(x_1, x_2) - \int_{g(x_2)}^{x_1} \left(\int_{x_2}^{h(y_1)} f(y_1, y_2, u, u_{y_1}, u_{y_2}) \, dy_2 \right) dy_1. \tag{1.68}$$

Let $u^1(x)$ and $u^2(x)$ be two solutions of the initial-value problem. Then from (1.46), (1.52), (1.54), and (1.55) we obtain the estimates

$$|u^1 - u^2| \leqslant M \int_{\mathcal{B}(x)} \Psi(x)\,dx, \qquad |p^1 - p^2| \leqslant M \left| \int_{\hat{x}_2}^{x_2} \Psi(x)\,dx_2 \right|,$$

$$|q^1 - q^2| \leqslant M \left| \int_{\hat{x}_1}^{x_1} \Psi(x)\,dx_1 \right|. \qquad (1.69)$$

Here we set

$$\Psi(x) = |u^1(x) - u^2(x)| + |u^1_{x_1}(x) - u^2_{x_1}(x)| + |u^1_{x_2}(x) - u^2_{x_2}(x)|. \qquad (1.70)$$

If we set $\mu = \max_{x \varepsilon \bar{\mathcal{B}}(\xi)} \Psi(x)$ and notice (1.56) and (1.57), then, for the point x at which the maximum is assumed, (1.69) becomes

$$\mu \leqslant 3M\mu |x_1 - \alpha_0|\,|x_2 - \beta_0|. \qquad (1.71)$$

If first we choose ξ sufficiently close to k so that $6M|x_1 - \alpha_0|\,|x_2 - \beta_0| \leqslant 1$ for all $x \varepsilon \bar{\mathcal{B}}(\xi)$, then from (1.71) it follows that $\mu \leqslant \frac{1}{2}\mu$; hence $\mu = 0$ in $\bar{\mathcal{B}}(\xi)$. By suitable repetition of this procedure we obtain the entire conclusion.

A completely analogous theorem with analogous proof exists for a system of equations of the form (1.46). If we put $x = (x_1, x_2)$, $u = (u_1(x), \ldots, u_n(x))$, $p = (p_1, \ldots, p_n)$, $q = (q_1, \ldots, q_n)$, and $f = (f_1, \ldots, f_n)$, then the system

$$u_{x_1 x_2} = f(x, u, p, q) \qquad (1.72)$$

which consists of n equations for the n unknown functions $u_1, \ldots, u_n$ will be considered. We set $p_i = \partial u_i / \partial x_1$ and $q_j = \partial u_j / \partial x_2$. Then on k the values of $u \varepsilon C^1$, p, $q \varepsilon C^0$ are prescribed so that the strip relation

$$\dot{u} = p\dot{x}_1 + q\dot{x}_2 \quad \text{with} \quad \cdot = \frac{d}{d\lambda} \qquad (1.73)$$

holds.

The following theorem now contains the desired lemma from Section 1.4 as a special case. A special quasi-linear system in $N = n + m$ unknown functions $u = (u_1, \ldots, u_N)$ is considered in the domain $\mathcal{D}$ of the x_1, x_2-plane. For abbreviation we set $u_{k x_1} = p_k$, $u_{k x_2} = q_k$. The system is of the form

$$l^i \equiv \sum_{k=1}^{N} a^i_k p_k + g^i = 0, \qquad i = 1, \ldots, n,$$

$$l^i \equiv \sum_{k=1}^{N} a^i_k q_k + g^i = 0, \qquad i = n+1, \ldots, N, \qquad (1.74)$$

where a^i_k, g^i are functions of x and u; and a^i_k, $g^i \varepsilon C^2$ for $x \varepsilon \mathcal{D}$, $-\infty < u_1, \ldots, u_N < \infty$. We consider an initial-value problem for (1.74). As initial curve k we select the curve (1.47) together with the hypotheses made there, and we suppose that it is contained in $\mathcal{D}$; on k we prescribe the values

$$u = \varphi(\lambda) \quad \text{on} \quad k \quad \text{where} \quad \varphi = (\varphi_1, \ldots, \varphi_N) \quad \text{and} \quad \varphi_i(\lambda) \varepsilon C^1. \qquad (1.75)$$

THEOREM 2.   Under the foregoing hypotheses, let the prescriptions φ be such that the determinant $|a_k^i(x,\varphi)| \neq 0$ on k where in a_k^i the functions φ were substituted in place of u.   Then the initial-value problem has exactly one solution $u(x) \in C^1$ in a suitable neighborhood $\mathfrak{U}(k)$ of k for which $u_{x_1 x_2} \in C^0$ in $\mathfrak{U}(k) - k$.

Proof.   We show that in a neighborhood of k this initial-value problem is equivalent to a special case of the problem (1.72), (1.73).   Whereas Theorem 1 was a global existence theorem, Theorem 2 is only a local one.

First Step.   Let $u(x)$ be a solution of (1.74) and (1.75) for which $u \in C^1$ in $\mathfrak{U}(k)$ and $u_{x_1 x_2} \in C^0$ in $\mathfrak{U}(k) - k$.   By differentiation in (1.74) we form $l_{x_2}^i$, $\bar{l}_{x_1}^i$.   Now $|a_k^i| \neq 0$ on k and by continuity in a neighborhood of k.

Therefore the differentiated equations may be uniquely solved for $u_{i x_1 x_2}$.   Then we obtain a system of the form (1.72) where $f \in C^1$ in this neighborhood.   On k in any case the values $u = \varphi \in C^1$ are prescribed.   Since $u \in C^1$ on k in particular, u satisfies the strip relation $\dot{u} = p\dot{x}_1 + q\dot{x}_2$.   Since k is strictly monotone this can be solved in the form $q = 1/\dot{x}_2(\dot{\varphi} - p\dot{x}_1)$. If we substitute this expression for q in $\bar{l}^i = 0$ of (1.74), then on k (1.74) can be interpreted as a linear system of equations for p, because the values of u are prescribed on k.   Since $|a_k^i| \neq 0$, this system may be solved uniquely for p, which is determined on k.   But by the equation $q = 1/\dot{x}_2(\dot{\varphi} - p\dot{x}_1)$ the $q = (q_1, \ldots, q_N)$ are uniquely determined on k, and a problem of the form (1.72) and (1.73) which has the solution u has been established in a suitable neighborhood of k.

Second Step.   Let $u(x)$ be a solution in a neighborhood $\mathfrak{V}(k)$ of k of the problem

$$u_{x_1 x_2} = f(x, u, p, q), \qquad u = \varphi \quad \text{on} \quad k \quad \text{where} \quad \dot{\varphi} = p\dot{x}_1 + q\dot{x}_2 \quad \text{on} \quad k; \qquad (1.76)$$

p and q are given on k so that $l^i = 0$ and $\bar{l}^i = 0$ there and f has the special form obtained in the first step.

Because of the special form of f and because $|a_k^i| \neq 0$ on k, u satisfies $l_{x_2}^i = 0$ and $\bar{l}_{x_1}^i = 0$ in $\mathfrak{V}(k)$.   Then $l^i = 0$ and $\bar{l}^i = 0$; thus u satisfies the system (1.74) and solves the problem (1.74) and (1.75); furthermore $u \in C^1$ in $\mathfrak{V}(k)$, $u_{x_1 x_2} \in C^0$ in $\mathfrak{V}(k) - k$.

After the equivalence in a neighborhood of k has been proved the extended version of Theorem 1 gives the desired conclusion.

Now the lemma of Section 1.4 is a special case of Theorem 2.   If, for instance, in (1.43) we set

$$x_1(\xi) = u_1(\xi), \qquad x_2(\xi) = u_2(\xi), \qquad u(\xi) = u_3(\xi), \qquad v(\xi) = u_4(\xi),$$

where $\xi = (\xi_1, \xi_2)$, then, after writing

$$\begin{aligned}
x_{1\xi_1} &= p_1, & x_{2\xi_1} &= p_2; & u_{\xi_1} &= p_3, & v_{\xi_1} &= p_4; \\
x_{1\xi_2} &= q_1, & x_{2\xi_2} &= q_2, & u_{\xi_2} &= q_3, & v_{\xi_2} &= q_4,
\end{aligned}$$

we obtain the system

$$
\begin{aligned}
\alpha_2 p_1 - \alpha_1 p_2 &&&= 0 \\
g p_1 &+ \alpha_1 \rho p_3 + \alpha_1 \sigma p_4 &&= 0 \\
\bar{\alpha}_2 q_1 - \bar{\alpha}_1 q_2 &&&= 0 \\
\bar{g} q_1 &+ \bar{\alpha}_1 \bar{\rho} q_3 + \bar{\alpha}_1 \bar{\sigma} q_4 &&= 0.
\end{aligned}
\tag{1.77}
$$

From (1.74) we obtain $N = 4$, $n = m = 2$. Calculation of $|a_k^i|$ in (1.77) gives

$$
|a_k^i| = -\alpha_1 \bar{\alpha}_1 \begin{vmatrix} \alpha_1 & \alpha_2 \\ \bar{\alpha}_1 & \bar{\alpha}_2 \end{vmatrix} \begin{vmatrix} \rho & \sigma \\ \bar{\rho} & \bar{\sigma} \end{vmatrix},
\tag{1.78}
$$

which is different from zero on k by (1.38) and (1.39). Thus the proof in Section 1.4 is finished.

1.6 Hyperbolic Systems in the Form of Conservation Theorems[7]

For linear and almost-linear equations we obtained global existence theorems but we did not succeed with this in the case of quasi-linear equations. There it was only possible to guarantee the existence of a solution in a neighborhood of k and nothing was said about the size of this neighborhood. By means of an example we now show that these statements cannot be improved in general. To this end we let the system (1.34) reduce to one equation and denote the variable x_2 as time t.

· *Example.* $u u_x + u_t = 0$, $u(x,0) = \varphi(x)$, $-\infty < x < \infty$, $0 \leqslant t < \infty$.

THEOREM. Let $\varphi(x) \in C^1$ in $-\infty < x < \infty$. If $\varphi'(x) \geqslant 0$ there, then there is exactly one solution $u(x,t) \in C^1$ in $-\infty < x < \infty$, $0 \leqslant t < \infty$. If $\varphi'(x)$ can assume negative values and if $-\kappa$ is the greatest lower bound of $\varphi'(x)$, then a solution does not exist at time $t = 1/\kappa$.

Proof. The uniquely determined (as an exercise, prove it) solution is given implicitly by

$$
f(x, t, u) \equiv u - \varphi(x - ut) = 0.
\tag{1.79}
$$

If $\partial f/\partial u = 1 + t\varphi' \neq 0$, then by the implicit-function theorem (1.79) may be solved for u and we have $u = u(x,t)$ for a solution. If $\partial f/\partial u = 0$, then u cannot have a continuous and not even a bounded first derivative u_x. This case occurs when $t = 1/\kappa$, which concludes the proof.

In this way it can be shown that in the case of quasi-linear equations in general we have to restrict ourselves to local results. The situation becomes more complicated if such quasi-linear equations arise from a natural phenomena. For example, as was indicated in Section I-1.6 the flows in gas dynamics are described by a quasi-linear system (I-1.27), which, according to Section II-1.3, is of hyperbolic or elliptic type respectively, depending

---

[7] A summarizing report has been given by P. Lax[7].

on whether the velocity $|v|$ of the flow is larger or smaller than the velocity of sound a. If we pose an initial-value problem for (I-1.27) where

$$v(x,0) = v_0(x), \qquad \rho(x,0) = \rho_0(x) \tag{1.80}$$

at time $t = 0$ and the prescriptions are to be such that (I-1.27) is of hyperbolic type at time $t = 0$, then it is clear that the flow fixed in this way will exist in some form for all later times $t > 0$. Of course it may happen that the flow at some time reaches a subsonic speed or that shocks or other changes occur on the flow. Therefore we will have to define suitable functions as weak solutions of (I-1.27) even if they are not continuous.

In general, the description of natural phenomena is done by systems of partial differential equations which have the form of conservation laws[8], that is, all terms in which spatial derivatives of the unknown function appear have the form of divergence expressions.

If $u = (u_1, \ldots, u_n)$ are the unknown functions of x and t, and if $a(u)$, $b(x,u)$ are vector-valued functions of u,x, respectively, $u: a = (a_1(u), \ldots, a_n(u))$ and $b = (b_1(x,u) \ldots, b_n(x,u))$, then we consider the initial-value problem

$$u_t + [a(u)]_x + b(x,u) = 0, \qquad u(x,0) = u_0(x), \qquad -\infty < x < \infty, \qquad 0 \leqslant t < \infty. \tag{1.81}$$

Let $w(x,t) \in C^1$ be an arbitrary vector which vanishes for sufficiently large $|x|$ and t. If $u \in C^1$ is a solution of (1.81), then

$$0 = -\int_0^\infty \int_{-\infty}^\infty (w, u_t + [a(u)]_x + b(x,u)) \, dx \, dt. \tag{1.82}$$

After integration by parts we find

$$0 = \int_0^\infty \int_{-\infty}^\infty \{(w_t,u) + (w_x,a(u)) - (w, b(x,u))\} dx \, dt + \int_{-\infty}^\infty (w(x,0),u_0(x)) \, dx. \tag{1.83}$$

The function $u(x,t)$ is called a weak solution of (1.81) if it is bounded and satisfies relation (1.83) for every admissible w.

If this $u \in C^1$, then these considerations can be traced back, and by the fundamental lemma of the calculus of variations, this u turns out to be a classical solution of (1.81).

Problem. According to (I-1.27), the equations of gas dynamics in one space dimension are of the form

$$v_{1t} + v_1 v_{1x} + \frac{p'}{\rho} \rho_x = 0, \qquad \rho_t + (\rho v_1)_x = 0. \tag{1.84}$$

Show that they can be written in the form (1.81) where $u_1 = \rho v_1$, $u_2 = \rho$; $a_1 = \dfrac{(u_1)^2}{u_2}$, $a_2 = u_1$; $b_1 = p_x$, $b_2 = 0$.

1.7   Riemann's Method

The topics to be treated here do not seem to occupy a central position in the modern theory of partial differential equations, so we restrict ourselves to the presentation of a special case that is of great importance.

The most general linear hyperbolic system in two unknown functions and two independent variables may be assumed in the normal form (1.3). As in the considerations of

Section II-1.4 we may pass analogously to the integrable normal form; this, however, would not be possible in the case of more than two unknown functions. Therefore we begin with the system

$$U^\tau_{x_\tau} + \sum_{j=1}^{2} A^\tau_j U^j + C^\tau = 0, \qquad \tau = 1, 2, \tag{1.85}$$

and assume A^τ_j, $C^\tau \in C^1$ in a domain $\mathfrak{D}$ of the $x = (x_1,x_2)$-plane. The characteristic curves c_τ are $x_\tau = \text{const}$. For (1.85) we pose an initial-value problem: $U^\tau = \Psi^\tau \in C^0$ on k. Here k can be represented in the form $x_2 = x_2(x_1)$ as well as in the form $x_1 = x_1(x_2)$. We consider the solution $U^\tau \in C^1$ of the initial-value problem at all points $y \in \mathfrak{D}$ for which the corresponding domain $\bar{\mathfrak{B}}(y)$ is contained in $\mathfrak{D}$. According to Figure 32, the boundary of

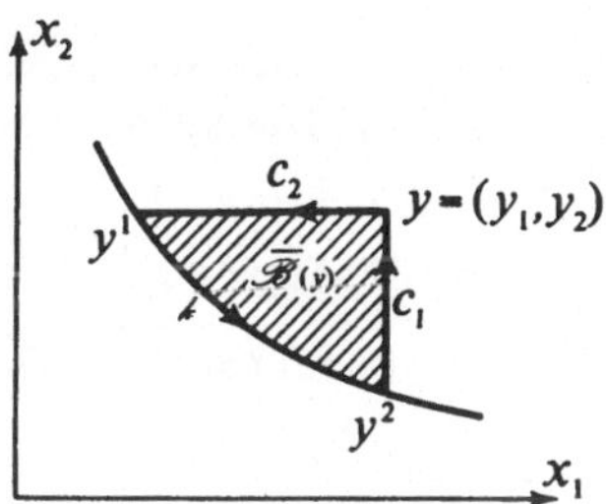

Figure 32

$\mathfrak{B}(y)$ is $k + c_1 + c_2$. Furthermore, let $V^\tau(x) \in C^1$ in $\bar{\mathfrak{B}}(y)$ be two auxiliary functions to be determined. By the Gauss integral theorem we have

$$\int_{\dot{\mathfrak{B}}(y)} [-V^2 U^2 \, dx_1 + V^1 U^1 \, dx_2] = \int_{\mathfrak{B}(y)} \{V^1 U^1_{x_1} + V^2 U^2_{x_2} + V^1_{x_1} U^1 + V^2_{x_2} U^2\} \, dx. \tag{1.86}$$

Now we compute $U^\tau_{x_\tau}$ from (1.85) and substitute these expressions in (1.86). Finally the $V^\tau \in C^1$ in $\bar{\mathfrak{B}}(y)$ shall be solutions of the *adjoint system* to (1.85):

$$V^\tau_{x_\tau} = \sum_{j=1}^{2} A^j_\tau V^j, \qquad \tau = 1, 2. \tag{1.87}$$

Then (1.86) gives the relation

$$\int_{\dot{\mathfrak{B}}(y)} - V^2 U^2 \, dx_1 + V^1 U^1 \, dx_2 + \int_{\mathfrak{B}(y)} (V^1 C^1 + V^2 C^2) \, dx = 0. \tag{1.88}$$

In order to transform the boundary integral we solve (1.85):

$$A^1_2 U^2 = -(U^1_{x_1} + A^1_1 U^1 + C^1), \qquad A^2_1 U^1 = -(U^2_{x_2} + A^2_2 U^2 + C^2). \tag{1.89}$$

If finally we introduce the new auxiliary functions $\tilde{V}^\tau(x) \in C^1$,[8]

$$V^1 = \tilde{V}^1 A^2_1 \quad \text{on} \quad c_1, \qquad V^2 = \tilde{V}^2 A^1_2 \quad \text{on} \quad c_2, \tag{1.90}$$

[8] This trick is necessary in order not to have to exclude eventual zeros of A^1_2, A^2_1. The case in which one or both quantities vanish identically is without interest. According to Section 1.2, $A^1_2 \equiv 0$ for instance, means that the solution of the initial-value problem can be found immediately by quadratures. The method of Riemann which is developed here does not give that much. The case $A^2_1 \equiv 0$ can be reduced to the above by renumbering.

then the boundary integrals in (1.88) on c_1 and c_2 become

$$\int_{c_1} V^1 U^1 \, dx_2 = \int_{c_1} \tilde{V}^1 A_1^2 U^1 \, dx_2 = - \int_{c_1} \tilde{V}^1 (U_{x_2}^2 + A_2^2 U^2 + C^2) \, dx_2$$

$$= - \tilde{V}^1 U^2 \Big|_{y^2}^{y} + \int_{c_1} U^2 (\tilde{V}_{x_2}^1 - A_2^2 \tilde{V}^1) \, dx_2 - \int_{c_1} \tilde{V}^1 C^2 \, dx_2. \tag{1.91}$$

$$\int_{c_2} - V^2 U^2 \, dx_1 = \int_{-c_2} V^2 U^2 \, dx_1$$

$$= - \tilde{V}^2 U^1 \Big|_{y^1}^{y} + \int_{-c_2} U^1 (\tilde{V}_{x_1}^2 - A_1^1 \tilde{V}^2) \, dx_1 - \int_{-c_2} \tilde{V}^2 C^1 \, dx_1. \tag{1.92}$$

Furthermore we require:

$$\tilde{V}_{x_2}^1 - A_2^2 \tilde{V}^1 = 0 \quad \text{on} \quad c_1, \tag{1.93a}$$

$$\tilde{V}_{x_1}^2 - A_1^1 \tilde{V}^2 = 0 \quad \text{on} \quad c_2. \tag{1.93b}$$

Thus the variable point y will enter in the auxiliary functions V^r and $\tilde{V}^r$ as a parameter, so that more precisely we write $V^r(x,y)$, and so forth. Using (1.90) and (1.93), on c_1 or c_2 we find

$$V^1(y_1, x_2, y) = C_1 A_1^2(y_1, x_2) \exp \left[\int_{\kappa_1}^{x_2} A_2^2(y_1, \xi) \, d\xi \right],$$

$$V^2(x_1, y_2, y) = C_2 A_2^1(x_1, y_2) \exp \left[\int_{\kappa_2}^{x_1} A_1^1(\eta, y_2) \, d\eta \right]. \tag{1.94}$$

Here κ_1 and κ_2 are fixed points on c_1 and c_2, respectively. Along with the representation of k it will be best to choose the initial points $\kappa_2 = x_1(y_2)$ and $\kappa_1 = x_2(y_1)$. By Section 1.3, equations (1.87) and (1.94) form a distinguished characteristic problem for the solutions V^r of the adjoint system. Finally from (1.88) we obtain the integral representation for the solution U^r of the initial-value problem:

$$\tilde{V}^2(y,y) U^1(y) + \tilde{V}^1(y,y) U^2(y) = \tilde{V}^2(y^1,y) \Psi^1(y^1) + \tilde{V}^1(y^2,y) \Psi^2(y^2) + \int_k (V^1 \Psi^1 \, dx_2 - V^2 \Psi^2 \, dx_1)$$

$$- \int_{c_1} \tilde{V}^1 C^2 \, dx_2 + \int_{c_2} \tilde{V}^2 C^1 \, dx_1 + \int_{\mathcal{B}(y)} (V^1 C^1 + V^2 C^2) \, dx \equiv H(V^1, V^2). \tag{1.95}$$

By fixing the free constants C_1 and C_2 in (1.94), from (1.95) we can in many ways obtain formulas for the solution $U^r(y)$ of the initial-value problem at the arbitrary point y. For reasons of elegance we should strive for formulas that are as symmetric as possible, though we look for the simplest ones possible. By fixing the constants of integration in (1.94) in two ways we arrive at two pairs of uniquely determined adjoint functions $V^r(x,y)$ and $\underline{V}^r(x,y)$ by requiring

$$\tilde{V}^1(x,y) = 0, \quad \tilde{V}^2(y,y) = 1 \quad \text{on} \quad c_1, \tag{1.96a}$$

$$\underline{\tilde{V}}^1(y,y) = 1, \quad \underline{\tilde{V}}^2(x,y) = 0 \quad \text{on} \quad c_2. \tag{1.96b}$$

Then (1.95) yields[9]

$$U^1(y) = \tilde{V}^2(y^1,y) \Psi^1(y^1) + \int_k (V^1 \Psi^1 \, dx_2 - V^2 \Psi^2 \, dx_1)$$

$$+ \int_{c_2} \tilde{V}^2 C^1 \, dx_1 \pm \int_{\mathcal{B}(y)} (V^1 C^1 + V^2 C^2) \, dx,$$

$$U^2(y) = \underline{\tilde{V}}^1(y^2,y) \Psi^2(y^2) + \int_k (\underline{V}^1 \Psi^1 \, dx_2 - \underline{V}^2 \Psi^2 \, dx_1)$$

$$- \int_{c_1} \underline{\tilde{V}}^1 C^2 \, dx_2 \pm \int_{\mathcal{B}(y)} (\underline{V}^1 C^1 + \underline{V}^2 C^2) \, dx. \tag{1.97}$$

---

[9] If $\mathcal{B}$ is traversed in the positive sense, then the plus or minus is to be used depending on whether k is decreasing or increasing. The second case, for example, occurs if y lies below k.

Since the adjoint functions do not know of the prescriptions Ψ^r on k and of the C^r it is, in particular, evident that the solutions of the initial-value problem depend continuously on the prescriptions and the C^r. The formula may be useful if for a system several initial-value problems with different $U^r = \Psi^r$ are to be solved. If the adjoint functions, which are independent of the initial-value problems are known, they can be solved by quadratures, using (1.97). Since only distinguished characteristic initial-value problems (see Section 1.3) were used, a corresponding formula to (1.97) can at once be found even for the initial-value problem for the system (1.1).[10] It will be noted, however, that in the case of more than two equations, because of the appearance of "inner" characteristics, the method becomes much more difficult.[11]

Problem 1. The adjoint system of the adjoint system is the original system (1.85) where $C^r \equiv 0$.

Problem 2. Give formulas that are as symmetric as possible.

1.8 An Example

As a special case we obtain the method of Riemann for a linear hyperbolic differential equation of the second order; it may be assumed in the form (1.24)

$$u_{x_1 x_2} + \sum_{k=1}^{2} a_k(x) u_{x_k} + a(x)u + f(x) = 0. \tag{1.98}$$

We prescribe the initial values $u = \psi_1$ and, say, $u_{x_1} = \psi_2$ on k. By putting $u = u_1, u_{x_1} = u_2$ and $U^1 = u_1$, $U^2 = a_2 u_1 + u_2$ we obtain the system (in normal form)

$$U^1_{x_1} + a_2 U^1 - U^2 = 0, \qquad U^2_{x_2} - (u_{2x_2} + a_1 a_2 - a) U^1 + a_1 U^2 + f = 0. \tag{1.99}$$

The corresponding adjoint system (1.87) reads

$$V^1_{x_1} = a_2 V^1 - (a_{2x_2} + a_1 a_2 - a) V^2, \qquad V^2_{x_2} = -V^1 + a_1 V^2. \tag{1.100}$$

Elimination of V^1 is possible, and for $V^2 = W$ it gives

$$W_{x_1 x_2} - (a_1 W)_{x_1} - (a_2 W)_{x_2} + aW = 0. \tag{1.101}$$

Because $A^1_2 = -1$, from (1.90) it follows that $V^2 = -\bar V^2$, and by the foregoing substitution $(1.93b)$ becomes

$$W_{x_1} - a_2 W = 0 \quad \text{on} \quad c_2. \tag{1.102}$$

For $\bar V^1$ we take $(1.96a)$: $\bar V^1(x,y) = 0$ on c_1. According to (1.90) this means $V^1(x,y) = 0$ on c_1, and from (1.100) we obtain the requirement

$$W_{x_2} - a_1 W = 0 \quad \text{on} \quad c_1. \tag{1.103}$$

[10] The presentation is similar to that in W. Haack and G. Hellwig[9].

[11] E. Holmgren[10], F. Rellich[11], G. Hellwig[12]. More important is the extension to equations with more than two independent variables; this leads to the theory of J. Hadamard. See J. Hadamard [13].

The function $W(x,y)$ is uniquely determined by (1.101), (1.102), (1.103), and $W(y,y) = -1$. The first equation in (1.97) gives the Riemann solution formula (here, $U^1 = u_1 = u$)

$$u(y) = -W(y^1,y)\psi_1(y^1) + \int_k (-W_{x_2} + a_1 W)\psi_1 \, dx_2 - W(a_2\psi_1 + \psi_2) \, dx_1 \pm \int_{\mathcal{B}(y)} Wf \, dx. \tag{1.104}$$

One can hardly find this very short formula in the literature. To arrive at the usual Riemann formula for ψ_i we have to write the corresponding u-values again, and on the right-hand side of (1.104) we must add and subtract the term

$$-\frac{1}{2}\int_k d(Wu) \equiv -\frac{1}{2}\int_k (W_{x_1}u + Wu_{x_1}) \, dx_1 + (W_{x_2}u + Wu_{x_2}) \, dx_2$$
$$\equiv -\tfrac{1}{2}W(y^2,y)u(y^2) + \tfrac{1}{2}W(y^1,y)u(y^1). \tag{1.105}$$

We obtain

$$u(y) = -\tfrac{1}{2}\{W(y^1,y)u(y^1) + W(y^2,y)u(y^2)\} + \int_k \{P(x,y) \, dx_2 - Q(x,y) \, dx_1\}$$
$$\pm \int_{\mathcal{B}(y)} W(x,y)f(x) \, dx, \tag{1.106}$$

where

$$P(x,y) = a_1(x)W(x,y) + \tfrac{1}{2}[W(x,y)u_{x_2}(x) - u(x)W_{x_2}(x,y)],$$
$$Q(x,y) = a_2(x)W(x,y) + \tfrac{1}{2}[W(x,y)u_{x_1}(x) - u(x)W_{x_1}(x,y)]. \tag{1.107}$$

Problem.* Solve the initial-value problem for the telegraph equation in one space dimension [see (I-1.24) and put $t = x_2$]

$$\alpha^2 u_{x_1x_1} - u_{x_2x_2} - 2\beta u_{x_2} = 0; \qquad u(x_1,0) = u_0(x_1), \qquad u_{x_2}(x_1,0) = u_1(x_1)$$

by means of formula (1.106). Here we put $\alpha^2 = c^2/\epsilon\mu$, $2\beta = 4\pi\sigma/\epsilon$.

2

BOUNDARY AND INITIAL-VALUE PROBLEMS FOR EQUATIONS OF HYPERBOLIC AND PARABOLIC TYPE IN TWO INDEPENDENT VARIABLES

2.1 Posing of the Problem

In most applications the pure initial-value problem is not the center of interest; usually, suitable boundary conditions are added which are distinguished in their great variety. The simplest examples were mentioned in Sections I-2.3 and I-4.4. Among the methods for treating such problems, we mentioned the separation method, with which the reader will certainly be familiar in the case of simple problems, and the calculus of Laplace transforms, which to a great extent takes into account the group or semigroup character of the solution operator. We shall justify this calculus to a modest extent.

We consider the second-order linear partial differential operator in the space variable x and time t:

$$\bar{D}u \equiv \bar{A}u + k(x)\{r_1 u_{tt} + r_2 u_t\}, \qquad 0 \leqslant t < \infty, \qquad l \leqslant x \leqslant m \tag{2.1}$$

where[12]

$$\bar{A}u \equiv -[p(x)u_x]_x + q(x)u$$

and make the following hypotheses:

$$p(x),\ k(x) \in C^3, \qquad q(x) \in C^1, \qquad p(x) > 0, \qquad k(x) > 0 \quad \text{for} \quad l \leqslant x \leqslant m \quad \text{with} \quad l < m,$$

and let the numbers r_1 and r_2 be such that $r_1 > 0$ (Case 1) or $r_1 = 0$, $r_2 > 0$ (Case 2).

We recognize immediately that for $l \leqslant x \leqslant m$ and $0 \leqslant t < \infty$, (2.1) is of hyperbolic type in Case 1 and of parabolic type in Case 2. Now we consider Problem I.

Problem I

$$\bar{D}u = k(x)f(x,t);$$
$$u(x,0) = u_0(x) \quad \text{(in Cases 1, 2)}, \qquad u_t(x,0) = u_1(x) \quad \text{(in Case 1)};$$
$$\bar{R}_i u \equiv a_{i1}u(l,t) + a_{i2}u_x(l,t) + a_{i3}u_t(l,t) + b_{i1}u(m,t) + b_{i2}u_x(m,t) + b_{i3}u_t(m,t)$$
$$= f_i(t), \qquad i = 1, 2, \tag{2.2}$$

[12] The more general expression $\bar{B}u = -a(x)u_{xx} + b(x)u_x + c(x)$, where $a(x) > 0$, can be put into the form $\bar{A}u$ by multiplication by $M(x) = (1/a)\exp[-\int(b/a)\,dx]$.

where $u_0(x)$, $u_1(x)$ and $f_i(t)$ are functions to be prescribed arbitrarily; $a_{i1}, \ldots, b_{i3}$ are constants. In the parabolic Case 2, the second initial condition disappears. The boundary conditions given in (2.2) are nearly always adequate in the applications. However, from now on we shall treat only boundary conditions without time derivatives: $a_{i3} = b_{i3} = 0$. Furthermore we assume

$$\begin{vmatrix} a_{12} & b_{12} \\ a_{22} & b_{22} \end{vmatrix} \neq 0, \tag{2.3}$$

a hypothesis from which we can free ourselves in many cases.[13] By *Liouville's transformation* we may simplify Problem I. If for x we introduce the new variable y, and for u the new unknown function w,

$$y = \frac{1}{\kappa} \int_l^x \sqrt{k/p}\, dx \quad \text{where} \quad \kappa = \int_l^m \sqrt{k/p}\, dx, \qquad w = \bar{g}u \quad \text{where} \quad \bar{g} = \sqrt[4]{kp}, \tag{2.4}$$

then an analogous problem in $w(y,t)$ arises, with the new quantities

$$\tilde{l} = 0, \qquad \tilde{m} = 1, \qquad \tilde{p} = \tilde{k} = 1, \qquad \tilde{q}(y) = \frac{\bar{g}_{yy}}{\bar{g}} + \kappa^2 \frac{q}{k}, \qquad \tilde{r}_1 = \kappa^2 r_1, \qquad \tilde{r}_2 = \kappa^2 r_2,$$

and the $\tilde{f}$ changed correspondingly. Also the boundary conditions and condition (2.3) remain preserved in the above form. Therefore we may assume these simplifications in (2.2) to begin with and put in (2.2):

$$\bar{A}u \equiv -u_{xx} + q(x)u, \qquad p = k = 1, \qquad l = 0, \qquad m = 1.$$

Problem. Verify these statements.

2.2   The Calculus of the Laplace Transform

Suppose $u(x,t)$ is a solution of Problem I, provided with the above simplifications, and with "sufficiently good" properties. The following considerations are formal though not superficial in nature, because the calculus must be enumerated before it can be justified.

---

[13] The essential but not the only hypothesis is

$$\text{rank} \begin{pmatrix} a_{11} & a_{12} & b_{11} & b_{12} \\ a_{21} & a_{22} & b_{21} & b_{22} \end{pmatrix} = 2.$$

Our further considerations also remain valid in the following cases

(1) $\begin{cases} \bar{R}_1 u \equiv a_{11}u(l,t) + b_{11}u(m,t) \\ \bar{R}_2 u \equiv a_{21}u(l,t) + b_{21}u(m,t) \end{cases}$ with $\begin{vmatrix} a_{11} & b_{11} \\ a_{21} & b_{21} \end{vmatrix} \neq 0,$

(2) $\begin{cases} \bar{R}_1 u \equiv a_{11}u(l,t) + b_{11}u(m,t) + u_x(m,t) \\ \bar{R}_2 u \equiv u(l,t) + b_{21}u(m,t), \end{cases}$

(3) $\begin{cases} \bar{R}_1 u \equiv a_{11}u(l,t) + u_x(l,t) + b_{11}u(m,t) \\ \bar{R}_2 u \equiv a_{21}u(l,t) + u(m,t), \end{cases}$

(4) $\begin{cases} \bar{R}_1 u \equiv u_x(l,t) - u_x(m,t) \\ \bar{R}_2 u \equiv u(l,t) - u(m,t) \end{cases}$

and in all cases which originate from the already mentioned cases by renumbering of $\bar{R}_1 u$ and $\bar{R}_2 u$ or suitable linear combinations of $\bar{R}_1 u$ and $\bar{R}_2 u$.

Let $s = \xi + i\eta$ be a complex parameter and let

$$\int_0^\infty e^{-st}u(x,t)\,dt = v(x,s)$$

be convergent for $s = s_0$. Then we say that v arose from u by application of the *Laplace operator*

$$L \equiv \int_0^\infty e^{-st}\{\cdots\}\,dt$$

and we write $Lu = v$. Then it follows that $v(x,s)$ is analytic in s in the right half-plane $\xi > \xi_0$ (ξ_0 being the real part of s_0), but this need not be proved by the reader. By integration by parts we formally find the rules

$$Lu_t = sv - u_0(x), \qquad Lu_{tt} = s^2 v - su_0(x) - u_1(x), \tag{2.5}$$

where the terms at the upper limit ∞ have been omitted. We still need the inverse operator L^{-1} which is obtained formally by way of the Fourier integral theorem as

$$u(x,t) = L^{-1}v \equiv \frac{1}{2\pi i}\int_{\xi_1-i\infty}^{\xi_1+i\infty} e^{st}v(x,s)\,ds, \qquad \xi_1 > \xi_0. \tag{2.6}$$

The formal deduction of (2.6) can be done as follows: The Fourier integral theorem for the function $\Phi(t)$, in its complex form, reads

$$\Phi(t) = \frac{1}{2\pi}\int_{-\infty}^\infty e^{i\eta t}\left\{\int_{-\infty}^\infty e^{-i\eta\tau}\Phi(\tau)\,d\tau\right\}d\eta.$$

If we apply this formula to the function

$$\Phi(t) = \begin{cases} e^{-\xi_1 t}u(x,t) & \text{for} \quad t \geqslant 0, \\ 0 & \text{for} \quad t < 0, \end{cases}$$

for fixed x and ξ_1, then for $t \geqslant 0$ and $s = \xi_1 + i\eta$ we obtain

$$e^{-\xi_1 t}u(x,t) = \frac{1}{2\pi}\int_{-\infty}^\infty e^{i\eta t}\left\{\int_0^\infty e^{-s\tau}u(x,\tau)\,d\tau\right\}d\eta$$

$$= \frac{1}{2\pi}\int_{-\infty}^\infty e^{i\eta t}v(x,s)\,d\eta.$$

Finally, (2.6) follows with $ds = i\,d\eta$. We remark that under suitable hypotheses L^{-1} has the semigroup property with respect to t. If for L^{-1} we write somewhat more explicitly $L^{-1}(t)$, then $L^{-1}(t_1 + t_2) = L^{-1}(t_1)L^{-1}(t_2)$.

If we apply the L-operator formally to (2.2), take account of the simplifications agreed upon, and interchange the operators $\bar{A}$ and $\bar{R}_i$ with L, we obtain

$$L\bar{A}u \equiv L\{-u_{xx} + q(x)u\} \equiv \int_0^\infty e^{-st}\{-u_{xx}(x,t) + q(x)u(x,t)\}\,dt$$

$$= -\left(\int_0^\infty e^{-st}u(x,t)\,dt\right)_{xx} + q(x)\int_0^\infty e^{-st}u(x,t)\,dt$$

$$= -v_{xx}(x,s) + q(x)v(x,s) = -v'' + q(x)v,$$

since we consider s as a parameter and there are no derivatives with respect to s. Furthermore we have

$$L\{a_{i1}u(0,t) + a_{i2}u_x(0,t)\} = \int_0^\infty e^{-st}a_{i1}u(0,t)\,dt + \lim_{x\to 0}\left(\int_0^\infty e^{-st}a_{i2}u(x,t)\,dt\right)_x$$

$$= a_{i1}v(0,s) + a_{i2}\lim_{x\to 0}v_x(x,s) = a_{i1}v(0,s) + a_{i2}v'(0,s).$$

Formally applying the L-operator to the whole problem (2.2), applying (2.5), and setting $Lf = g$ and $Lf_i = g_i$ gives the transformed Problem II.

Problem II

$$Dv \equiv Av + \{r_1s^2 + r_2s\}v = \{r_1s + r_2\}u_0(x) + r_1u_1(x) + g(x,s),$$
$$R_iv \equiv a_{i1}v(0,s) + a_{i2}v'(0,s) + b_{i1}v(1,s) + b_{i2}v'(1,s) = g_i(s), \tag{2.7}$$

where $Av \equiv -v'' + q(x)v$. This is a boundary-value problem for an ordinary differential equation and it can be solved explicitly if a fundamental system is known. If $v(x,s)$ is the solution, then to justify the calculus we must prove that $u = L^{-1}v$ is a solution of Problem I.

First it is useful to decompose Problem I into subproblems. If $f \equiv f_i \equiv 0$, then the *initial-boundary value problem* I_1 arises; if $u_0 \equiv u_1 \equiv 0$, $f_i \equiv 0$, then the *inhomogeneous problem* I_2 arises, and if, finally, $u_0 \equiv u_1 \equiv 0$, $f \equiv 0$, then we have the *equilibrium problem* I_3. If these problems are solved, then obviously the sum of their solutions is a solution of Problem I. Problem I_1 is the most difficult and I_3 is the simplest. Actually, I_3 could be reduced to I_1 and I_2, although this is not to be recommended. Indeed, if $h(x,t)$ is a function which satisfies $\bar{R}_ih = f_i$ and if u is the solution of I_3, then $w = u - h$ satisfies the problem

$$\bar{D}w = -\bar{D}h; \qquad w(x,0) = -h(x,0), \qquad w_t(x,0) = -h_t(x,0); \qquad \bar{R}_iw = 0, \tag{2.8}$$

which may be decomposed into I_1 and I_2. The converse then gives the desired result.

2.3   Solution of the Transformed Problem II[14]

In $0 \leqslant x \leqslant 1$ we consider the equation

$$D_\rho v \equiv -v'' + q(x)v + \rho^2 v = 0 \quad \text{where} \quad \rho^2 = r_1s^2 + r_2s, \qquad s = \xi + i\eta. \tag{2.9}$$

Sometimes it will be useful to work in the complex ρ-plane with $\rho = \zeta + i\mu$. In the s-plane we denote (2.9) correspondingly by D_sv. The multiple-valuedness arising from the inverse transformation $\rho = \sqrt{r_1s^2 + r_2s}$ is without consequence; it can be removed by the agreement that in case of a real positive radicand, ρ itself should be real and positive.

Let $v^k = v^k(x,\rho)$ or $v^k = v^k(x,s)$, $k = 1, 2$ be a *fundamental system* of (2.9). The v^k may be chosen so that they are entire functions in s.

---

[14] This section deals exclusively with questions about ordinary differential equations. For the convenience of the reader we prove them in Section 2.5.

LEMMA 1.   Let $q(x)$ be continuously differentiable in $0 \leqslant x \leqslant 1$.   For a suitably normed fundamental system $v^k(x,\rho)$ the following representation holds in a domain $\mathcal{B}_1$: $|\rho| \geqslant R$, $\varsigma \geqslant \varsigma_0$ (with a constant ς_0) of the ρ-plane for sufficiently large R:

$$v^k(x,\rho) = e^{\rho \varphi^k x} \left\{ 1 + \frac{[\varphi^k Q(x)]}{\rho} \right\},$$

$$\{v^k(x,\rho)\}' = \rho \varphi^k \, e^{\rho \varphi^k x} \left\{ 1 + \frac{[\varphi^k Q(x)]}{\rho} \right\}, \tag{2.10}$$

where $\varphi^1 = -\varphi^2 = 1$ and

$$Q(x) = \frac{1}{2} \int_0^x q(\tau) \, d\tau.$$

By $[\alpha]$ we understand Birkhoff's expression $[\alpha] = \alpha + E/\rho$, where $E = E(x,\rho)$ is continuous in both variables and uniformly bounded for large $|\rho|$.   The same representations with other E-quantities also hold in $\mathcal{B}_2$: $|\rho| \geqslant R$, $\varsigma \leqslant \varsigma_0$.

We consider the operator D_ρ which has the domain of definition

$$\mathfrak{D}_{D_\rho}: v \in C^2 \quad \text{in} \quad 0 \leqslant x \leqslant 1, \qquad R_i v = 0.$$

As is well known, the solution of the problem $D_\rho v = j$, $R_i v = 0$ for continuous $j(x)$ and arbitrary complex ρ is given by the *Green's resolvent* G, in the form

$$v = D_\rho^{-1} j \equiv \int_0^1 G(x, y, \rho) j(y) \, dy, \tag{2.11}$$

provided that G exists.   Conversely, in this case every $v \in \mathfrak{D}_{D_\rho}$ may be written in the form (2.11) where we only have to set $j = D_\rho v$.

To solve Problem II we shall decompose it into three subproblems and then put the solution together from these.   From (2.7) and (2.11), where again s is chosen as a parameter, we obtain

$$v(x,s) = \int_0^1 G(x, y, s)\{(r_1 s + r_2)u_0(y) + r_1 u_1(y) + g(y,s)\} \, dy + \sum_{k=1}^{2} g_k(s) G_k(x,s). \tag{2.12}$$

The last term is the solution of Problem II$_3$.   This expression is found by elementary determination of the constants c_1, c_2 in the general solution $v = c_1 v^1 + c_2 v^2$ for Problem II$_3$.

By means of the fundamental system v^k, the quantities G and G_k can be constructed explicitly.   We know that

$$G(x, y, s) = -\frac{1}{2\Delta(s)} \begin{vmatrix} v^1 & v^2 & \gamma \\ R_1 v^1 & R_1 v^2 & R_1 \gamma \\ R_2 v^1 & R_2 v^2 & R_2 \gamma \end{vmatrix}, \qquad G_k(x,s) = \frac{\Delta_k(x,s)}{\Delta(s)}, \tag{2.13}$$

where

$$\Delta(s) = \begin{vmatrix} R_1 v^1 & R_1 v^2 \\ R_2 v^1 & R_2 v^2 \end{vmatrix}, \qquad \Delta_1(x,s) = \begin{vmatrix} v^1 & v^2 \\ R_2 v^1 & R_2 v^2 \end{vmatrix}, \qquad \Delta_2(x,s) = -\begin{vmatrix} v^1 & v^2 \\ R_1 v^1 & R_1 v^2 \end{vmatrix},$$

$$\gamma(x, y, s) = \pm \frac{1}{\delta(x,s)} \begin{vmatrix} v^1(x,s) & v^2(x,s) \\ v^1(y,s) & v^2(y,s) \end{vmatrix}, \qquad \begin{matrix} + & \text{for} & x \geqslant y, \\ - & \text{for} & x \leqslant y, \end{matrix} \tag{2.14}$$

$$\delta(x,s) = \begin{vmatrix} v^{1\prime}(x,s) & v^{2\prime}(x,s) \\ v^1(x,s) & v^2(x,s) \end{vmatrix} = \delta(0,s).$$

Since the v^k are entire functions in s, G turns out to be a meromorphic function in s because of (2.13) and (2.14); that is, the only singularities of G are poles. Obviously G exists for those s-values for which $\Delta(s) \neq 0$. The Wronskian determinant $\delta(x,s)$ in (2.14) does not depend on x, as is well known.

The most convenient method for finding the expression for G given in (2.13) is as follows: As is well known, all solutions of $D_s v = j$ are determined from a fundamental system v^1, v^2 of $D_s v = 0$ by the formula

$$v(x,s) = c_1' v^1(x,s) + c_2' v^2(x,s) - \frac{1}{\delta(0,s)} \int_0^x (v^1(x,s)v^2(y,s) - v^2(x,s)v^1(y,s))j(y)\,dy$$

whose validity can be verified immediately. Here c_1' and c_2' are arbitrary constants. Quite analogously we can represent all solutions of $D_s v = j$ also in the form

$$v(x,s) = c_1'' v^1(x,s) + c_2'' v^2(x,s) + \frac{1}{\delta(0,s)} \int_x^1 (v^1(x,s)v^2(y,s) - v^2(x,s)v^1(y,s))j(y)\,dy$$

with new constants c_1'' and c_2''. By adding the two formulas and using new constants c_1 and c_2, we find

$$v(x,s) = c_1 v^1(x,s) + c_2 v^2(x,s) - \frac{1}{2} \int_0^1 \gamma(x,\,y,\,s)j(y)\,dy, \tag{2.14a}$$

where for $\gamma(x,\,y,\,s)$ we have expression (2.14). Now the constants c_1 and c_2 have to be determined so that v satisfies exactly the boundary conditions $R_i v = 0$. By using the formula for γ and decomposing the integral $\int_0^1 \cdots$ into $\int_0^x \cdots + \int_x^1 \cdots$, after some calculation, we find

$$R_1 v = c_1 R_1 v^1 + c_2 R_1 v^2 - \frac{1}{2} \int_0^1 (R_1\gamma)j\,dy,$$

$$R_2 v = c_1 R_2 v^1 + c_2 R_2 v^2 - \frac{1}{2} \int_0^1 (R_2\gamma)j\,dy.$$

The conditions $R_i v = 0$ now give two inhomogeneous equations in two unknowns c_1 and c_2. If these are determined and their values substituted in (2.14a), then, after some easy changes, (2.14a) appears in the form

$$v(x,s) = \int_0^1 G(x,\,y,\,s)j(y)\,dy$$

where $G(x,\,y,\,s)$ coincides exactly with the expression given in (2.13).

LEMMA 2.   Under hypothesis (2.3) there is a half-plane $\xi > \xi_*$ which does not contain the zeros s_1, s_2, $\ldots$ of $\Delta(s)$ and in which G is analytic in s.

The zeros are cut out by small circles (punctured s-plane) and are mapped into the ρ-plane by $\rho_i^2 = r_1 s_i^2 + r_2 s_i$ and removed there in a similar manner (punctured ρ-plane).

LEMMA 3.   From (2.3), in the punctured ρ-plane for $|\rho| \geqslant r$ and $0 \leqslant x \leqslant 1$ for suitably large r, we have the estimates

$$|D_\rho^{-1}j| \leqslant \frac{M}{|\rho|}, \qquad \left|\frac{d}{dx} D_\rho^{-1}j\right| \leqslant M \quad \text{for} \quad \text{continuous } j(x), \tag{2.15}$$

$$|D_\rho^{-1}j| \leqslant \frac{M}{|\rho|^2}, \qquad \left|\frac{d}{dx} D_\rho^{-1}j\right| \leqslant \frac{M}{|\rho|} \quad \text{for } j, j', \text{ continuous} \tag{2.16}$$

where for fixed j the constants M depend only on the radii of the circles which had been cut out. The hypotheses indicate that in general the estimates (2.16) can be obtained from (2.15) by integration by parts.

Before carrying these estimates over to the s-plane we distinguish the Cases 1 and 2 by putting Case 2 in parentheses.

For sufficiently large $|s|$ in the punctured s-plane, we find for (2.15):

$$\text{Case 1: } |D_s^{-1}j| \leqslant \frac{N}{|s|}, \qquad \left|\frac{d}{dx} D_s^{-1}j\right| \leqslant N;$$

$$\left(\text{Case 2: } |D_s^{-1}j| \leqslant \frac{N}{\sqrt{|s|}}, \qquad \left|\frac{d}{dx} D_s^{-1}j\right| \leqslant N\right); \tag{2.17}$$

and analogously for (2.16)

$$\text{Case 1: } |D_s^{-1}j| \leqslant \frac{N}{|s|^2}, \qquad \left|\frac{d}{dx} D_s^{-1}j\right| \leqslant \frac{N}{|s|};$$

$$\left(\text{Case 2: } |D_s^{-1}j| \leqslant \frac{N}{|s|}, \qquad \left|\frac{d}{dx} D_s^{-1}j\right| \leqslant \frac{N}{\sqrt{|s|}}\right). \tag{2.18}$$

2.4   Justification of the Calculus

This shall be done for Problem I_1 where we use the simplification of Problem I_1 obtained by means of the Liouville transformation. According to (2.12), the solution of the transformed Problem I_1 is given by

$$v(x,s) = D_s^{-1}h \equiv \int_0^1 G(x, y, s)h(y,s)\,dy \qquad h(x,s) = (r_1 s + r_2)u_0(x) + r_1 u_1(x). \tag{2.19}$$

THEOREM.   Let $q(x) \in C^1$, $u_0(x) \in C^3$, $u_1(x) \in C^2$ in $0 \leqslant x \leqslant 1$ and let u_0, u_1 satisfy the boundary conditions $\bar{R}_i u_k = 0$ for (2.3) and $a_{i3} = b_{i3} = 0$, $i = 1, 2$; $k = 0, 1$. Then, for suitable $\xi_1 > 0$,

$$u(x,t) = L^{-1}v \equiv \frac{1}{2\pi i} \int_{\xi_1 - i\infty}^{\xi_1 + i\infty} e^{st} \left\{\int_0^1 G(x, y, s)h(y,s)\,dy\right\} ds \tag{2.20}$$

$$= \lim_{\beta \to \infty} \sum_{|s_\nu| < \beta} e^{s_\nu t} R_\nu(x,t)$$

is a weak solution of I_1 where $e^{s_\nu t}R_\nu(x,t)$ is the residue of $e^{st}v(x,s)$ at the pole $s = s_\nu$ of $G(x, y, s)$. The numbers $\xi_1 > \xi_0$ and ξ_0 are chosen so that, according to Lemma 2, G becomes analytic in $\xi > \xi_0$, and thus so does v. More precisely, we have for Case 1: $u \in C^1$ in $0 \leqslant x \leqslant 1$, $0 \leqslant t < \infty$, and u satisfies the integrated equation

$$\int_0^x \int_0^t \bar{D}u\,dx\,dt = 0$$

with the initial conditions $u(x,0) = u_0(x)$, $u_t(x,0) = u_1(x)$, and the boundary conditions $\bar{R}_i u = 0$. For Case 2: u, $u_x \in C_0$ in $0 \leqslant x \leqslant 1$, $0 \leqslant t < \infty$ and $u = u(x,t)$ satisfies the integrated equation, with $u(x,0) = u_0(x)$ and $\bar{R}_i u = 0$.

Remark. (1) The assumptions $a_{i3} = b_{i3} = 0$ and the validity of the boundary conditions for the initial conditions $u_0(x)$, $u_1(x)$ are superfluous. Even if these are not satisfied, the u in (2.20) is still a solution in the classical sense, but then the proof becomes rather complicated.[15] In the cases where derivatives with respect to time occur in the boundary conditions the calculus turns out to be superior to the separation methods. (2) By using the theory of Lebesgue integration we prove a little more than was stated in the theorem.

Proof. First Step. It suffices to consider Case 1. By Lemma 2, $D_s^{-1}h$ exists in $\xi > \xi_0$ and is analytic in s. The initial prescriptions $u_i(x)$, $i = 0$, 1 are trivially solutions of

$$D_s v = D_s u_i \quad \text{where} \quad D_s u_i = A u_i + (r_1 s^2 + r_2 s) u_i. \tag{2.21}$$

By hypothesis $u_i \in \mathfrak{D}_D$, (see Section 2.3). Hence we have the representation

$$u_i(x) = D_s^{-1} D_s u_i = D_s^{-1} A u_i + (r_1 s^2 + r_2 s) D_s^{-1} u_i, \qquad i = 0, 1. \tag{2.22}$$

From (2.22) it follows that

$$
\begin{aligned}
(r_1 s + r_2) D_s^{-1} u_0 &= \frac{u_0}{s} - \frac{1}{s} D_s^{-1} A u_0, \\[2mm]
r_1 D_s^{-1} u_1 &= \frac{u_1}{s^2} - \frac{1}{s^2} D_s^{-1}(A u_1 + r_2 s u_1)
\end{aligned}
\tag{2.23}
$$

for $i = 0$, 1. By consideration of (2.19) we have

$$v(x,s) = D_s^{-1}h = \frac{u_0(x)}{s} + \frac{u_1(x)}{s^2} + w(x,s), \tag{2.24}$$

where

$$w(x,s) = -\frac{1}{s} D_s^{-1} z \quad \text{and} \quad z(x,s) = A u_0 + \frac{1}{s} A u_1 + r_2 u_1. \tag{2.25}$$

Relation (2.24) is an appropriate representation for the investigation of $u = L^{-1}v$. Different paths of integration which are needed for the complex integral are shown in Figures 33, 34, and 35. By our integral along the path of integration in Figure 33 we understand

$$\int_{\xi_1 - i\infty}^{\xi_1 + i\infty} \{\cdots\} \, ds \equiv \lim_{R_1, R_2 \to \infty} \int_{\xi_1 - iR_1}^{\xi_1 + iR_2} \{\cdots\} \, ds,$$

and the integrals for the paths of integration in Figures 34 and 35 are denoted by $\oint$ and $\oint$, respectively. Here R always has to be chosen so large that we may use the estimates in the s-plane from Lemma 3.

Second Step. First we consider the integral

$$\frac{1}{2\pi i} \oint e^{st} w(x,s) \, ds.$$

[15] See G. Hellwig[14,15]. The proof given above uses ideas of W. Mächler[16].

In any case z in (2.25) is continuous in x. Because of (2.17), w satisfies an estimate of the form $|w(x,s)| \leqslant c/|s|^2$ for sufficiently large $|s|$. Hence the integrals over $W_1 + W_2 + W_3$ converge to zero as $R \to \infty$. Thus we have

$$\lim_{R \to \infty} \frac{1}{2\pi i} \oint e^{st} w(x,s)\, ds = \frac{1}{2\pi i} \int_{\xi_1 - i\infty}^{\xi_1 + i\infty} e^{st} w(x,s)\, ds. \tag{2.26}$$

The last integral is uniformly convergent in $0 \leqslant x \leqslant 1,\ 0 \leqslant t \leqslant t_1 < \infty$, because we have the estimate

$$\left| \int_{\xi_1 + iA}^{\xi_1 + i\infty} e^{st} w(x,s)\, ds \right| \leqslant C\, e^{\xi_1 t_1} \int_A^\infty \frac{d\eta}{\xi_1^2 + \eta^2} < \epsilon \tag{2.27}$$

for $A > A_0(\epsilon)$. Therefore we may pass to the limit $t \to 0$ under the integral sign and obtain

$$\lim_{t \to 0} \frac{1}{2\pi i} \int_{\xi_1 - i\infty}^{\xi_1 + i\infty} e^{st} w(x,s)\, ds = \frac{1}{2\pi i} \int_{\xi_1 - i\infty}^{\xi_1 + i\infty} w(x,s)\, ds = \lim_{R \to \infty} \oint w(x,s)\, ds = 0. \tag{2.28}$$

The last term is correct, since

$$\lim_{R \to \infty} \int_{W_4} w(x,s)\, ds = 0$$

by the estimate given for w. The zero in (2.28) results from Cauchy's theorem: $\oint w(x,s)\, ds = 0$, since w is analytic in s in the enclosed domain of Figure 35; this is because, by Lemma

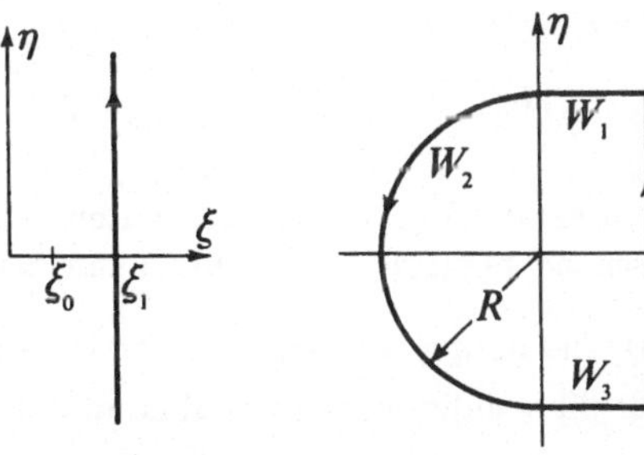

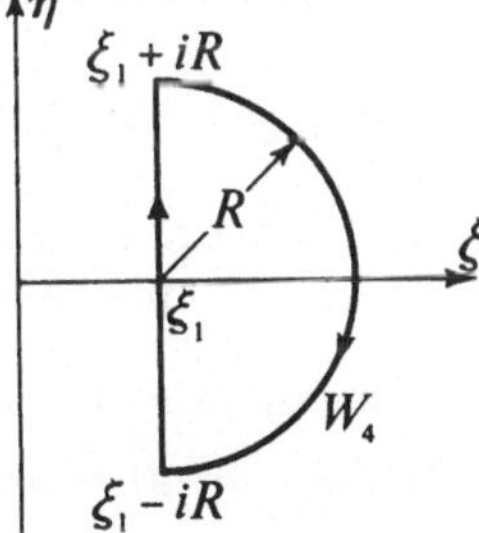

Figure 33   Figure 34   Figure 35

2, the poles of the Green's resolvent do not enter the half-plane $\xi > \xi_0$. Furthermore, by the residue theorem,

$$\frac{1}{2\pi i} \oint \frac{e^{st}}{s}\, ds = 1, \qquad \frac{1}{2\pi i} \oint \frac{e^{st}}{s^2}\, ds = t, \qquad \frac{1}{2\pi i} \oint \frac{e^{st}}{s^3}\, ds = \frac{t^2}{2}, \tag{2.29}$$

since $s = 0$ is the only pole and the *Laurent* expansions for the integrands are of the form

$$\frac{1}{s} + t + \cdots, \qquad \frac{1}{s^2} + \frac{t}{s} + \cdots, \qquad \frac{1}{s^3} + \frac{t}{s^2} + \frac{t^2}{2} \cdot \frac{1}{s} + \cdots. \tag{2.30}$$

Using easy estimates we find that the integrals over $W_1 + W_2 + W_3$ tend to zero as

$R \to \infty$. This happens uniformly for the first integral in $0 < t_0 \leqslant t \leqslant t_1 < \infty$,[16] and for the two others in $0 \leqslant t \leqslant t_1 < \infty$.

On account of (2.29) it is suggestive to define the function $u(x,t)$ first by

$$u(x,t) = \lim_{R \to \infty} \frac{1}{2\pi i} \oint e^{st} v(x,s)\, ds = u_0(x) + u_1(x)t + \frac{1}{2\pi i} \int_{\xi_1 - i\infty}^{\xi_1 + i\infty} e^{st} w(x,s)\, ds. \quad (2.31)$$

Then $\lim_{t \to 0} u(x,t) = u_0(x)$. By the foregoing considerations, for $t > 0$ we can write (2.31) in the form

$$u(x,t) = L^{-1}v = \frac{1}{2\pi i} \int_{\xi_1 - i\infty}^{\xi_1 + i\infty} e^{st} v(x,s)\, ds \quad \text{where} \quad \lim_{t \to 0} u(x,t) = u_0(x). \quad (2.32)$$

But we should note that in (2.32) the limit $t \to 0$ must not be taken under the integral sign since in general

$$\frac{1}{2\pi i} \int_{\xi_1 - i\infty}^{\xi_1 + i\infty} v(x,s)\, ds$$

does not make sense.

Third Step. In (2.25) $w(x,s)$ even satisfies an estimate of the form $|w(x,s)| \leqslant C/|s|^3$ for sufficiently large $|s|$. In fact in (2.25) we have Au_0, $u_1 \in C^1$ so that we can use the estimate (2.18) for these terms. The remaining term $-(1/s^2)D_s^{-1}Au_1$ also satisfies the given estimate since $Au_1 \in C^0$ and because then the estimate (2.17) holds. Then we immediately obtain the uniform convergence of the integral

$$\int_{\xi_1 - i\infty}^{\xi_1 + i\infty} s\, e^{st} w(x,t)\, ds \quad \text{and analogously that of} \quad \int_{\xi_1 - i\infty}^{\xi_1 + i\infty} e^{st} w'(x,s)\, ds \quad (2.33)$$

in $0 \leqslant x \leqslant 1$, $0 \leqslant t \leqslant t_1 < \infty$. Again, the integrals in (2.33) can be written in the form $\lim_{R \to \infty} \oint \{\cdots\}\, ds$. Quite analogously to the second step it follows from Cauchy's theorem that $\int_{\xi_1 - i\infty}^{\xi_1 + i\infty} sw(x,s)\, ds = 0$. Thus using (2.26), and noting (2.31) and (2.32), we prove that u_x, $u_t \in C^0$ and $\lim_{t \to 0} u_t(x,t) = u_1(x)$. This settles the validity of the initial conditions.

---

[16] From (2.29) we have

$$\lim_{R \to \infty} \int_{W_1 + W_2 + W_3} e^{st} \frac{1}{s}\, ds + \int_{\xi_1 - i\infty}^{\xi_1 + i\infty} e^{st} \frac{1}{s}\, ds = 1.$$

It suffices to show that the second integral is uniformly convergent in $0 < t_0 \leqslant t \leqslant t_1 < \infty$. By separation into real and imaginary parts we are essentially led to an integral of the form

$$\int_{-\infty}^{\infty} \frac{\eta \sin \eta t}{\xi_1^2 + \eta^2}\, d\eta = 2 \int_0^{\infty} \frac{\eta \sin \eta t}{\xi_1^2 + \eta^2}\, d\eta.$$

To see the uniform convergence in the last integral, apply the second mean-value theorem of the calculus of integration to

$$\int_0^{A} \frac{\eta \sin \eta t}{\xi_1^2 + \eta^2}\, d\eta$$

and then let $A \to \infty$.

Fourth Step. To show the validity of the integrated equation, first we have

$$\int_0^x \int_0^t Du \, dx \, dt = - \int_0^t \left\{ u_x - u_x \Big|_{x=0} \right\} dt + \int_0^x q(x) \left(\int_0^t u \, dt \right) dx$$
$$+ \int_0^x \left\{ r_1 \left(u_t - u_t \Big|_{t=0} \right) + r_2 \left(u - u \Big|_{t=0} \right) \right\} dx. \quad (2.34)$$

By integrating the second term in (2.31) with respect to t, then (by the uniform convergence) exchanging the order of integration and taking (2.28) into consideration, we find

$$\int_0^t u(x,t) \, dt = u_0(x)t + u_1(x) \frac{t^2}{2} + \frac{1}{2\pi i} \int_{\xi_1-i\infty}^{\xi_1+i\infty} \left[w(x,s) \int_0^t e^{st} \, dt \right] ds$$
$$= u_0(x)t + u_1(x) \frac{t^2}{2} + \frac{1}{2\pi i} \int_{\xi_1-i\infty}^{\xi_1+i\infty} \frac{e^{st}}{s} w(x,s) \, ds, \quad (2.35)$$

and quite analogously

$$\int_0^t u_x(x,t) \, dt = u_0'(x)t + u_1'(x) \frac{t^2}{2} + \frac{1}{2\pi i} \int_{\xi_1-i\infty}^{\xi_1+i\infty} \frac{e^{st}}{s} w'(x,s) \, ds. \quad (2.36)$$

If we now replace u, u_x, and u_t by the expressions resulting from (2.31), and

$$\int_0^t u \, dt, \qquad \int_0^t u_x \, dt$$

by (2.35), (2.36), then (2.34) becomes

$$\int_0^x \int_0^t Du \, dx \, dt = t \left\{ -u_0'(x) + u_0'(0) + \int_0^x [q(x)u_0(x) + r_2u_1(x)] \, dx \right\}$$
$$+ \frac{t^2}{2} \left\{ -u_1'(x) + u_1'(0) + \int_0^x q(x)u_1(x) \, dx \right\}$$
$$+ \frac{1}{2\pi i} \int_{\xi_1-i\infty}^{\xi_1+i\infty} \frac{e^{st}}{s} \left\{ w'(x,s) + w'(0,s) + \int_0^x [r_1s^2 + r_2s + q(x)]w(x,s) \, dx \right\} ds. \quad (2.37)$$

We have to show that the right-hand side in (2.37) equals zero. Now $v(x,s)$ satisfies the transformed equation

$$-v'' + q(x)v + (r_1s^2 + r_2s)v - \{(r_1s + r_2)u_0(x) + r_1u_1(x)\} = 0. \quad (2.38)$$

The representation for v found in (2.24) is substituted in (2.38) and then integrated over x from 0 to x and we obtain

$$\frac{1}{s} \left\{ -u_0'(x) + u_0'(0) + \int_0^x [q(x)u_0(x) + r_2u_1(x)] \, dx \right\}$$
$$+ \frac{1}{s^2} \left\{ -u_1'(x) + u_1'(0) + \int_0^x q(x)u_1(x) \, dx \right\} - w'(x,s) + w'(0,s)$$
$$+ \int_0^x [r_1s^2 + r_2s + q(x)]w(x,s) \, dx = 0. \quad (2.39)$$

If we multiply (2.39) by $(1/2\pi i s) \, e^{st}$ and integrate with respect to s from $\xi_1 - i\infty$ to $\xi_1 + i\infty$, then (2.37) follows immediately by use of (2.29), and

$$\int_0^x \int_0^t Du \, dx \, dt = 0$$

is finally proved.

The validity of the boundary conditions $R_i u = 0$ follows from (2.31) in this way. First, by hypothesis we have $R_i u_k = 0$, $k = 0, 1$. Furthermore, w has the representation

$w = -(1/s)D_s^{-1}z$, by (2.25), and thus it satisfies the transformed boundary conditions. But because of the uniform convergence, differentiation with respect to x and the passages to the limit $x \to 0$, $x \to 1$ in the last expression in (2.31) may be carried out under the integral sign.

The second expression in (2.20) results from the representation (2.31)

$$u(x,t) = \lim_{R \to \infty} \frac{1}{2\pi i} \oint e^{st}v(x,s)\, ds$$

by simply evaluating the integral by means of the calculus of residues. Thus the theorem is proved.

Addendum 1. In addition to

$$u(x,t) = \lim_{R \to \infty} \frac{1}{2\pi i} \oint e^{st}v(x,s)\, ds = \frac{1}{2\pi i} \int_{\xi_1 - i\infty}^{\xi_1 + i\infty} e^{st}v(x,s)\, ds$$

we easily verify the further representations

$$u_t(x,t) = \lim_{R \to \infty} \frac{1}{2\pi i} \oint s\, e^{st}v(x,s)\, ds, \qquad u_x(x,t) = \lim_{R \to \infty} \frac{1}{2\pi i} \oint e^{st}w'(x,s)\, ds, \qquad (2.40)$$

where the integrals in (2.40) converge uniformly in $0 \leqslant x \leqslant 1, 0 \leqslant t \leqslant t_1 < \infty$. Furthermore if we denote the residue of $se^{st}v(x,s)$ at the pole $s = s_\nu$, by

$$e^{s_\nu t}R_\nu^{(1)}(x,t),$$

then besides (2.20)

$$u_t(x,t) = \lim_{\beta \to \infty} \sum_{|s_\nu| < \beta} e^{s_\nu t}R_\nu^{(1)}(x,t),$$

$$u_x(x,t) = \lim_{\beta \to \infty} \sum_{|s_\nu| < \beta} e^{s_\nu t}\frac{\partial R_\nu(x,t)}{\partial x} \qquad (2.41)$$

holds uniformly in $0 \leqslant x \leqslant 1, 0 \leqslant t \leqslant t_1 < \infty$ and from this we obtain the representations

$$u_0(x) = \lim_{\beta \to \infty} \sum_{|s_\nu| < \beta} R_\nu(x,0), \qquad u_0'(x) = \lim_{\beta \to \infty} \sum_{|s_\nu| < \beta} \frac{\partial R_\nu(x,0)}{\partial x},$$

$$u_1(x) = \lim_{\beta \to \infty} \sum_{|s_\nu| < \beta} R_\nu^{(1)}(x,0), \qquad (2.42)$$

which are uniformly convergent too. Thus, sharp representation theorems have been found.

Addendum 2. It had been proved that the right-hand side of (2.34) equals zero. We reorder, take notice of $u(x,0) = u_0(x)$ and $u_t(x,0) = u_1(x)$, and obtain

$$-\int_0^t u_x\, dt + r_1 \int_0^t u_t\, dx = -\int_0^x q(x)\left(\int_0^t u\, dt\right) dx - \int_0^t u_x(0,t)\, dt$$

$$- \int_0^x \{-r_1 u_1(x) + r_2(u - u_0(x))\}\, dx. \qquad (2.43)$$

Since the right-hand side is continuously differentiable in x,t, so is the left-hand side:

$$\left[\left(-\int_0^t u_x\, dt\right)_x + r_1 u_t\right]_t + q(x)u + r_2 u_t = 0,$$

$$\left[-u_x + r_1\left(\int_0^x u_t\, dx\right)_t\right]_x + q(x)u + r_2 u_t = 0. \qquad (2.44)$$

If in (2.31) we formed u_{tt} by—so far not permitted—differentiating twice under the integral sign, then the integral

$$\int_{\xi_1-i\infty}^{\xi_1+i\infty} s^2\, e^{st} w(x,s)\, ds$$

would arise. If we set $s^2 w = \bar{w}$, then $\bar{w}(x,s)$ is of the order $1/s$ and the integral

$$\int_{-\infty}^{\infty} |\bar{w}(x,\ \xi_1 + i\eta)|^2\, d\eta \tag{2.45}$$

converges uniformly in $0 \leqslant x \leqslant 1$. If all integrals are taken in the sense of *Lebesgue*, then the existence of u_{tt} can be obtained almost everywhere. Then by (2.44)

$$\left(-\int_0^t u_x\, dt\right)_{xt} = -u_{xx}$$

exists almost everywhere and we have

$$\check{D}u \equiv -u_{xx} + r_1 u_{tt} + r_2 u_t + q(x)u = 0 \quad \text{in} \quad l \leqslant x \leqslant m, \qquad 0 \leqslant t < \infty, \tag{2.46}$$

except for a two-dimensional point set of measure zero.

The theorem given here contains numerous problems of physics and engineering. By using comprehensive tables[17] we can often obtain the solution explicitly. The tables contain the $u = L^{-1}v$ belonging to the v. Of the literature we mention G. Doetsch[18] and, in particular for numerical calculations, R. V. Churchill[19].

2.5 Auxiliary Considerations

Proof of Lemma 1 from Section 2.3. A fundamental system of equation (2.9):

$$D_\rho v \equiv -v'' + q(x)v + \rho^2 v = 0 \quad \text{where} \quad \rho = \zeta + i\mu \quad \text{and} \quad 0 \leqslant x \leqslant 1 \tag{2.47}$$

will be determined as follows: Let $v^1(x,\rho)$ be a solution of (2.47) with the initial conditions $v^1(0,\rho) = 1$, $v^{1\prime}(0,\rho) = \rho$. Furthermore let $v^2(x,\rho)$ be a solution of (2.47) with the initial conditions $v^2(0,\rho) = 1$, $v^{2\prime}(0,\rho) = -\rho$. Such initial-value problems can be solved uniquely by using known theorems. Then v^1, v^2 form a fundamental system since their Wronskian determinant turns out to be $\delta(0,\rho) \equiv v^{1\prime}v^2 - v^{2\prime}v^1 = 2\rho \neq 0$, because Lemma 1 requires the investigation only for such ρ for which $|\rho| \geqslant R$. We see immediately that $\delta(x,\rho) = \delta(0,\rho)$ so that the Wronskian is always a constant. For if we form $-v^2 D_\rho v^1 + v^1 D_\rho v^2 = 0$, where the zero follows from $D_\rho v^1 = 0$, $D_\rho v^2 = 0$, then the left-hand side gives exactly $\delta'(x,s)$, which proves $\delta(x,s) = \text{const}$.

For our domain $\mathcal{B}_1$ in Lemma 1 we choose $\mathcal{B}_1$: $|\rho| \geqslant 2M$, $\zeta \geqslant 0$ where $M = \max\limits_{0 \leqslant x \leqslant 1} |q(x)|$. The function $v^1(x,\rho)$ satisfies the relation

$$v^1(x,\rho) = e^{\rho x} + \frac{1}{2\rho} \int_0^x \left\{ e^{\rho(x-y)} - e^{-\rho(x-y)} \right\} q(y) v^1(y,\rho)\, dy \tag{2.48}$$

as is verified immediately by substitution into (2.47) and the initial conditions. From (2.48) it follows immediately that

$$v^1(x,\rho) = e^{\rho x} \left\{ 1 + \frac{1}{2\rho} \int_0^x (1 - e^{-2\rho(x-y)}) q(y) v^1(y,\rho)\, e^{-\rho y}\, dy \right\}. \tag{2.48a}$$

We set $\max\limits_{0\leqslant x\leqslant 1} |v^1(x,\rho)\, e^{-\rho x}| = M_\rho$ for $\rho \in \mathcal{B}_1$, and then from (2.48a) we obtain

$$|v^1(x,\rho)\, e^{-\rho x}| \leqslant 1 + \frac{2MM_\rho}{2|\rho|} = 1 + \frac{MM_\rho}{|\rho|} \leqslant 1 + \frac{M_\rho}{2}.$$

If we consider this inequality at the point x where the left-hand side assumes its maximum, then we obtain $M_\rho \leqslant 1 + M_\rho/2$ or $M_\rho \leqslant 2$, which proves that $|v^1(x,\rho)| \leqslant 2|e^{\rho x}|$ for all $\rho \in \mathcal{B}_1$ and all x in $0 \leqslant x \leqslant 1$. In (2.48) we now change notation:

$$v^1(y,\rho) = e^{\rho y} + \frac{1}{2\rho} \int_0^y \{e^{\rho(y-z)} - e^{-\rho(y-z)}\} q(z) v^1(z,\rho)\, dz$$

and substitute this representation in the right-hand side of (2.48a). Then we obtain

$$v^1(x,\rho) = e^{\rho x} \left\{ 1 + \frac{1}{2\rho} \int_0^x q(y)(1 - e^{-2\rho(x-y)})\, e^{-\rho y} \left[e^{\rho y} \right.\right.$$
$$\left.\left. + \frac{1}{2\rho} \int_0^y (e^{\rho(y-z)} - e^{-\rho(y-z)}) q(z) v^1(z,\rho)\, dz \right] dy \right\}.$$

Multiplying out and setting

$$Q(x) = \frac{1}{2} \int_0^x q(y)\, dy$$

then yields

$$v^1(x,\rho) = e^{\rho x} \left\{ 1 + \frac{Q(x)}{\rho} - \frac{1}{2\rho} \int_0^x q(y)\, e^{-2\rho(x-y)}\, dy + \frac{E_1(x,\rho)}{\rho^2} \right\}.$$

Here we set

$$E_1(x,\rho) = \frac{1}{4} \int_0^x q(y) \left\{ \int_0^y q(z) v^1(z,\rho)\, e^{-\rho z} [1 - e^{-2\rho(y-z)} - e^{-2\rho(x-y)} + e^{-2\rho(x-z)}]\, dz \right\} dy.$$

Because $1 \geqslant x \geqslant y \geqslant z \geqslant 0$, the continuity of $E_1(x,\rho)$ in both variables and $|E_1(x,\rho)| \leqslant C_1$ for all $\rho \in \mathcal{B}_1$ and all x in $0 \leqslant x \leqslant 1$ follow from this representation. Integration by parts finally gives

$$-\frac{1}{2\rho} \int_0^x \rho(y)\, e^{-2\rho(x-y)}\, dy = -\frac{1}{4\rho^2} e^{-2\rho(x-y)} q(y) \Big|_{y=0}^{y=x} + \frac{1}{4\rho^2} \int_0^x q'(y)\, e^{-2\rho(x-y)}\, dy$$

so that in conclusion we obtain for $\rho \in \mathcal{B}_1$ the representation

$$v^1(x,\rho) = e^{\rho x} \left\{ 1 + \frac{Q(x)}{\rho} + \frac{E(x,\rho)}{\rho^2} \right\} = e^{\rho x} \left\{ 1 + \frac{[Q(x)]}{\rho} \right\}. \tag{2.49}$$

By analogous considerations for $v^2(x,\rho)$ the first formula in (2.10) is thus proved.

From (2.48) and (2.49) it now follows that

$$v^{1\prime}(x,\rho) = \rho\, e^{\rho x} + \frac{1}{2} \int_0^x \{e^{\rho(x-y)} + e^{-\rho(x-y)}\} q(y) v^1(y,\rho)\, dy$$

$$= \rho\, e^{\rho x} \left\{ 1 + \frac{1}{2\rho} \int_0^x q(y)(e^{\rho(x-y)} + e^{-\rho(x-y)})\, e^{-\rho x} e^{\rho y} \left[1 + \frac{Q(y)}{\rho} + \frac{E(y,\rho)}{\rho^2} \right] dy \right\}$$

$$= \rho\, e^{\rho x} \left\{ 1 + \frac{1}{2\rho} \int_0^x q(y)(1 + e^{-2\rho(x-y)})\, dy + \frac{E_2(x,\rho)}{\rho^2} \right\}$$

$$= \rho\, e^{\rho x} \left\{ 1 + \frac{1}{2\rho} \int_0^x q(y)\, dy + \frac{E_3(x,\rho)}{\rho^2} \right\} = \rho\, e^{\rho x} \left\{ 1 + \frac{[Q(x)]}{\rho} \right\}.$$

If we carry out analogous considerations for $v^2(x,\rho)$, then formula (2.10) is proved completely. After similar arguments for the domain $\mathcal{B}_2$: $|\rho| \geq 2M$, $\zeta \leq 0$, Lemma 1 is proved. The proof makes it obvious that the special choice $\zeta_0 = 0$ in Lemma 1 is not essential. For a different choice of the constant ζ_0 the only change is that the estimates in the proof become a little more complicated.

Proof of Lemma 2 from 2.3. We carry out our arguments in the domain $\mathcal{B}_1(|\rho| \geq R,$ $\zeta \geq \zeta_0$, ζ_0 a negative constant). For $v^1(x,\rho)$, $v^2(x,\rho)$ we use the representations (2.10) but we do not use the particular initial prescriptions which were mentioned in the proof of Lemma 1. Using (2.10) we find

$$
\begin{aligned}
R_i v^k &\equiv a_{i1}v^k(0,\rho) + a_{i2}v^{k'}(0,\rho) + b_{i1}v^k(1,\rho) + b_{i2}v^{k'}(1,\rho) \\
&= a_{i1}\left\{1 + \frac{[0]}{\rho}\right\} + a_{i2}\rho\varphi^k\left\{1 + \frac{[0]}{\rho}\right\} + b_{i1}e^{\rho\varphi^k}\left\{1 + \frac{[\varphi^k Q(1)]}{\rho}\right\} \\
&\qquad\qquad + b_{i2}\rho\varphi^k e^{\rho\varphi^k}\left\{1 + \frac{[\varphi^k Q(1)]}{\rho}\right\} \\
&= \rho[a_{i2}\varphi^k] + \rho\, e^{\rho\varphi^k}[b_{i2}\varphi^k].
\end{aligned}
\tag{2.50}
$$

Here of course $[a_{i2}\varphi^k] = a_{i2}\varphi^k + E(\rho)/\rho$. Although these E-quantities may differ from one formula to the other we shall not distinguish them in our notation. From (2.50) it follows that

$$
\Delta(\rho) \equiv \begin{vmatrix} R_1 v^1 & R_1 v^2 \\ R_2 v^1 & R_2 v^2 \end{vmatrix} = \rho^2 \begin{vmatrix} [a_{12}] + e^\rho[b_{12}] & [-a_{12}] + e^{-\rho}[-b_{12}] \\ [a_{22}] + e^\rho[b_{22}] & [-a_{22}] + e^{-\rho}[-b_{22}] \end{vmatrix}.
\tag{2.51}
$$

If in (2.3) we put

$$
A \equiv \begin{vmatrix} a_{12} & b_{12} \\ a_{22} & b_{22} \end{vmatrix} \neq 0
$$

then calculating the determinant in (2.51) gives

$$
\Delta(\rho) = \rho^2 e^\rho \left\{ A(1 - e^{-2\rho}) + \frac{E_0(\rho) + E_1(\rho)\, e^{-\rho} + E_2(\rho)\, e^{-2\rho}}{\rho} \right\},
\tag{2.52}
$$

where the $E_i(\rho)$ are uniformly bounded for $\rho \in \mathcal{B}_1$. By analogous calculations we would find the representation

$$
\Delta(\rho) = \rho^2 e^{-\rho} \left\{ A(e^{2\rho} - 1) + \frac{\tilde{E}_0(\rho) + \tilde{E}_1(\rho)\, e^\rho + \tilde{E}_2(\rho)\, e^{2\rho}}{\rho} \right\}
\tag{2.52a}
$$

in $\mathcal{B}_2(|\rho| \geq R, \zeta \leq \zeta_1, \zeta_1$ a positive constant). The zeros of $1 - e^{-2\rho}$, as well as the zeros of $e^{2\rho} - 1$, are given by $\rho_j = j\pi i$ where $j = 0, \pm1, \pm2, \ldots$. If we introduce the functions $\Phi(\rho) = A(1 - e^{-2\rho})$ and

$$
\Psi(\rho) = \frac{E_0(\rho) + E_1(\rho)\, e^{-\rho} + E_2(\rho)\, e^{-2\rho}}{\rho},
$$

then from the periodicity of $\Phi(\rho)$ it follows that the zeros ρ_j can be surrounded by circles Γ_j: $|\rho - \rho_j| = \sigma$ of sufficiently small radius σ so that in the whole ρ-plane, with the exception of these circles, the inequality $|\Phi(\rho)| \geq m(\sigma) > 0$ holds, where m depends only on the radius of these small circles. By the punctured ρ-plane we understand the ρ-plane from which these circles about the zeros ρ_j were removed. For sufficiently large $|\rho|$, say $|\rho| \geqslant R$ we can achieve $|(\Phi\rho)| > |\Psi(\rho)|$ on $\dot{\Gamma}_j$. This inequality is valid for all but finitely many of the $\dot{\Gamma}_j$. If we apply Rouché's theorem[17] to the functions $\Phi(\rho)$ and $\Psi(\rho)$ we see that $\Phi(\rho) + \Psi(\rho)$ has exactly one zero in infinitely many Γ_j: $|\rho - \rho_j| < \sigma$. From (2.52) it then follows that $\Delta(\rho)$ has infinitely many zeros which, except for a finite number, are simple, and which lie in the circles Γ_j. We must now carry the zeros of $\Delta(\rho)$ over to the s-plane by means of the substitution $\rho^2 = r_1 s^2 + r_2 s$ and then show that these zeros of $\Delta(s)$ do not enter a suitable right half-plane $\xi > \xi_* > 0$ of the $s = \xi + i\eta$-plane. We begin with the centers $\rho_j = j\pi i$ of the circles. Let their images in the s-plane be denoted by s_j so that $\rho_j^2 = r_1 s_j^2 + r_2 s_j$. For $s_j = \xi_j + i\eta_j$ this gives the relations

$$-\pi^2 j^2 = r_1(\xi_j^2 - \eta_j^2) + r_2 \xi_j, \qquad 0 = (2r_1 \xi_j + r_2)\eta_j.$$

For sufficiently large j these s_j do not enter the half-plane $\xi > \xi_* > 0$ since the second equation requires $\eta_j = 0$ or $\xi_j = -r_2/2r_1$. By choosing ξ_* sufficiently large we can exclude the second case. Then $-\pi^2 j^2 = r_1 \xi_j^2 + r_2 \xi_j$ remains, which certainly is wrong for sufficiently large $\xi > \xi_* > 0$. Since for sufficiently large $|\rho|$ the zeros of $\Delta(\rho)$ are arbitrarily close to ρ_j, it has also been shown that all the zeros of $\Delta(s)$ do not enter a suitable right half-plane of the s-plane because the finitely many zeros of $\Delta(s)$ which were not considered can be banned from this half-plane by a suitable choice of the half-plane. Thus Lemma 2 is proved.

Proof of Lemma 3 from Section 2.3.   According to (2.13), $G(x, y, \rho)$ is of the form

$$
\begin{aligned}
G(x, y, \rho) &= -\frac{1}{2\Delta(\rho)}
\begin{vmatrix}
v^1(x,\rho) & v^2(x,\rho) & \gamma(x, y, \rho) \\
R_1 v^1 & R_1 v^2 & R_1 \gamma \\
R_2 v^1 & R_2 v^2 & R_2 \gamma
\end{vmatrix} \\[2mm]
&= -\frac{1}{\Delta(\rho)}
\begin{vmatrix}
v^1 & v^2 & \gamma/2 \\
R_1 v^1 & R_1 v^2 & R_1 \gamma/2 \\
R_2 v^1 & R_2 v^2 & R_2 \gamma/2
\end{vmatrix}.
\end{aligned}
\tag{2.53}
$$

To begin with we shall do our considerations in the ρ-plane and restrict ourselves to the domain $\mathcal{B}_1$: $|\rho| \geqslant R$, $\zeta \geqslant \zeta_0$. To determine G in the last representation of (2.53) we multiply the first column of the determinant by

$$-\frac{1}{2}\frac{v^2(y,\rho)}{\delta(0,\rho)},$$

the second column by

$$-\frac{1}{2}\frac{v^1(y,\rho)}{\delta(0,\rho)},$$

and add these to the last column. Then $G(x, y, \rho)$ appears in the form

$$
G(x, y, \rho) = -\frac{1}{\Delta(\rho)}
\begin{vmatrix}
v^1 & v^2 & \gamma_0 \\
R_1 v^1 & R_1 v^2 & \gamma_1 \\
R_2 v^1 & R_2 v^2 & \gamma_2
\end{vmatrix} = -\gamma_0 + \gamma_1 G_1 + \gamma_2 G_2
\tag{2.54}
$$

---

[17] This theorem states: Let $\varphi(\rho)$, $\psi(\rho)$ be analytic in the simply connected domain $\mathfrak{D}$. Let the closed curve $\dot{\Gamma}$ be contained in $\mathfrak{D}$ on which $\varphi(\rho) \neq 0$ and $|\varphi(\rho)| > |\psi(\rho)|$. Then the two functions $\varphi(\rho)$ and $\varphi(\rho) + \psi(\rho)$ have the same number of zeros in the subdomain of $\mathfrak{D}$ which is enclosed by $\dot{\Gamma}$.

where (after an easy calculation)

$$G_1(x,\rho) = \frac{1}{\Delta(\rho)} \begin{vmatrix} v^1(x,\rho) & v^2(x,\rho) \\ R_2 v^1 & R_2 v^2 \end{vmatrix}, \qquad G_2(x,\rho) = -\frac{1}{\Delta(\rho)} \begin{vmatrix} v^1(x,\rho) & v^2(x,\rho) \\ R_1 v^1 & R_1 v^2 \end{vmatrix},$$

$$\gamma_0(x, y, \rho) = \begin{cases} -\dfrac{v^1(y,\rho)v^2(x,\rho)}{\delta(0,\rho)} & \text{for } x \geqslant y, \\[2ex] -\dfrac{v^1(x,\rho)v^2(y,\rho)}{\delta(0,\rho)} & \text{for } x \leqslant y, \end{cases} \tag{2.55}$$

$$\gamma_i(y,\rho) = -\frac{1}{\delta(0,\rho)} \{ (a_{i1}v^1(0,\rho) + a_{i2}v^{1'}(0,\rho))v^2(y) + (b_{i1}v^2(1,\rho) + b_{i2}v^{2'}(1,\rho))v^1(y) \}.$$

Using formula (2.10) we obtain the representations

$$\delta(0,\rho) = 2\rho \left\{ 1 + \frac{[0]}{\rho} \right\},$$

$$\gamma_0(x, y, \rho) = -\frac{1}{2\rho} e^{-\rho|x-y|} \left\{ 1 + \frac{\operatorname{sgn}(x-y)}{\rho} [Q(y) - Q(x)] \right\}, \tag{2.56}$$

$$\gamma_i(y,\rho) = -\left[\frac{a_{i2}}{2}\right] e^{-\rho y} + \left[\frac{b_{i2}}{2}\right] e^{-\rho(1-y)},$$

where sgn $(x - y)$ denotes the algebraic sign of $x - y$. In an analogous way we obtain

$$G_1(x,\rho) = -\frac{\rho}{\Delta(\rho)} \{ [a_{22}] e^{\rho x} + [a_{22}] e^{-\rho x} + [b_{22}] e^{\rho(x-1)} + [b_{22}] e^{-\rho(x-1)} \},$$

$$G_2(x,\rho) = \frac{\rho}{\Delta(\rho)} \{ [a_{12}] e^{\rho x} + [a_{12}] e^{-\rho x} + [b_{12}] e^{\rho(x-1)} + [b_{12}] e^{-\rho(x-1)} \}. \tag{2.57}$$

From (2.56) and (2.57), and with suitable E-quantities, we now have

$$\gamma_1(y,\rho)G_1(x,\rho) + \gamma_2(y,\rho)G_2(x,\rho) = \frac{\rho}{\Delta(\rho)} \left\{ \left[\frac{A}{2}\right] e^{-\rho(1-x+y)} + \left[\frac{A}{2}\right] e^{-\rho(x+y-1)} \right.$$

$$+ \left[\frac{A}{2}\right] e^{-\rho(1-x-y)} + \left[\frac{A}{2}\right] e^{-\rho(x-y+1)} + \frac{E}{\rho} e^{-\rho(y-x)} + \frac{E}{\rho} e^{-\rho(x+y)} + \frac{E}{\rho} e^{-\rho(2-x-y)} + \left. \frac{E}{\rho} e^{-\rho(x-y)} \right\}.$$

From (2.52) in $\mathfrak{B}_1$ we get the representation

$$\frac{1}{\Delta(\rho)} = \frac{e^{-\rho}[1]}{\rho^2 A(1 - e^{-2\rho})}.$$

Using this representation in the previous formula, we finally obtain

$$\gamma_1 G_1 + \gamma_2 G_2 = \frac{[\frac{1}{2}]}{\rho\{1 - e^{-2\rho}\}} \left\{ [1] e^{-\rho(2-x+y)} + [1] e^{-\rho(x+y)} + [1] e^{-\rho(2-x-y)} \right.$$

$$+ [1] e^{-\rho(x-y+2)} + \frac{E}{\rho} e^{-\rho(1+y-x)} + \frac{E}{\rho} e^{-\rho(1+x+y)} + \frac{E}{\rho} e^{-\rho(3-x-y)} + \left. \frac{E}{\rho} e^{-\rho(1+x-y)} \right\}.$$

From this representation, and from the representation for γ_0 in (2.56), we can now read off the validity of the first inequalities in (2.15) and (2.16) in the domain $\mathfrak{B}_1$. The remaining two inequalities in Lemma 3 would be obtained in an analogous way. By similar considerations in $\mathfrak{B}_2$ we would obtain the conclusion of Lemma 3. But inequalities (2.17) and (2.18) can be inferred from (2.15) and (2.16) in such a simple way that we shall not carry it out here.

2.6   The Formal Calculus of Laplace Transforms

It is not enough merely to justify this calculus, we must also learn its formal application. Here we treat the equilibrium Problem I_3 with constant coefficients and with simple boundary conditions. Obviously it can be put into the form

$$
\begin{aligned}
-u_{xx} + qu + r_1 u_{tt} + r_2 u_t &= 0, \\
u(x,0) = 0, \qquad u_t(x,0) &= 0, \\
u(l,t) = f_1(t), \qquad u(m,t) &= f_2(t),
\end{aligned}
\tag{2.58}
$$

where $l \leqslant x \leqslant m$, $0 \leqslant t < \infty$. As is already indicated by the initial conditions, we consider the hyperbolic Case 1 with $r_1 > 0$. q, r_1 and r_2 are constants. Problem (2.58) can now be further simplified without loss of generality.

Instead of $u(x,t)$ we introduce a new unknown function $U(x,t)$ where $u(x,t) = U(x,t)\, e^{\alpha t}$. Then for $U(x,t)$ we obtain the problem

$$
\begin{aligned}
-U_{xx} + (q + r_1\alpha^2 + r_2\alpha)U + r_1 U_{tt} + (r_1 2\alpha + r_2)U_t &= 0, \\
U(x,0) = 0, \qquad U_t(x,0) &= 0, \\
U(l,t) = f_1(t)e^{-\alpha t}, \qquad U(m,t) &= f_2(t)e^{-\alpha t}.
\end{aligned}
\tag{2.59}
$$

If we now put $\alpha = -r_2/2r_1$, then the coefficient of U_t vanishes. By introducing the new independent variable $\bar{x} = [1/(m - l)](x - l)$ we can finally achieve $0 \leqslant \bar{x} \leqslant 1$. By further setting putting $\bar{t} = \beta t$ and choosing β suitably, we can achieve $r_1 = 1$. Thus from the beginning we may restrict ourselves to the problem

$$
\begin{aligned}
-u_{xx} + qu + u_{tt} &= 0, \\
u(x,0) = 0, \qquad u_t(x,0) &= 0, \\
u(0,t) = f_1(t), \qquad u(1,t) &= f_2(t),
\end{aligned}
\tag{2.60}
$$

where $0 \leqslant x \leqslant 1$, $0 \leqslant t < \infty$. We suppose that the solution of this problem is given in the form $u(x,t) \equiv u_1(x,t) + u_2(x,t)$, where u_1 satisfies problem (2.60) with the boundary conditions $u_1(0,t) = f_1(t)$, $u_1(1,t) = 0$, and u_2 satisfies problem (2.60) with the boundary conditions $u_2(0,t) = 0$, $u_2(1,t) = f_2(t)$. If the first problem is solved for $u_1(x,t)$, then the solution $u_2(x,t)$ obviously is obtained by substituting $1 - x$ for x in u_1 and f_2 for f_1. Thus it finally remains to solve problem (2.60) with $f_2(t) \equiv 0$.

By setting $v = Lu$ and $g_1 = Lf_1$, the transformed Problem II_3

$$
\begin{aligned}
-v'' + Q^2 v = 0, \qquad Q^2 &= s^2 + q, \\
v(0,s) = g_1(s), \qquad v(1,s) &= 0
\end{aligned}
\tag{2.61}
$$

arises, where $0 \leqslant x \leqslant 1$ and $' = d/dx$. The quantity $Q = \sqrt{s^2 + q}$ is defined so that for sufficiently large real positive s, Q itself is real and positive. Problem II_3 can easily be solved explicitly. First, all solutions of the ordinary differential equation (2.61) are determined in the form

$$
v(x,s) = c_1 e^{Qx} + c_2 e^{-Qx}.
\tag{2.62}
$$

The constants c_1 and c_2 are easily determined from the boundary conditions:

$$
c_1 = \frac{e^{-Q}g_1(s)}{e^{-Q} - e^{Q}}, \qquad c_2 = \frac{e^{Q}g_1(s)}{e^{-Q} - e^{Q}}.
\tag{2.63}
$$

Thus as solution of (2.61) we obtain

$$v(x,s) = \frac{e^{-Qx} - e^{-Q(2-x)}}{1 - e^{-2Q}}\, g_1(s)$$

$$= g_1(s)\{e^{-Qx} - e^{-Q(2-x)}\} \sum_{j=1}^{\infty} e^{-2jQ}$$

$$= \sum_{j=0}^{\infty} e^{-(2j+x)Q} g_1(s) - \sum_{k=0}^{\infty} e^{-(2k-x)Q} g_1(s). \tag{2.64}$$

The formal solution of our problem will now be given by

$$u(x,t) = L^{-1}v = \sum_{j=0}^{\infty} L^{-1}\{e^{-(2j+x)Q} g_1(s)\} - \sum_{k=1}^{\infty} L^{-1}\{e^{-(2k-x)Q} g_1(s)\}. \tag{2.65}$$

Thus it only remains to determine $L^{-1}\{\cdots\}$. For this we need two simple formal calculation rules, which we derive now.

Rule 1.   If $g = Lf$, then $e^{-\alpha s}g = LF$ where $\alpha \geq 0$ and

$$F(t) = \begin{cases} f(t - \alpha) & \text{for} \quad t \geq \alpha, \\ 0 & \text{for} \quad t < \alpha. \end{cases}$$

The formal proof starts with

$$g(s) = \int_0^{\infty} e^{-st}f(t)\, dt \quad \text{or} \quad g = Lf.$$

Furthermore we have

$$e^{-\alpha s}g(s) = \int_0^{\infty} e^{-s(\alpha+t)} f(t)\, dt = \int_{\alpha}^{\infty} e^{-st}f(t - \alpha)\, dt$$

$$= \int_0^{\infty} e^{-st}F(t)\, dt. \tag{2.66}$$

Rule 2.   If $g_1 = Lf_1$, $g_2 = Lf_2$, then

$$g_1 g_2 = LF \quad \text{where} \quad F(t) = \int_0^t f_1(t - \tau)f_2(\tau)\, d\tau.$$

The formal proof begins with

$$g_1(s)g_2(s) = \int_0^{\infty} e^{-s\xi}f_1(\xi)\, d\xi \int_0^{\infty} e^{-s\eta} f_2(\eta)\, d\eta$$

$$= \int_0^{\infty} \int_0^{\infty} e^{-s(\xi+\eta)} f_1(\xi)f_2(\eta)\, d\xi\, d\eta. \tag{2.67}$$

If we put $\xi + \eta = t$, $\eta = \tau$, then in the new independent variables t,τ the quarter-plane $0 \leq \xi,\, \eta < \infty$ is exhausted by $0 \leq \tau \leq t$, $0 \leq t < \infty$. After the transformation (the functional determinant here is 1) we obtain

$$g_1(s)g_2(s) = \int_0^{\infty} e^{-st} \left\{\int_0^t f_1(t - \tau)f_2(\tau)\, d\tau\right\} dt = LF. \tag{2.68}$$

For the determination of $L^{-1}\{\cdots\}$ in (2.65) we may use the tables of G. Doetsch[17] and select a formula which is as close to our expressions as possible. Of course much practice is necessary for this. We copy formula 43 on p. 109 in the tables of G. Doetsch:

$$\frac{e^{-b\sqrt{s^2+a^2}}}{\sqrt{s^2+a^2}} = L \begin{cases} J_0(a\sqrt{t^2-b^2}) & \text{for} \quad t \geqslant b, \\ 0 & \text{for} \quad t < b, \end{cases} \tag{2.69}$$

where J_0 is the Bessel function of order zero. We put the formula in our notation and get

$$\frac{e^{-bQ}}{Q} = LG \quad \text{where} \quad G(t) = \begin{cases} J_0(\sqrt{q(t^2-b^2)}) & \text{for} \quad t \geqslant b, \\ 0 & \text{for} \quad t < b. \end{cases} \tag{2.70}$$

If we multiply (2.70) by $g_1(s)$ and write the right-hand side in detail, then

$$\frac{e^{-bQ}}{Q} g_1(s) = g_1(s) \int_b^\infty e^{-st} J_0(\sqrt{q(t^2-b^2)})\, dt. \tag{2.71}$$

We differentiate this equation with respect to b, and note $J_0(0) = 1$, $J_0'(x) = -J_1(x)$. The result is

$$e^{-bQ}g_1(s) = e^{-sb}g_1(s) - b\sqrt{q}\, g_1(s) \int_b^\infty e^{-st} \frac{J_1(\sqrt{q(t^2-b^2)})}{\sqrt{t^2-b^2}}\, dt. \tag{2.72}$$

If we set

$$h = LH \quad \text{where} \quad H(t) = \begin{cases} \dfrac{J_1(\sqrt{q(t^2-b^2)})}{\sqrt{t^2-b^2}} & \text{for} \quad t \geqslant b, \\ 0 & \text{for} \quad t < b, \end{cases} \tag{2.73}$$

then (2.72) becomes

$$e^{-bQ}g_1(s) = e^{-sb}g_1(s) - b\sqrt{q}\, g_1(s)h(s). \tag{2.74}$$

If we recall putting $g_1 = Lf_1$ and use Rules 1 and 2, then from (2.74) we obtain

$$L^{-1}(e^{-bQ}g_1(s)) = \begin{cases} f_1(t-b) & \text{for} \quad t \geqslant b \\ 0 & \text{for} \quad t < b \end{cases}$$

$$- b\sqrt{q} \begin{cases} \displaystyle\int_b^t \frac{f_1(t-\tau)J_1(\sqrt{q(\tau^2-b^2)})}{\sqrt{\tau^2-b^2}}\, d\tau & \text{for} \quad t \geqslant b, \\ 0 & \text{for} \quad t < b. \end{cases} \tag{2.75}$$

Thus the formal solution (2.65) can be determined immediately. We only have to replace b once by $2j + x$ and the other time by $2k - x$. Finally we obtain

$$u(x,t) = \sum_{j=0}^\infty f_1(t-(2j+x)) - \sum_{k=1}^\infty f_1(t-(2k-x))$$

$$- \sum_{j=0}^\infty (2j+x) \int_{2j+x}^t f_1(t-\tau) \frac{\sqrt{q}\, J_1(\sqrt{q(\tau^2-(2j+x)^2)})}{\sqrt{\tau^2-(2j+x)^2}}\, d\tau \tag{2.76}$$

$$+ \sum_{k=1}^\infty (2k-x) \int_{2k-x}^t f_1(t-\tau) \frac{\sqrt{q}\, J_1(\sqrt{q(\tau^2-(2k-x)^2)})}{\sqrt{\tau^2-(2k-x)^2}}\, d\tau.$$

Here we always have finite sums, since the only summands different from zero are those for which $t \geqslant 2j + x$ or $t \geqslant 2k - x$, respectively, is satisfied. If finally $q < 0$ then $\sqrt{q}$ is purely imaginary. But then J_1 is also purely imaginary so that the solution always turns out to be real. The physical interpretation of this solution is not the purpose of this book.

The reader is invited to write down the formal solution also for the general problem (2.58) by tracing back all the transformations.

We now treat the equilibrium problem I_3 for the parabolic Case 2 with $r_1 = 0$, $r_2 > 0$, constant coefficients and simple boundary conditions. It can be brought into the form

$$-u_{xx} + qu + r_2 u_t = 0,$$
$$u(x,0) = 0,$$
$$u(l,t) = f_1(t), \qquad u(m,t) = f_2(t), \tag{2.77}$$

where $l \leqslant x \leqslant m$, $0 \leqslant t < \infty$. By setting

$$u(x,t) = U(x,t)\, e^{\alpha t}$$

we can achieve a problem of the form (2.77) for U for which the coefficient of U is zero. For this it suffices to choose $\alpha = -q/r_2$. If we use the further simplifications which were already described in the hyperbolic case, then it finally suffices to consider a problem of the form

$$-u_{xx} + u_t = 0, \qquad u(x,0) = 0,$$
$$u(0,t) = f_1(t), \qquad u(1,t) = 0, \tag{2.78}$$

where $0 \leqslant x \leqslant 1$, $0 \leqslant t < \infty$. After we set $v = Lu$, $g_1 = Lf_1$, the transformed problem

$$-v'' + sv = 0,$$
$$v(0,s) = g_1(s), \qquad v(1,s) = 0 \tag{2.79}$$

arises with the solution

$$v(x,s) = g_1(s)\, \frac{\sinh \sqrt{s}\, (1 - x)}{\sinh \sqrt{s}}.$$

In the tables mentioned we find on p. 119, formula 45:

$$\frac{\sinh \sqrt{s}\, (1 - x)}{\sinh \sqrt{s}} = L\left\{ -\frac{\partial}{\partial x}\, \vartheta_3\left(\frac{x}{2}, t\right) \right\}, \tag{2.80}$$

where $\vartheta_3(x,t)$ is a theta function which is defined by[18]

$$\vartheta_3(x,t) = \frac{1}{\sqrt{\pi t}} \sum_{k=-\infty}^{+\infty} e^{-(1/t)(x+k)^2} = 1 + 2 \sum_{k=1}^{\infty} e^{-\pi^2 k^2 t} \cos 2\pi k x. \tag{2.81}$$

Hence by Rule 2 we find

$$u(x,t) = L^{-1}v = -\int_0^t f_1(t - \tau)\, \frac{\partial}{\partial x}\, \vartheta_3\left(\frac{x}{2}, \tau\right) d\tau \tag{2.82}$$

as a formal solution.

[18] The equality of the two formulas for the theta function in (2.81) represents a profound theorem from the theory of theta functions and cannot be verified immediately.

3

EQUATIONS OF ELLIPTIC TYPE

According to Section I-3.1 we have to consider boundary-value problems, but we must not expect that the existence problems can be settled easily. In Section II-1.5 it was shown that we cannot obtain a normal form even in the case of two independent variables since it is necessary to have an existence proof for *Beltrami's* differential equation for this. Hence we have to renounce any normal form. On the other hand, without additional difficulties we can carry out our considerations immediately for n variables.

3.1   Estimates for Potentials

In the following it frequently occurs that functions which are to be estimated are defined by improper integrals. Let $\mathfrak{D}$ be a normal domain in R_n with $x = (x_1, \ldots, x_n)$.

THEOREM.   The formula

$$\int_{\mathfrak{D}} |x - y|^{\sigma-n}\, dy \leqslant \frac{\omega_n}{\sigma} \left(\frac{nV}{\omega_n}\right)^{\sigma/n}$$

holds for $n \geqslant 2$, $0 < \sigma < n$ and all $x \in R_n$ where V is the volume of $\mathfrak{D}$.[19]

Proof. About the point x we construct a ball S which has the same volume as $\mathfrak{D}$, $V = \omega_n R^n/n$ and radius $R = (nV/\omega_n)^{1/n}$. By $S' = \mathfrak{D} \cap S$ we understand the intersection of $\mathfrak{D}$ and S; it may be empty. Then we have

$$\int_{S} |x - y|^{\sigma-n}\, dy = \int_{S'} |x - y|^{\sigma-n}\, dy + \int_{S-S'} |x - y|^{\sigma-n}\, dy$$
$$\geqslant \int_{S'} |x - y|^{\sigma-n}\, dy + R^{\sigma-n}(V - V_{S'}), \quad (3.1)$$

$$\int_{\mathfrak{D}} |x - y|^{\sigma-n}\, dy = \int_{S'} |x - y|^{\sigma-n}\, dy + \int_{\mathfrak{D}-S'} |x - y|^{\sigma-n}\, dy$$
$$\leqslant \int_{S'} |x - y|^{\sigma-n}\, dy + R^{\sigma-n}(V - V_{S'}) \quad (3.2)$$

---

[19] For $\sigma \geqslant n$ we have an integral with a continuous integrand which is without interest for us.

where $V_{S'}$ is the volume of S'. Using (3.1) the integral over S' in (3.2) can be estimated from above. Thus we find

$$\int_{\mathfrak{D}} |x - y|^{\sigma-n}\, dy \leqslant \int_{S} |x - y|^{\sigma-n}\, dy = \int_0^R \rho^{\sigma-n} \left(\int_{|y-x|=\rho} dS \right) d\rho$$

$$= \int_0^R \rho^{\sigma-n}\rho^{n-1} \left(\int_{|\nu|=1} d\omega \right) d\rho = \omega_n \frac{R^\sigma}{\sigma}, \qquad (3.3)$$

which is the desired result.

3.2   A Solution of $\Delta_n u = f(x)$

The boundary-value problem from Section I-3.4

$$\Delta_n u = f(x) \quad \text{in} \quad \mathfrak{D}, \qquad u = \varphi \quad \text{on} \quad \dot{\mathfrak{D}} \qquad (3.4)$$

can be decomposed into subproblems in two different ways.

1. Let $v(x) \in C^2$ in $\mathfrak{D}$, $\in C^0$ in $\overline{\mathfrak{D}}$ be an arbitrary solution of $\Delta_n v = f$. If we put $u = v + w$, then $w(x)$ has to satisfy the simpler problem $\Delta_n w = 0$, $w = \varphi - v$ on $\dot{\mathfrak{D}}$. Then $u = v + w$ is a solution of (3.4).

2. Let $v(x) \in C^2$ in $\mathfrak{D}$, $\in C^0$ in $\overline{\mathfrak{D}}$ be an arbitrary function which satisfies $v = \varphi$ on $\dot{\mathfrak{D}}$. If we put $u = v + w$, then $w(x)$ must satisfy the problem $\Delta_n w = f - \Delta_n v = \tilde{f}$, $w = 0$ on $\dot{\mathfrak{D}}$. Then again $u = v + w$ is a solution of (3.4).

We now determine $v(x)$ for subproblem 1.

THEOREM.   (1) Let $f(x) \in C^1$ in $\overline{\mathfrak{D}}$. Then

$$u(x) = - \int_{\mathfrak{D}} s(x,y) f(y)\, dy \qquad (3.5)$$

$\in C^1$ in $\overline{\mathfrak{D}}$, $\in C^2$ in $\mathfrak{D}$, and satisfies $\Delta_n u = f(x)$ where $s(x,y)$ is the singularity function from Section I-3.2.   (2) If $f(x) \in C^0$ in $\mathfrak{D}$ and bounded in $\mathfrak{D}$, then $u(x) \in C^1$ in $\overline{\mathfrak{D}}$.

Addendum.   The condition $f(x) \in C^1$ in $\overline{\mathfrak{D}}$ may be weakened to: $f(x)$ is Hölder-continuous in $\overline{\mathfrak{D}}$. The function $f(x)$ is said to be Hölder-continuous if for any two points x^1, x^2 the Hölder condition

$$|f(x^1) - f(x^2)| \leqslant H|x^1 - x^2|^\alpha \qquad (3.6)$$

is satisfied with fixed numbers H and α, $0 < \alpha \leqslant 1$, independent of x^1, x^2.

If $f(x) \in C^1$ in $\overline{\mathfrak{D}}$, then, by the mean-value theorem of the differential calculus, in every convex subdomain of $\overline{\mathfrak{D}}$ we have a relation of the form (3.6) with $\alpha = 1$.

Remark.   The continuity of $f(x)$ is not sufficient for the $u(x)$ from (3.5) to possess second derivatives. By using the results of Cl. Müller[20] one can prove: In the unit sphere $|x| \leqslant 1$ in R_3 there exists a continuous function $f(x)$ such that the function $u(x)$ of (3.5), with $\overline{\mathfrak{D}}: |x| \leqslant 1$, is not twice differentiable at an everywhere-dense sequence of points in the sphere $|x| \leqslant 1$.

[20] See Reference[11], Part I.

Proof: We restrict ourselves to $n \geqslant 3$. Then

$$s(x,y) = \frac{1}{(n-2)\omega_n} |x-y|^{2-n},$$

and the integral given in (3.5) exists by Section 3.1.

First Step. We prove (2) and consider the auxiliary function

$$h_\delta(x,y) = \begin{cases} s(x,y) \quad \text{for} \quad |x-y| \geqslant \delta \\ \dfrac{\delta^{2-n}}{(n-2)\omega_n} \left[2 - \dfrac{2}{n} - \dfrac{n-2}{n} \left(\dfrac{|x-y|}{\delta} \right)^n \right] \quad \text{for} \quad |x-y| \leqslant \delta, \end{cases} \qquad (3.7)$$

which is εC^1 for $x,\, y \in R_n$ and every $\delta > 0$. Then

$$u_\delta(x) = - \int_{\mathfrak{D}} h_\delta(x,y) f(y)\, dy, \qquad (3.8)$$

εC^1 for $x \in R_n$. From (3.5) and (3.8) we have

$$|u(x) - u_\delta(x)| \leqslant \int_{|x-y| \leqslant \delta} \{ s(x,y) + |h_\delta(x,y)| \} |f(y)|\, dy \leqslant \text{const } \delta^2. \qquad (3.9)$$

The last estimate can easily be obtained from Section 3.1 if we identify $\mathfrak{D}$ with $|x-y| \leqslant \delta$ and V with $(\omega_n/n)\delta^n$. If we set $\delta = 1/m$ and $u_{1/m}(x) = u^m(x)$, then $|u(x) - u^m(x)| < \epsilon$ for $m > N(\epsilon)$ and the sequence $\{u^m(x)\}$ converges uniformly to $u(x)$ in $\mathfrak{D}$ so that $u(x) \varepsilon C^0$ in $\mathfrak{D}$. But also the sequence $u^m_{x_i}(x)$ converges uniformly in $\mathfrak{D}$ to

$$- \int_{\mathfrak{D}} s_{x_i}(x,y) f(y)\, dy.$$

Since in (3.8) we may interchange differentiation and integration we find quite analogously

$$\left| - \int_{\mathfrak{D}} s_{x_i}(x,y) f(y)\, dy - u_{\delta x_i}(x) \right| \leqslant \int_{|x-y| \leqslant \delta} \{ |s_{x_i}(x,y)| + |h_{\delta x_i}(x,y)| \} |f(y)|\, dy \qquad (3.10)$$
$$\leqslant \text{const } \delta,$$

and this, in the foregoing notation, implies the uniform convergence of $\{u^m_{x_i}(x)\}$ to that integral. Thus $u(x) \varepsilon C^1$ in $\mathfrak{D}$ and

$$u_{x_i}(x) = - \int_{\mathfrak{D}} s_{x_i}(x,y) f(y)\, dy \qquad (3.11)$$

is proved.

Second Step. We prove (1). We have $|x-y|^{2-n}_{x_i} = -|x-y|^{2-n}_{y_i}$ so that we may write (3.11) as

$$u_{x_i}(x) = \int_{\mathfrak{D}} s_{y_i}(x,y) f(y)\, dy = \int_{\dot{\mathfrak{D}}} \nu_i(y) s(x,y) f(y)\, dS - \int_{\mathfrak{D}} s(x,y) f_{y_i}(y)\, dy; \qquad (3.12)$$

the latter is given by the Gauss theorem in Section I-1.2 in spite of the singularity of $s_{y_i}(x,y)$ at $x = y$. We argue as follows: For fixed $x \in \mathfrak{D}$ construct a sphere S_ϵ with x as center and radius $\epsilon > 0$, and which is wholly contained in $\mathfrak{D}$. Then

$$\int_{\mathfrak{D}} s_{y_i}(x,y) f(y)\, dy = \lim_{\epsilon \to 0} \int_{\mathfrak{D} - \bar{S}_\epsilon} s_{y_i}(x,y) f(y)\, dy.$$

Since the integrand is sufficiently regular in $\mathfrak{D} - \bar{S}_\epsilon$ we have, by the Gauss theorem,

$$\int_{\mathfrak{D}-\bar{S}_\epsilon} s_{\nu_i}(x,y)f(y)\, dy = \int_{\dot{S}_\epsilon} \nu_i(y)s(x,y)f(y)\, dS + \int_{\dot{\mathfrak{D}}} \nu_i(y)s(x,y)f(y)\, dS - \int_{\mathfrak{D}-\bar{S}_\epsilon} s(x,y)f_{\nu_i}(y)\, dy.$$

Since $|s(x,y)| \leqslant \text{const } |x - y|^{2-n}$ the surface integral over $\dot{S}_\epsilon$ may be estimated by

$$\text{const} \int_{\dot{S}_\epsilon} |x - y|^{2-n}\, dS = \text{const} \int_{|x-y|=\epsilon} |x - y|^{2-n}\, dS = \text{const } \epsilon^{2-n}\epsilon^{n-1}\omega_n$$
$$= \text{const } \omega_n \epsilon$$

and by passing to the limit $\epsilon \to 0$ we obtain (3.12). Using the first step, another differentiation with respect to x_i may be carried out under the integral sign. The integral over $\dot{\mathfrak{D}}$ belongs to C^∞ relative to x since $x \in \mathfrak{D}$ but $y \in \dot{\mathfrak{D}}$. Hence we have $u(x) \in C^2$ in $\mathfrak{D}$ and furthermore

$$u_{x_i x_i}(x) = \int_{\dot{\mathfrak{D}}} \nu_i(y)s_{x_i}(x,y)f(y)\, dS - \int_{\mathfrak{D}} s_{x_i}(x,y)[f(y) - f(x)]_{\nu_i}\, dy. \tag{3.13}$$

If in the last term in (3.13) we integrate first over $\mathfrak{D} - \bar{S}_\epsilon$, then by applying the Gauss theorem and then letting $\epsilon \to 0$ we obtain for it, just as before, the equation

$$\int_{\mathfrak{D}} s_{x_i}(x,y)[f(y) - f(x)]_{\nu_i}\, dy = \int_{\dot{\mathfrak{D}}} \nu_i(y)s_{x_i}(x,y)[f(y) - f(x)]\, dS$$
$$- \int_{\mathfrak{D}} s_{x_i \nu_i}(x,y)[f(y) - f(x)]\, dy, \tag{3.14}$$

particularly since the integrand of the surface integral over $\dot{S}_\epsilon$ admits the following estimate (because of the Hölder continuity of $f(x)$):

$$|\nu_i(y)s_{x_i}(x,y)[f(y) - f(x)]| \leqslant \text{const } |x - y|^{1-n} H\, |x - y|^\alpha$$
$$\leqslant H \text{ const } |x - y|^{1+\alpha-n}. \tag{3.15}$$

Existence must also be proved for the last integral in (3.14); this can be done by using the estimate

$$|s_{x_i \nu_i}(x,y)[f(y) - f(x)]| \leqslant \text{const } |x - y|^{-n} H|x - y|^\alpha = H \text{ const } |x - y|^{\alpha-n}.$$

For $f(x) \in C^1$ in $\dot{\mathfrak{D}}$ we obtain such an estimate with $\alpha = 1$ for a suitable sphere in $\mathfrak{D}$ with center x which is sufficient for existence of the last integral in (3.14). From (3.13) and (3.14) it follows that

$$u_{x_i x_i}(x) = f(x) \int_{\dot{\mathfrak{D}}} s_{x_i}(x,y)\nu_i(y)\, dS + \int_{\mathfrak{D}} s_{x_i \nu_i}(x,y)[f(y) - f(x)]\, dy. \tag{3.16}$$

Noting $s_{x_i} = -s_{y_i}$, $s_{x_i \nu_i} = -s_{x_i x_i}$, and $\Delta_n s = 0$, we find

$$\Delta_n u \equiv \sum_{i=1}^{n} u_{x_i x_i}(x) = -f(x) \int_{\dot{\mathfrak{D}}} s_\nu(x,y)\, dS. \tag{3.17}$$

The integral in (3.17) is evaluated by a little trick. The function $v(x) \equiv 1$ is $\in C^2$ in $\dot{\mathfrak{D}}$, and it satisfies $\Delta_n v = 0$. If we apply the theorem from Section I-3.3 to this v by putting $\Phi(a,x) \equiv 0$ and $a = x$ in the fundamental solution, then (I-3.8) gives

$$v(x) \equiv 1 = - \int_{\dot{\mathfrak{D}}} s_\nu(x,y)\, dS, \tag{3.18}$$

and, by (3.17), the theorem follows.

The addendum, however, has not been proved; to do so we merely would have to approximate the Hölder-continuous $f(y)$ by suitable functions which are εC^1 in $\mathfrak{D}$, since in some of the calculations—though not in the final formula (3.16)—terms of the form $f_{y_i}(y)$ appear.

The foregoing considerations are intended to give the reader the necessary routine in integrating by parts using the Gauss theorem; hereafter, these details are omitted.

3.3   Formulation of the General Boundary-Value Problem

We consider the most general second-order linear partial differential equation

$$Du = f \quad \text{where} \quad Du \equiv \sum_{i,k=1}^{n} a_{ik}(x)u_{x_i x_k} + \sum_{i=1}^{n} a_i(x)u_{x_i} + [c(x) - \kappa]u \tag{3.19}$$

in the normal domain $\mathfrak{D}$ of R_n. We make the following *hypotheses:*

1. Let the coefficients $a_{ik}(x)$, $a_i(x)$, $c(x)$, and $f(x)$ be three times, twice, once, and once, respectively, continuously differentiable in $\mathfrak{D}$. Let κ be a number which is to be chosen suitably. Further let Du be uniformly elliptic in $\mathfrak{D}$, that is,

$$\sum_{i,k=1}^{n} a_{ik}(x)\xi_i\xi_k \geqslant c_0 \sum_{i=1}^{n} (\xi_i)^2 \tag{3.20}$$

with fixed $c_0 > 0$ for all numbers $\xi_1, \ldots, \xi_n$, and for all $x \in \mathfrak{D}$.

2. Let the boundary $\dot{\mathfrak{D}}$ of $\mathfrak{D}$ be as follows: Every point $\dot{x} \in \dot{\mathfrak{D}}$ has a neighborhood $\mathfrak{U}$ which can be mapped one-to-one and three times continuously differentiable in both directions onto the interior of a ball S where $\dot{x}$ is mapped onto the center of the ball, and the intersection $\mathfrak{U} \cap \dot{\mathfrak{D}}$ onto the intersection of S with a hyperplane.

The goal of our investigations is the following profound existence theorem.

THEOREM 1.   The boundary-value problem $Du = f$ in $\mathfrak{D}$, where $u = 0$ on $\dot{\mathfrak{D}}$, has exactly one solution $u(x) \in C^0$ in $\mathfrak{D}$, εC^2 in $\mathfrak{D}$ for sufficiently large κ.

Remark. If in (3.19) we set $a(x) \equiv c(x) - \kappa$, then the uniqueness theorem for our boundary-value problem in Section III-1.1 was proved for $a(x) \leqslant 0$ in $\mathfrak{D}$. This condition can certainly be fulfilled by choosing $\kappa \geqslant \max_{x \in \dot{\mathfrak{D}}} |c(x)|$. In the existence theorem we shall not look for a best possible value of κ, since there will be enough other difficulties. We shall show that in particular

$$\kappa \geqslant \frac{1}{c_0} \max_{x \in \dot{\mathfrak{D}}} \left\{ \sum_{i=1}^{n} \left(\sum_{k=1}^{n} [a_{ik}(x)]_{x_k} \right)^2 + \sum_{i=1}^{n} a_i^2(x) + c_0|c(x)| \right\} \tag{3.21}$$

is sufficient.

However, we remark explicitly that in the particularly important, simpler case $a_{ik} = $ const, $a_i = 0$, Theorem 1 gives existence to the same extent as we proved the uniqueness in Section III-1.1. Theorem 2 is a simple corollary to Theorem 1.

THEOREM 2.   The boundary-value problem

$$Du \equiv \sum_{i,k=1}^{n} a_{ik}u_{x_ix_k} + c(x)u = f(x) \quad \text{where} \quad u = 0 \quad \text{on} \quad \dot{\mathfrak{D}} \quad \text{and} \quad a_{ik} = \text{const},$$

$c(x) \leqslant 0$ in $\mathfrak{D}$ has one and only one solution $u(x)\ \varepsilon\ C^0$ in $\bar{\mathfrak{D}}$, εC^2 in $\mathfrak{D}$.

Proof.   This boundary-value problem can be rewritten as follows:

$$Du \equiv \sum_{i,k=1}^{n} a_{ik}u_{x_ix_k} + [\tilde{c}(x) - \kappa]u = f(x), \qquad u = 0 \quad \text{on} \quad \dot{\mathfrak{D}},$$

where

$$\tilde{c}(x) = c(x) + \max_{x\varepsilon\bar{\mathfrak{D}}} |c(x)|, \qquad \kappa = \max_{x\varepsilon\bar{\mathfrak{D}}} |c(x)|.$$

By Theorem 1 and (3.21) we merely have to prove

$$\kappa = \max_{x\varepsilon\bar{\mathfrak{D}}} |c(x)| \geqslant \max_{x\varepsilon\bar{\mathfrak{D}}} |\tilde{c}(x)| = \max_{x\varepsilon\bar{\mathfrak{D}}} |c(x) + \max_{x\varepsilon\bar{\mathfrak{D}}} |c(x)||,$$

and this is obtained as follows:

$$0 \leqslant c(x) + \max |c(x)| \leqslant \max |c(x)|.$$

Thus $|c(x) + \max |c(x)|| \leqslant \max |c(x)|$, $\max |c(x) + \max |c(x)|| \leqslant \max |c(x)|$, which proves everything.

The properties of $u(x)$ mentioned in Theorem 1 are not sufficient to apply *Green's formulas*, but under the hypotheses 1 and 2 we have Theorem 3.

THEOREM 3.   For sufficiently large κ, the boundary-value problem $Du = f$ in $\mathfrak{D}$, where $u = 0$ on $\dot{\mathfrak{D}}$, has exactly one solution $u(x)\ \varepsilon\ C^0$ in $\bar{\mathfrak{D}}$, εC^2 in $\mathfrak{D}$ which satisfies $|u_{x_i}| \leqslant \text{const}$ in $\mathfrak{D}$.

3.4   Outline of Proof and Notations

First we outline the main ideas which lead to the proof of Theorem 1.   Let $u(x)$ be the desired solution of the boundary-value problem.   Then for every $\Phi(x)\ \varepsilon\ C^0$ in $\bar{\mathfrak{D}}$ we have

$$\int_{\mathfrak{D}} \Phi(x)\, Du\, dx = \int_{\mathfrak{D}} \Phi(x)f(x)\, dx. \tag{3.22}$$

If furthermore $\Phi(x)\ \varepsilon\ C^2$ and if it vanishes outside some compact[21] $\bar{\mathcal{E}}$ contained in $\mathfrak{D}$ then by integration by parts and an application of Gauss's theorem it follows that

$$\int_{\mathfrak{D}} u(x)\, D^*\Phi\, dx = \int_{\mathfrak{D}} \Phi(x)f(x)\, dx; \tag{3.23}$$

---

[21] Compact in R_n: closed, bounded point set.   Compact in $\mathfrak{D}$: closed point set contained in $\mathfrak{D}$ (remember that $\mathfrak{D}$ is bounded).

since Φ vanishes in a neighborhood of $\dot{\mathfrak{D}}$, all integrals over $\dot{\mathfrak{D}}$ vanish. Here D^* denotes the operator formally adjoint to D:

$$D^*u \equiv \sum_{i,k=1}^{n} [a_{ik}(x)u]_{x_i x_k} - \sum_{i=1}^{n} [a_i(x)u]_{x_i} + [c(x) - \kappa]u. \qquad (3.24)$$

Furthermore the solution $u(x)$, assumed to be known, of our boundary-value problem can be approximated in mean square by a sequence $\{\Phi_m(x)\}$ of such functions:

$$\lim_{m \to \infty} \int_{\mathfrak{D}} [u(x) - \Phi_m(x)]^2 \, dx = 0 \qquad (3.25)$$

such that each sequence $\{\Phi_{mx_i}(x)\}$, $i = 1, 2, \ldots, n$ converges in mean square. This, of course, does not follow immediately; we must know something about the behavior of the derivatives $u_{x_i}(x)$ near $\dot{\mathfrak{D}}$.

Now it is suggestive to take a new point of view and call any function $u(x)$ which satisfies relation (3.23) for all such $\Phi(x)$, a *weak solution of the partial differential equation $Du = f$*. If furthermore this $u(x)$ can be approximated by these Φ_m-functions in the way described, then we shall call it a *weak solution of the boundary-value problem* (briefly: a *W-solution*).

The proof now proceeds as follows: First we prove the existence of such a *W*-solution and then apply the profound Weyl's lemma which allows us to show that the *W*-solution belongs to C^2 in $\mathfrak{D}$. This immediately implies that the *W*-solution satisfies the differential equation $Du = f$ in $\mathfrak{D}$. Finally it remains to show that the *W*-solution $u(x)$ which was first defined in $\mathfrak{D}$ can be extended continuously to $\bar{\mathfrak{D}}$ and that it is zero on $\dot{\mathfrak{D}}$.[22]

Notations. 1. Let $\mathfrak{D}$ be an arbitrary normal domain of R_n. Then $C_0^\infty(\mathfrak{D})$ denotes the set of all infinitely differentiable functions which vanish outside some compact subset of $\mathfrak{D}$.[23]

2. Let $H_0(\mathfrak{D})$ be the set of all functions which are square-integrable over $\mathfrak{D}$ in the sense of Lebesgue with the inner product:

$$(u,v)_0^{\mathfrak{D}} = \int_{\mathfrak{D}} u(x)v(x) \, dx, \quad \text{and} \quad \text{the norm} \quad \|u\|_0^{\mathfrak{D}} = [(u,u)_0^{\mathfrak{D}}]^{\frac{1}{2}}. \qquad (3.26)$$

The set $H_0(\mathfrak{D})$ is complete in this norm, that is, for every sequence

$$\{u_m(x)\} \subset H_0(\mathfrak{D}) \quad \text{where} \quad \lim_{l,m \to \infty} \|u_l - u_m\|_0^{\mathfrak{D}} = 0^{24}$$

there exists a $u(x) \in H_0(\mathfrak{D})$ such that $\lim\limits_{m \to \infty} \|u - u_m\|_0^{\mathfrak{D}} = 0$; for this is exactly the conclusion of the well-known *Riesz-Fischer* theorem.

---

[22] The procedure of proof described here probably goes back to the direct methods in the calculus of variations with which it was possible to treat certain existence problems of this kind successfully. The book of R. Courant and D. Hilbert [4] reports about this. During recent years, however, these methods have been generalized so much that an immediate connection with the calculus of variations no longer exists. Essential contributions were made by Browder, Friedrichs, Gårding, Lax, Morrey, Nirenberg, Višik, and others. The reports mentioned at the end of this book should be consulted for literature.

[23] The compact subset will vary from function to function.

[24] This means: For any $\epsilon > 0$ there is an $N(\epsilon)$ such that $\|u_l - u_m\|_0^{\mathfrak{D}} < \epsilon$ holds for all $l, m > N(\epsilon)$.

3. Let $\overset{0}{H}_1(\mathfrak{D})$ be the set of all functions $u(x) \; \varepsilon \; H_0(\mathfrak{D})$ for which there exist sequences

$$\{\Phi_m(x)\} \subset C_0^\infty(\mathfrak{D})$$

with the properties

$$\lim_{m \to \infty} \|\Phi_m(x) - u(x)\|_0^{\mathfrak{D}} = 0, \tag{3.27a}$$

$$\lim_{l,m \to \infty} \|\Phi_{m x_i}(x) - \Phi_{l x_i}(x)\|_0^{\mathfrak{D}} = 0, \tag{3.27b}$$

$i = 1, 2, \ldots, n$. Then because of (3.27b) to every $u(x) \; \varepsilon \; \overset{0}{H}_1(\mathfrak{D})$ there exist uniquely determined elements $D_1' u, \ldots, D_n' u$ for which

$$\lim_{m \to \infty} \|\Phi_{m x_i}(x) - D_i' u\|_0^{\mathfrak{D}} = 0. \tag{3.28}$$

These elements are called generalized first derivatives of $u(x)$.

If furthermore $u(x) \; \varepsilon \; \overset{0}{H}_1(\mathfrak{D})$ is continuously differentiable, then $D_i' u = u_{x_i}(x)$. Indeed, using the Gauss theorem and integrating by parts (where again integrals over $\dot{\mathfrak{D}}$ vanish), we have

$$\begin{aligned}
(\Phi, u_{x_i})_0^{\mathfrak{D}} &= -(\Phi_{x_i}, u)_0^{\mathfrak{D}} = \lim_{m \to \infty} - (\Phi_{x_i}, \Phi_m)_0^{\mathfrak{D}} \\
&= \lim_{m \to \infty} (\Phi, \Phi_{m x_i})_0^{\mathfrak{D}} = (\Phi, D_i' u)_0^{\mathfrak{D}}
\end{aligned} \tag{3.29}$$

for all $\Phi(x) \; \varepsilon \; C_0^\infty(\mathfrak{D})$. Written out this means

$$\int_{\mathfrak{D}} \Phi(x) \{u_{x_i}(x) - D_i' u\} \, dx = 0 \tag{3.30}$$

for all $\Phi(x) \; \varepsilon \; C_0^\infty(\mathfrak{D})$ and $u_{x_i}(x) = D_i' u(x)$ except possibly on a set of measure zero.

In $\overset{0}{H}_1(\mathfrak{D})$ we define an inner product $(u,v)_1^{\mathfrak{D}}$ and a norm $\|u\|_1^{\mathfrak{D}}$ by

$$(u,v)_1^{\mathfrak{D}} = (u,v)_0^{\mathfrak{D}} + \sum_{i=1}^{n} (D_i' u, D_i' v)_0^{\mathfrak{D}}, \qquad \|u\|_1^{\mathfrak{D}} = [(u,u)_1^{\mathfrak{D}}]^{1/2}. \tag{3.31}$$

The class $\overset{0}{H}_1(\mathfrak{D})$ is complete with respect to this norm.

3.5 Existence of a *W*-Solution

First Step. For functions $\Phi(x)$, $\Psi(x) \; \varepsilon \; C_0^\infty(\mathfrak{D})$ the following form makes sense:

$$B[\Phi, \Psi] \equiv \int_{\mathfrak{D}} \left\{ \sum_{i,k=1}^{n} [a_{ik}(x)\Psi(x)]_{x_k} \Phi_{x_i}(x) - \sum_{i=1}^{n} a_i(x) \Phi_{x_i}(x)\Psi(x) \right.$$
$$\left. - [c(x) - \kappa]\Phi(x)\Psi(x) \right\} dx, \tag{3.32}$$

which by completion can also be defined for functions $\Phi(x)$, $\Psi(x) \; \varepsilon \; \overset{0}{H}_1(\mathfrak{D})$. By this we mean the following: If $\Phi(x)$, $\Psi(x) \; \varepsilon \; \overset{0}{H}_1(\mathfrak{D})$, then by (3.27) there are sequences $\{\Phi_m(x)\}$, $\{\Psi_m(x)\} \subset C_0^\infty(\mathfrak{D})$ such that [with the notation introduced in (3.31)]

$$\lim_{m \to \infty} \|\Phi_m(x) - \Phi(x)\|_1^{\mathfrak{D}} = 0, \qquad \lim_{m \to \infty} \|\Psi_m(x) - \Psi(x)\|_1^{\mathfrak{D}} = 0. \tag{3.33}$$

Hence the sequence of numbers $B[\Phi_m,\Psi_m]$ converges, for we have

$$|B[\Phi_l,\Psi_l] - B[\Phi_m,\Psi_m]| \leqslant |B[\Phi_l,\Psi_l] - B[\Phi_l,\Psi_m]| + |B[\Phi_l,\Psi_m] - B[\Phi_m,\Psi_m]|$$
$$\leqslant \text{const } \|\Phi_l\|_1^{\mathfrak{D}} \|\Psi_l - \Psi_m\|_1^{\mathfrak{D}} + \text{const } \|\Psi_m\|_1^{\mathfrak{D}} \|\Phi_l - \Phi_m\|_1^{\mathfrak{D}}. \qquad (3.34)$$

Here we used the well-known *Schwarz inequality*

$$\left(\int_{\mathfrak{D}} u(x)v(x)\,dx\right)^2 \leqslant \int_{\mathfrak{D}} u^2(x)\,dx \int_{\mathfrak{D}} v^2(x)\,dx \quad \text{or} \quad |(u,v)_0^{\mathfrak{D}}| \leqslant \|u\|_0^{\mathfrak{D}} \|v\|_0^{\mathfrak{D}}. \qquad (3.35)$$

The right-hand side in (3.34) tends to zero as $l, m \to \infty$. Thus $\lim\limits_{m\to\infty} B[\Phi_m,\Psi_m]$ exists, and this limit is independent of the special choice of the sequences $\{\Phi_m\}$, $\{\Psi_m\}$. Hence it is permissible to define $B[\Phi,\Psi]$ for $\Phi(x)$, $\Psi(x) \in \overset{0}{H}_1(\mathfrak{D})$ by $\lim\limits_{m\to\infty} B[\Phi_m,\Psi_m]$, which is called a definition by completion. The third step will give further properties of this form $B[\Phi,\Psi]$. The next step is intended to show the reader how this form is connected with our problem.

Second Step. If we could show the existence of a function $u(x) \in \overset{0}{H}_1(\mathfrak{D})$ which satisfies

$$B[u,\Phi] = -(f,\Phi)_0^{\mathfrak{D}} \qquad (3.36)$$

for all $\Phi(x) \in \overset{0}{H}_1(\mathfrak{D})$, then (3.36) would *a priori* hold for all $\Phi(x) \in C_0^{\infty}(\mathfrak{D})$. Here $f(x)$ is the given right-hand side of the differential equation (3.19). By integration by parts and the Gauss theorem, the left-hand side in (3.36) becomes

$$B[u,\Phi] = -\int_{\mathfrak{D}} u(x)\Big\{ \sum_{i,k=1}^{n} [a_{ik}(x)\Phi(x)]_{x_i x_k} - \sum_{i=1}^{n} [a_i(x)\Phi(x)]_{x_i} + [c(x) - \kappa]\Phi(x)\Big\}\,dx. \qquad (3.37)$$

With the usual notation, the first term for instance is immediately obtained from

$$\int_{\mathfrak{D}} (a_{ik}\Phi)_{x_k} u_{x_i}\,dx = \int_{\dot{\mathfrak{D}}} (a_{ik}\Phi)_{x_k} u\nu_i\,dS - \int_{\mathfrak{D}} u(a_{ik}\Phi)_{x_i x_k}\,dx, \qquad (3.38)$$

where the integral over $\dot{\mathfrak{D}}$ vanishes because $\Phi(x) \in C_0^{\infty}(\mathfrak{D})$. From (3.37), (3.36), and (3.24) there results

$$(u,D^*\Phi)_0^{\mathfrak{D}} = (f,\Phi)_0^{\mathfrak{D}} \qquad (3.39)$$

for all $\Phi(x) \in C_0^{\infty}(\mathfrak{D})$. So $u(x)$ would be a weak solution of the differential equation, and hence $u(x) \in \overset{0}{H}_1(\mathfrak{D})$; by (3.27) it would even follow that $u(x)$ is a desired W-solution.

Third Step. The existence of a function $u \in \overset{0}{H}_1(\mathfrak{D})$ for which (3.36) holds for all $\Phi \in \overset{0}{H}_1(\mathfrak{D})$ is guaranteed by the following lemma.

LEMMA. If $B[\Phi,\Psi]$ has the following four properties:

$$B[\Phi, \Psi + X] = B[\Phi,\Psi] + B[\Phi,X], \qquad B[\Phi + \Psi, X] = B[\Phi,X] + B[\Psi,X], \qquad (3.40)$$
$$B[\Phi,c\Psi] = cB[\Phi,\Psi], \qquad B[c\Phi,\Psi] = cB[\Phi,\Psi], \qquad (3.41)$$
$$|B[\Phi,\Psi]| \leqslant \text{const } \|\Phi\|_1^{\mathfrak{D}} \|\Psi\|_1^{\mathfrak{D}}, \qquad (3.42)$$
$$B[\Phi,\Phi] \geqslant \gamma(\|\Phi\|_1^{\mathfrak{D}})^2 \qquad (3.43)$$

with a fixed $\gamma > 0$ for all Φ, Ψ, $X \in \overset{0}{H}_1(\mathfrak{D})$ and all real numbers c, then there is a function $u \in \overset{0}{H}_1(\mathfrak{D})$ for which

$$B[u,\Phi] = - \int_{\mathfrak{D}} f(x)\Phi(x)\, dx$$

for all $\Phi \in \overset{0}{H}_1(\mathfrak{D})$.

This lemma is a special case of a theorem from functional analysis which in Section V-1.3 is formulated as Lemma 6 and proved.

It only remains to show that the form $B[\Phi,\Psi]$ has the required properties. Because the first three are obvious we only have to give a proof for (3.43). From (3.32) and from the uniform ellipticity (3.20) we infer the inequality

$$B[\Phi,\Phi] \geqslant c_0 \int_{\mathfrak{D}} \sum_{i=1}^{n} (\Phi_{x_i})^2\, dx + R + \kappa \int_{\mathfrak{D}} \Phi^2\, dx \quad \text{for all } \Phi \in C_0^{\infty}(\mathfrak{D}), \tag{3.44}$$

where we set

$$R = \int_{\mathfrak{D}} \{ \sum_{i,k=1}^{n} (a_{ik})_{x_k}\Phi_{x_i}\Phi - \sum_{i=1}^{n} a_i\Phi_{x_i}\Phi - c\Phi^2 \}\, dx. \tag{3.45}$$

For arbitrary real numbers A, B, α, where $\alpha > 0$, we have

$$|AB| \leqslant \frac{\alpha}{2} A^2 + \frac{1}{2\alpha} B^2, \tag{3.46}$$

which is obtained immediately from $0 \leqslant [\sqrt{\alpha}\,|A| - (1/\sqrt{\alpha})\,|B|]^2$. If, in particular, we set $\alpha = c_0/2$ we then find

$$|\Phi_{x_i}(\sum_{k=1}^{n} (a_{ik})_{x_k}\Phi)| \leqslant \frac{c_0}{4} \Phi_{x_i}^2 + \frac{1}{c_0} (\sum_{k=1}^{n} (a_{ik})_{x_k})^2 \Phi^2,$$

$$|\Phi_{x_i}a_i\Phi| \leqslant \frac{c_0}{4} \Phi_{x_i}^2 + \frac{1}{c_0} a_i^2\Phi^2.$$

This gives the estimate

$$|R| \leqslant \frac{c_0}{2} \int_{\mathfrak{D}} \sum_{i=1}^{n} (\Phi_x)^2\, dx + N \int_{\mathfrak{D}} \Phi^2\, dx \tag{3.47}$$

for R, where

$$N = \frac{1}{c_0} \max_{x \in \bar{\mathfrak{D}}} \{ \sum_{i=1}^{n} (\sum_{k=1}^{n} (a_{ik})_{x_k})^2 + \sum_{i=1}^{n} a_i^2 + c_0|c| \}.$$

If we consider the choice of κ in (3.21), then from (3.44) the inequality

$$B[\Phi,\Phi] \geqslant \frac{c_0}{2} \int_{\mathfrak{D}} \sum_{i=1}^{n} (\Phi_{x_i})^2\, dx \quad \text{for all } \Phi \in C_0^{\infty}(\mathfrak{D}) \tag{3.48}$$

follows. For the further evaluation we use the relation

$$\int_{\mathfrak{D}} \Phi^2(x)\, dx \leqslant \eta \int_{\mathfrak{D}} \sum_{i=1}^{n} (\Phi_{x_i})^2\, dx \quad \text{for all } \Phi \in C_0^{\infty}(\mathfrak{D}) \tag{3.49}$$

with a number η depending only on $\mathfrak{D}$. Thus from (3.48) the inequality

$$B[\Phi,\Phi] \geqslant \frac{c_0}{4} \int_{\mathfrak{D}} \sum_{i=1}^{n} \Phi_{x_i}^2 \, dx + \frac{c_0}{4\eta} \int_{\mathfrak{D}} \Phi^2 \, dx \geqslant \gamma(\|\Phi\|_1^{\mathfrak{D}})^2 \tag{3.50}$$

results for all $\Phi \, \varepsilon \, C_0^{\infty}(\mathfrak{D})$, where $\gamma = \min\,(c_0/4, c_0/4\eta)$. If in (3.50) we pass to $\Phi \, \varepsilon \, \overset{0}{H}_1(\mathfrak{D})$ (by completion), we obtain (3.43). The validity of (3.49) can be seen as follows: By the Gauss theorem,

$$\int_{\mathfrak{D}} \Phi^2 \, dx = - \int_{\mathfrak{D}} x_1(\Phi^2)_{x_1} \, dx,$$

because $\qquad\qquad O = \int_{\dot{\mathfrak{D}}} x_1 \Phi^2(x)\nu_1 \, dS = \int_{\mathfrak{D}} (\Phi^2(x) + x_1(\Phi^2(x))_{x_1}) \, dx.$

If μ denotes the maximum of $|x_1|$ in $\mathfrak{D}$, then by (3.46)

$$\mu|(\Phi^2)_{x_1}| = 2\mu \, |\Phi| \, |\Phi_{x_1}| \leqslant \tfrac{1}{2}\Phi^2 + 2\mu^2(\Phi_{x_1})^2,$$

and we obtain

$$\int_{\mathfrak{D}} \Phi^2 \, dx \leqslant \frac{1}{2} \int_{\mathfrak{D}} \Phi^2 \, dx + 2\mu^2 \int_{\mathfrak{D}} \Phi_{x_1}^2 \, dx. \tag{3.51}$$

Hence

$$\int_{\mathfrak{D}} \Phi^2 \, dx \leqslant 4\mu^2 \int_{\mathfrak{D}} \Phi_{x_1}^2 \, dx \leqslant 4\mu^2 \int_{\mathfrak{D}} \sum_{i=1}^{n} \Phi_{x_i}^2 \, dx. \tag{3.52}$$

3.6   Differentiability of the *W*-Solution

We show that the *W*-solution $u(x)$ which had been secured in Section 3.5 is twice continuously differentiable in any compact $\bar{\mathcal{E}} \subset \mathfrak{D}$ and that it satisfies $Du = f$ in $\mathfrak{D}$. To do so we use a version of Weyl's lemma.

WEYL'S LEMMA.   In the domain $\mathfrak{D}$ in R_n consider the elliptic differential operator

$$Kv \equiv \sum_{i,k=1}^{n} \alpha_{ik}(x)v_{x_i x_k} + \sum_{i=1}^{n} \alpha_i(x)v_{x_i} + \gamma(x)v, \tag{3.53}$$

where $\sum\limits_{i,k=1}^{n} \alpha_{ik}(x)\xi_i\xi_k > 0$ for all $\xi = (\xi_1, \xi_2, \ldots, \xi_n) \neq 0$ and $x \, \varepsilon \, \mathfrak{D}$. For the coefficients assume $\alpha_{ik}(x) \, \varepsilon \, C^3$, $\alpha_i(x) \, \varepsilon \, C^2$, $\gamma(x) \, \varepsilon \, C^1$ in $\mathfrak{D}$. Further let $\eta(x)$ be a Hölder-continuous function in $\mathfrak{D}$.

If $w(x)$ is locally integrable in $\mathfrak{D}$ $\left(\int_{\mathcal{E}} |w(x)| \, dx < \infty \text{ for every compact } \bar{\mathcal{E}} \subset \mathfrak{D} \right)$ and if

$$\int_{\mathfrak{D}} w(y)K\Phi \, dy = \int_{\mathfrak{D}} \Phi(y)\eta(y) \, dy \tag{3.54}$$

for all $\Phi(x) \, \varepsilon \, C_0^{\infty}(\mathfrak{D})$, then $w(x)$ coincides almost everywhere with a function $\bar{w} \, \varepsilon \, C^2$ in $\mathfrak{D}$.

The proof of a sharper version of this lemma will be given in Section 4.

Let $u(x)$ be the W-solution constructed in Section 3.5. Since $u(x) \in \overset{0}{H}_1(\mathfrak{D})$, there exists a sequence $\{\Phi_m(x)\} \subset C_0^\infty(\mathfrak{D})$ for which

$$B[u,\Phi] = \lim_{m \to \infty} B[\Phi_m,\Phi] \quad \text{for all } \Phi(x) \in C_0^\infty(\mathfrak{D}). \tag{3.55}$$

But by integration by parts from (3.32) we have

$$B[\Phi_m,\Phi] = - \int_{\mathfrak{D}} \Phi_m(x) D^*\Phi \, dx; \tag{3.56}$$

thus

$$B[u,\Phi] = - \int_{\mathfrak{D}} u(x) D^*\Phi \, dx \quad \text{for all } \Phi(x) \in C_0^\infty(\mathfrak{D}).$$

Because of relation (3.36), we finally have

$$\int_{\mathfrak{D}} u(x) D^*\Phi \, dx = \int_{\mathfrak{D}} \Phi(x) f(x) \, dx \quad \text{for all } \Phi(x) \in C_0^\infty(\mathfrak{D}). \tag{3.57}$$

Hence, by Weyl's lemma, the W-solution $u(x)$ coincides almost everywhere in $\mathfrak{D}$ with a function $\bar{u}(x) \in C^2$.

We show that $\bar{u}(x)$ is a W-solution. Clearly $\bar{u}(x)$ satisfies relation (3.57), and after integration by parts we have $B[\bar{u},\Phi] = -(f,\Phi)_0^{\mathfrak{D}}$. It merely remains to show that $\bar{u}(x) \in \overset{0}{H}_1(\mathfrak{D})$. But this is trivial; for $u(x) \in \overset{0}{H}_1(\mathfrak{D})$, and from the validity of relations (3.27) it follows that

$$\|\Phi_m(x) - \bar{u}(x)\|_0^{\mathfrak{D}} \to 0, \qquad \|\Phi_{mx_i}(x) - \Phi_{lx_i}(x)\|_0^{\mathfrak{D}} \to 0 \quad \text{as} \quad l, m \to \infty. \tag{3.58}$$

Thus the existence of a W-solution $u(x) \in C^2$ in $\mathfrak{D}$ is proved. Integration by parts in (3.57) now gives

$$\int_{\mathfrak{D}} \Phi(x) \, Du \, dx = \int_{\mathfrak{D}} \Phi(x) f(x) \, dx \quad \text{for all } \Phi(x) \in C_0^\infty(\mathfrak{D}), \tag{3.59}$$

which in turn implies the validity of the differential equation $Du = f$ in $\mathfrak{D}$. Thus Theorem 1 in Section 3.3 is proved, except for the statements that $u(x) \in C^0$ in $\bar{\mathfrak{D}}$ and $u = 0$ on $\dot{\mathfrak{D}}$.

3.7 Continuous Assumption of the Boundary Values[25]

In order to show that the solution $u(x)$ of $Du = f$ which is twice continuously differentiable in $\mathfrak{D}$ can be extended continuously to $\mathfrak{D} + \dot{\mathfrak{D}}$ and that it assumes the value zero on $\dot{\mathfrak{D}}$, fix an arbitrary point $\dot{x} \in \dot{\mathfrak{D}}$. By the hypotheses on $\dot{\mathfrak{D}}$ in Section 3.3 we can now find a neighborhood $\mathfrak{U}$ about $\dot{x}$ which may be mapped one-to-one and three times continuously differentiable in both directions onto a ball $S: |z - \dot{z}| < d$ where $\dot{x}$ is mapped onto the center of the ball, and the intersection $\mathfrak{U} \cap \dot{\mathfrak{D}}$ onto the intersection of S with a hyperplane. We may assume that this hyperplane is given by $z_n = 0$ and that under our transformation $\mathfrak{U} \cap \mathfrak{D}$ goes over into a domain $\mathcal{E}_1$ which consists of all points z for which $|z - \dot{z}| < d$ and $z_n > 0$. In general by $\mathcal{E}_\lambda$ we understand the set of all z for which $|z - \dot{z}| < \lambda d$ and

[25] The method and presentation of the proof given here are due to E. Wienholtz and have not been published (except in the German original of this book).

$z_n > 0$. Finally we may assume that the functional determinant of the mapping is bounded from above and is bounded from below by some positive number.

If in $\mathcal{E}_1$ we define a function $v(z)$ by $v(z) = u(x)$ where x is the inverse image of z, then obviously our problem is to show that

$$\lim_{z \to \dot{z},\ z\varepsilon\mathcal{E}_1} v(z) = 0.$$

We have $v(z) \varepsilon C^2$ in $\mathcal{E}_1$ and we check that $v(z)$ satisfies a uniformly elliptic differential equation

$$\check{D}v \equiv \sum_{i,k=1}^{n} \bar{a}_{ik}(z)v_{z_iz_k} + \sum_{i=1}^{n} \bar{a}_i(z)v_{z_i} + \bar{c}(z)v = \bar{f}(z) \tag{3.60}$$

in $\mathcal{E}_1$ whose coefficients $\bar{a}_{ik}$, $\bar{a}_i$, $\bar{c}$, and $\bar{f}$ are at least twice, once, once, and once, respectively, continuously differentiable in $\bar{\mathcal{E}}_1$ because of the differentiability properties of the transformation.

Since we had $u(x) \varepsilon \overset{0}{H}_1(\mathfrak{D})$, $\int_{\mathcal{E}_1} \{v^2(z) + \sum_{i=1}^{n} [v_{z_i}(z)]^2\}\, dz$ exists and $v(z)$ may be approximated in $\mathcal{E}_1$ by three times continuously differentiable functions $\Phi_m(z)$ which vanish in a neighborhood of $z_n = 0$ such that in analogy to (3.27) and (3.28) we have

$$\lim_{m \to \infty} \int_{\mathcal{E}_1} [\Phi_m(z) - v(z)]^2\, dz = 0, \tag{3.61a}$$

$$\lim_{m \to \infty} \int_{\mathcal{E}_1} [\Phi_{mz_i}(z) - v_{z_i}(z)]^2\, dz = 0, \tag{3.61b}$$

$i = 1, 2, \ldots, n$. This follows immediately from the differentiability properties of the transformation and from the fact that $v(z) \varepsilon C^2$ in $\mathcal{E}_1$. From (3.61) we infer the inequality, important for the sequel,

$$\int_{C_h} [v(z)]^2\, dz \leqslant h^2 \int_{C_h} [v_{z_n}(z)]^2\, dz \tag{3.62}$$

if the domain of integration C_h is a cylinder of height h which is wholly contained in $\mathcal{E}_1$ and which has part of the hyperplane $z_n = 0$ for its base.

To see this we note that the approximating functions $\Phi(z)$ vanish on $z_n = 0$; hence

$$\Phi(z) = \int_0^{z_n} \Phi_{z_n}(z_1, \ldots, z_{n-1}, z_n)\, dz_n \quad \text{for} \quad z \varepsilon \mathcal{E}_1.$$

By Schwarz's inequality

$$\Phi^2(z) \leqslant h \int_0^h \Phi_{z_n}^2\, dz_n \quad \text{for} \quad z \varepsilon C_h$$

and from this the inequality (3.62) is obtained by integration over C_h.

If we assume the existence of

$$\lim_{z \to \dot{z},\ z\varepsilon\mathcal{E}_1} v(z),$$

say $v(\dot{z})$, then we easily see that $v(\dot{z}) = 0$. Indeed, for sufficiently small $h > 0$, the cylinder

$$C_h: \begin{matrix} (z_1 - \dot{z}_1)^2 + \cdots + (z_{n-1} - \dot{z}_{n-1})^2 \leqslant h^{2/(n-1)}, \\ 0 < z_n \leqslant h, \end{matrix} \tag{3.63}$$

is contained in $\mathcal{E}_1$, its volume $V(C_h)$ has the value const $\cdot\, h^2$, and by (3.62) the estimate

$$\frac{\text{const}}{V(C_h)} \int_{C_h} v^2(z)\, dz \leqslant \int_{C_h} [v_{z_n}(z)]^2\, dz \tag{3.64}$$

is valid. By our assumption we have $|v(z)| \geqslant \frac{1}{2}|v(\dot{z})|$ for all $z \in C_h$ for h sufficiently small. Hence by (3.64) for such h we obtain

$$\frac{\text{const}}{4}\, v^2(\dot{z}) \leqslant \int_{C_h} [v_{z_n}(z)]^2\, dz. \tag{3.65}$$

If we let $h \to 0$, then $v(\dot{z}) = 0$.

Now we shall show that

$$\lim_{z \to \dot{z},\ z \varepsilon \mathcal{E}_1} v(z)$$

exists. For this it suffices to prove Hölder continuity of $v(z)$ in $\mathcal{E}_{1/2}$ with a fixed exponent α and a fixed coefficient H:

$$|v(z^1) - v(z^2)| \leqslant H|z^1 - z^2|^\alpha, \qquad 0 < \alpha < 1, \quad \text{for} \quad \text{all } z^1,\, z^2 \varepsilon \mathcal{E}_{1/2}.$$

As an essential tool we introduce a pseudo-fundamental solution $\gamma(y,z)$, $y \varepsilon \mathcal{E}_1$, $z \varepsilon \bar{\mathcal{E}}_1$, $z \neq y$, corresponding to the differential equation (3.60) in $\mathcal{E}_1$ which has the value zero when z lies on the hyperplane $z_n = 0$. In Section 3.8 we study the precise properties of $\gamma(y,z)$ and prove its existence, but the reader will find all the needed properties listed here.

In particular, for fixed $y \varepsilon \mathcal{E}_1$, $\gamma(y,z)$ is arbitrarily often differentiable in z for $z \varepsilon \bar{\mathcal{E}}_1$ and $z \neq y$. Furthermore, for fixed $z \varepsilon \bar{\mathcal{E}}_1$, $\gamma(y,z)$ and every derivative with respect to z is twice continuously differentiable in y for

$$y \varepsilon \mathcal{E}_1, \qquad y \neq z.$$

Finally, for all $y \varepsilon \mathcal{E}_1$, $z \varepsilon \bar{\mathcal{E}}_1$, $y \neq z$ we have the estimates

$$|\gamma(y,z)| \leqslant \frac{\text{const}}{|y - z|^{n-3/2}}, \qquad |\gamma_{z_i}(y,z)| \leqslant \frac{\text{const}}{|y - z|^{n-1}},$$

$$|\gamma_{z_i z_k}(y,z)| \leqslant \frac{\text{const}}{|y - z|^n}, \qquad |\gamma_{y_i}(y,z)| \leqslant \frac{\text{const}}{|y - z|^{n-1}}, \tag{3.66}$$

$$|\gamma_{y_j z_i}(y,z)| \leqslant \frac{\text{const}}{|y - z|^n}, \qquad i,\, j,\, k = 1,\, \ldots,\, n;$$

$$\Big|\sum_{i,k=1}^{n} \bar{a}_{ik}(z)\gamma_{z_i z_k}(y,z)\Big| \leqslant \frac{\text{const}}{|y - z|^{n-1}},$$

$$\Big|\sum_{i,k=1}^{n} \bar{a}_{ik}(z)\gamma_{z_i z_k y_j}(y,z)\Big| \leqslant \frac{\text{const}}{|y - z|^n}, \qquad j = 1,\, \ldots,\, n. \tag{3.67}$$

The particularly important property of $\gamma(y,z)$, which also justifies the name pseudo-fundamental solution, is the fact that for all functions $\Phi(z)$ which are twice continuously differentiable in $\mathcal{E}_1$ and which vanish identically in a neighborhood of $\dot{\mathcal{E}}_1$ we have the relation

$$\Phi(y) = -\int_{\mathcal{E}_1} \gamma(y,z)\mathcal{D}\Phi(z)\, dz + \int_{\mathcal{E}_1} \Phi(z)\mathcal{D}^*\gamma(y,z)\, dz, \tag{3.68}$$

where $\mathcal{D}^*$ is the operator formally adjoint to $\mathcal{D}$ [see (3.24)].

Fix a $y_0 \in \mathcal{E}_{1/2}$ and let $\zeta(z)$ denote a twice continuously differentiable function in $z_n \geqslant 0$ which has the value 1 in $\mathcal{E}_{1/2}$, zero outside of $\bar{\mathcal{E}}_{3/4}$, and which satisfies $0 \leqslant \zeta(z) \leqslant 1$. Because of the validity of the differential equation (3.60) for small $\rho > 0$ we have

$$\int_{\substack{z \in \mathcal{E}_1 \\ z_n \geqslant \rho}} \zeta(z)\gamma(y_0,z)\, \bar{D}v(z)\, dz = \int_{\substack{z \in \mathcal{E}_1 \\ z_n \geqslant \rho}} \zeta(z)\gamma(y_0,z)\bar{f}(z)\, dz. \tag{3.69}$$

Of course these integrals exist, since $\gamma(y_0,z)$ was multiplied by functions which are bounded in a neighborhood of $z = y_0$ and since for $\gamma(y_0,z)$ we have estimates of the form (3.66).

To be able to apply (3.68) to the left-hand side of (3.69), introduce a function $v^1(z) \in C^2$ in $\mathcal{E}_1$ with the properties

$$\begin{aligned}
&(1)\ \ v^1(z) = v(z) \quad \text{in a neighborhood of} \quad z = y_0,\\
&(2)\ \ v^1(z) \equiv 0 \quad \text{outside some compact set contained in } \mathcal{E}_{1/2}.
\end{aligned} \tag{3.70}$$

Then the left-hand side of (3.69) is[26]

$$\int_{z \in \mathcal{E}_1} \gamma(y_0,z)\, \bar{D}v^1(z)\, dz + \int_{\substack{z \in \mathcal{E}_1 \\ z_n \geqslant \rho}} \zeta(z)\gamma(y_0,z)\bar{D}[v(z) - v^1(z)]\, dz. \tag{3.71}$$

We apply (3.68) to the first integral. The integrand of the second integral has no singularities. If we transform the second integral in (3.71) by integration by parts using the Gauss theorem and denote the outer normal on the boundary by $\nu = (\nu_1, \ldots, \nu_n)$, then

$$\int_{z_n = \rho} \Big\{ \zeta\gamma \sum_{i,k=1}^{n} \bar{a}_{ik} v_{z_i}\nu_k - \sum_{i,k=1}^{n} (\bar{a}_{ik}\zeta\gamma)_{z_k} v\nu_i + \zeta\gamma \sum_{i=1}^{n} \bar{a}_i v\nu_i \Big\}\, dS$$
$$+ \int_{\substack{z \in \mathcal{E}_1 \\ z_n \geqslant \rho}} (v - v^1)\bar{D}^*(\zeta\gamma)\, dz. \tag{3.72}$$

We used the fact that for sufficiently small $\rho > 0$, $v^1(z)$ vanishes identically in a neighborhood of $z_n = \rho$ and that no further boundary integrals occur because of the factor $\zeta(z)$.

From (3.69) and the above we have

$$-v(y_0) + \int_{z_n = \rho} \Big\{ \zeta\gamma \sum_{i,k=1}^{n} \bar{a}_{ik} v_{z_i}\nu_k - \sum_{i,k=1}^{n} (\bar{a}_{ik}\zeta\gamma)_{z_k} v\nu_i + \zeta\gamma \sum_{i=1}^{n} \bar{a}_i v\nu_i \Big\}\, dS$$
$$+ \int_{\substack{z \in \mathcal{E}_1 \\ z_n \geqslant \rho}} v\, \bar{D}^*(\zeta\gamma)\, dz = \int_{\substack{z \in \mathcal{E}_1 \\ z_n \geqslant \rho}} \zeta\gamma\bar{f}\, dz. \tag{3.73}$$

If we let $\rho \to 0$, then the integrals over domains converge because $v(z)$ is integrable on $\mathcal{E}_1$ and the coefficient of $v(z)$ is bounded in a neighborhood of $z_n = 0$. Thus the integral $\int_{z_n = \rho} \{\cdots\}\, dS$ must converge; let its limit be p. In the sequel we show that $p = 0$.

Therefore (3.73) gives

$$v(y) = \int_{z \in \mathcal{E}_1} \Big\{ \sum_{i,k=1}^{n} (\bar{a}_{ik}\zeta\gamma)_{z_i z_k} - \sum_{i=1}^{n} (\bar{a}_i\zeta\gamma)_{z_i} + \bar{c}\zeta\gamma \Big\} v\, dz - \int_{z \in \mathcal{E}_1} \zeta\gamma\bar{f}\, dz \tag{3.74}$$

---

[26] The first integral originally had ζ in the integrand and the same domain of integration as the second integral. Because of (3.70) we only have to integrate over $\mathcal{E}_{1/2}$ if ρ is sufficiently small. There $\zeta = 1$. Then because of (3.70) we also may choose $\mathcal{E}_1$ for the domain of integration.

for $y = y_0$; but, since $y_0 \in \mathcal{E}_{\frac{1}{2}}$ was arbitrary, (3.74) holds for all $y \in \mathcal{E}_{\frac{1}{2}}$. If we use the estimates (3.66) and (3.67), then from (3.74) the inequality

$$|v(y)| \leqslant \int_{z \in \mathcal{E}_1} \frac{c_1}{|y - z|^{n-1}}\, |v(z)|\, dz + c_2, \qquad y \in \mathcal{E}_{\frac{1}{2}} \tag{3.75}$$

follows, where c_1, c_2 are two constants. From this we prove the boundedness of $v(z)$ by iteration. In order not to end up with the "logarithmic case" we weaken the estimate, setting $\lambda = \frac{1}{2}\sqrt{2}$, to

$$|v(y)| \leqslant \int_{z \in \mathcal{E}_1} \frac{c_1}{|y - z|^{n-\lambda}}\, |v(z)|\, dz + c_2, \qquad y \in \mathcal{E}_{\frac{1}{2}}. \tag{3.76}$$

If instead of $\zeta(z)$ we had chosen a function which has the value 1 on $\bar{\mathcal{E}}_{\frac{1}{4}}$ and which vanishes outside $\bar{\mathcal{E}}_{\frac{1}{2}}$, then with new constants c_3, c_4 we would have obtained the inequality

$$|v(y)| \leqslant \int_{z \in \mathcal{E}_{\frac{1}{2}}} \frac{c_3}{|y - z|^{n-\lambda}}\, |v(z)|\, dz + c_4, \qquad y \in \mathcal{E}_{\frac{1}{4}}. \tag{3.77}$$

For the following presentation it is more convenient in (3.76) to denote the variable of integration by x, and the point y by z, that is,

$$|v(z)| \leqslant \int_{x \in \mathcal{E}_1} \frac{c_1}{|z - x|^{n-\lambda}}\, |v(x)|\, dx + c_2, \qquad z \in \mathcal{E}_{\frac{1}{2}}. \tag{3.78}$$

For the first step of the iteration procedure, we apply (3.78) to the integrand of (3.77) and obtain

$$|v(y)| \leqslant \int_{z \in \mathcal{E}_{\frac{1}{2}}} \frac{c_3}{|y - z|^{n-\lambda}} \left(\int_{x \in \mathcal{E}_1} \frac{c_1}{|z - x|^{n-\lambda}}\, |v(x)|\, dx \right) dz$$
$$+ \int_{z \in \mathcal{E}_{\frac{1}{2}}} \frac{c_2 c_3}{|y - z|^{n-\lambda}}\, dz + c_4 \quad \text{for} \quad y \in \mathcal{E}_{\frac{1}{4}}.$$

By interchanging the order of integration and using Section 3.1,

$$|v(y)| \leqslant \int_{x \in \mathcal{E}_1} \left(\int_{z \in \mathcal{E}_{\frac{1}{2}}} \frac{c_1 c_3}{|y - z|^{n-\lambda} |z - x|^{n-\lambda}}\, dz \right) |v(x)|\, dx + \text{const}$$

follows for all $y \in \mathcal{E}_{\frac{1}{4}}$. The inner integral may be estimated by Theorem 4 of Section 4.1 so that with two new constants c_5, c_6 we obtain the inequality

$$|v(y)| \leqslant \int_{x \in \mathcal{E}_1} \frac{c_5}{|y - x|^{n-2\lambda}}\, |v(x)|\, dx + c_6, \qquad y \in \mathcal{E}_{\frac{1}{4}}.$$

Now we can proceed by entering this result in the estimate

$$|v(z)| \leqslant \int_{y \in \mathcal{E}_{\frac{1}{4}}} \frac{c_7}{|z - y|^{n-\lambda}}\, |v(y)|\, dy + c_8, \qquad z \in \mathcal{E}_{\frac{1}{8}}$$

which is valid for certain constants c_7 and c_8. As before, we obtain

$$|v(z)| \leqslant \int_{x \in \mathcal{E}_1} \frac{c_9}{|z - x|^{n-3\lambda}}\, |v(x)|\, dx + c_{10} \quad \text{for} \quad z \in \mathcal{E}_{\frac{1}{8}}.$$

By Theorem 4 of Section 4.1, after finitely many steps we will come to some $\mathcal{E}_\alpha$ such that

$$|v(z)| \leqslant \int_{x \in \mathcal{E}_1} \text{const}\, |v(x)|\, dx + \text{const}, \qquad z \in \mathcal{E}_\alpha.$$

Thus $v(z)$ is bounded in $\mathcal{E}_\alpha$. But then without loss of generality we may assume that $v(z)$ is bounded in $\mathcal{E}_1$ and, because of the properties of $\gamma(y,z)$ we mentioned, the integrands in (3.74) satisfy the hypotheses of Theorem 1 of Section 4.1. Consequently $v(z)$ is Hölder-continuous in $\mathcal{E}_{1/2}$ with fixed Hölder-coefficient and fixed exponent, which was to be shown.

It remains to prove that in (3.73) the integrals over $z_n = \rho$ have the limit zero as $\rho \to 0$. If we put $h(\rho) = \int_{z_n = \rho} \{\cdots\} \, dS$ in (3.73), then we have shown that

$$\lim_{\rho \to 0,\, \rho > 0} h(\rho) = p$$

exists, and

$$\lim_{\rho \to 0,\, \rho > 0} \frac{2}{\rho} \int_{\rho/2}^{\rho} h(t) \, dt = p$$

by the mean-value theorem of the integral calculus. Now from (3.73) we have

$$\left| \frac{2}{\rho} \int_{\rho/2}^{\rho} h(t) \, dt \right| = \left| \frac{2}{\rho} \int_{\rho/2}^{\rho} \left(\int_{z_n = t} \{\cdots\} \, dS \right) dt \right| = \left| \frac{2}{\rho} \int_{\rho/2 \leqslant z_n \leqslant \rho} \{\cdots\} \, dz \right|$$

$$= \left| \frac{2}{\rho} \int_{\rho/2 \leqslant z_n \leqslant \rho} \{ \zeta\gamma \sum_{i,k=1}^{n} \bar{a}_{ik} v_{z_i} \nu_k - \sum_{i,k=1}^{n} (\bar{a}_{ik}\zeta\gamma)_{z_k} v \nu_i + \zeta\gamma \sum_{i=1}^{n} \bar{a}_i v \nu_i \} \, dz \right|. \tag{3.79}$$

But now $\zeta(z) \equiv 0$ outside of $\bar{\mathcal{E}}_{2/3}$. Hence the domain of integration $\rho/2 \leqslant z_n \leqslant \rho$ may be replaced by $z \, \varepsilon \, C_\rho$, $z_n \geqslant \rho/2$, where C_ρ denotes the cylinder

$$C_\rho: \quad \begin{aligned} &(z_1 - \dot{z}_1)^2 + (z_2 - \dot{z}_2)^2 + \cdots + (z_{n-1} - \dot{z}_{n-1})^2 \leqslant (\tfrac{2}{3}d)^2, \\ &0 < z_n \leqslant \rho \end{aligned}$$

($\dot{z}$ and d were introduced at the beginning of Section 3.7), then we have[27]

$$\left| \frac{2}{\rho} \int_{\rho/2}^{\rho} h(t) \, dt \right| \leqslant \frac{2}{\rho} \int_{\substack{z \varepsilon C_\rho \\ z_n \geqslant \rho/2}} |\{\cdots\}| \, dz \leqslant \frac{\text{const}}{\rho} \int_{z \varepsilon C_\rho} \{ \sum_{i=1}^{n} |\gamma(y_0,z)| \, |v_{z_i}(z)| + |v(z)| \} \, dz$$

$$\leqslant \frac{\text{const}}{\rho} \int_{z \varepsilon C_\rho} \sum_{i=1}^{n} |\gamma(y_0,z)| \, |v_{z_i}(z)| \, dz + \frac{\text{const}}{\rho} \int_{z \varepsilon C_\rho} |v(z)| \, dz.$$

These integrals exist since, in particular, $|v(z)|$ and $|v_{z_i}(z)|$ are integrable over C_ρ. In the first integral we use $|\gamma(y_0,z)| \leqslant \text{const} \cdot \rho$, since $\gamma(y_0,z) = 0$ when $z_n = 0$; to the second integral we apply Schwarz's inequality.

Then

$$\left| \frac{2}{\rho} \int_{\rho/2}^{\rho} h(t) \, dt \right| \leqslant \text{const} \int_{C_\rho} \sum_{i=1}^{n} |v_{z_i}(z)| \, dz + \frac{\text{const}}{\rho} \sqrt{\text{const}\,\rho \int_{C_\rho} v^2(z) \, dz},$$

and if here we apply (3.62) to the last term we obtain

$$\left| \frac{2}{\rho} \int_{\rho/2}^{\rho} h(t) \, dt \right| \leqslant \text{const} \int_{C_\rho} \sum_{i=1}^{n} |v_{z_i}(z)| \, dz + \text{const} \sqrt{\rho}.$$

From this we see that indeed $p = 0$.

We now come to the proof of Theorem 3 in Section 3.3. It can be skipped during the first study of the book.

---

[27] The coefficient of $v(z)$ was estimated by a constant; this is possible since y_0 has a fixed distance from $z_n = 0$.

Proof of Theorem 3. As we shall show presently, $v(z)$ has bounded first derivatives in $\mathcal{E}_{1/4}$. But then the derivatives $u_{z_i}(x)$ are bounded in $\mathcal{V} \cap \mathcal{D}$ where $\mathcal{V}$ is a neighborhood of $\dot{x}$. Since we already know that $u_{z_i}(x)$ is continuous in $\mathcal{D}$ and since $\dot{x}$ is an arbitrary point on $\dot{\mathcal{D}}$, to every $x \in \bar{\mathcal{D}}$ there is a neighborhood $\mathcal{V}$ such that $u(x)$ has bounded first derivatives in $\mathcal{V} \cap \mathcal{D}$. By the covering theorem of Heine and Borel, finitely many of these neighborhoods suffice to cover $\bar{\mathcal{D}}$. Let M be the maximum of the corresponding bounds for the $|u_{z_i}|$. Then $|u_{z_i}(x)| \leqslant M$ for $x \in \mathcal{D}$.

Now let $\zeta(z)$ be a function which is twice continuously differentiable in $z_n \geqslant 0$, has the value 1 in $\bar{\mathcal{E}}_{1/4}$ and the value 0 outside of $\bar{\mathcal{E}}_{1/3}$, and which satisfies $0 \leqslant \zeta(z) \leqslant 1$; furthermore let

$$H(y,z) = \sum_{i,k=1}^{n} [\bar{a}_{ik}(z)\zeta(z)\gamma(y,z)]_{z_i z_k} - \sum_{i=1}^{n} [\bar{a}_i(z)\zeta(z)\gamma(y,z)]_{z_i} + \bar{c}(z)\zeta(z)\gamma(y,z).$$

Then, analogously to (3.74), the equation

$$v(y) = \int_{\mathcal{E}_{1/2}} H(y,z)v(z) \, dz - \int_{\mathcal{E}_{1/2}} \zeta(z)\gamma(y,z)\bar{f}(z) \, dz \quad \text{for} \quad y \in \mathcal{E}_{1/4}$$

is valid. By Theorem 2 from Section 4.1, the first derivatives of the last integral are uniformly Hölder-continuous in $\mathcal{E}_{1/4}$ and are therefore bounded in $\mathcal{E}_{1/4}$.

By Theorem 3 from Section 4.1 we have

$$\frac{\partial}{\partial y_j} \int_{\mathcal{E}_{1/2}} H(y,z)v(z) \, dz = \int_{0_{1/2}} H_{y_j}(y,z)[v(z) - v(y)] \, dz + v(y) \frac{\partial}{\partial y_j} \int_{\mathcal{E}_{1/2}} H(y,z) \, dz,$$

since we know already that $v(z)$ is uniformly Hölder-continuous in $\mathcal{E}_{1/2}$ and we shall show presently that $\int_{\mathcal{E}_{1/2}} H(y,z) \, dz$ is Hölder-continuously differentiable.   Further we shall prove the estimate

$$\left| \frac{\partial}{\partial y_j} \int_{\mathcal{E}_{1/2}} H(y,z) \, dz \right| \leqslant \text{const} \, |y_n|^{-\alpha}$$

where y_n is the nth component of $y = (y_1, \dots, y_n)$ and α the Hölder exponent of $v(y)$ in $\mathcal{E}_{1/2}$.

Since $v(y_1, \dots, y_{n-1}, 0) = 0$, we have

$$|v(y)| = |v(y) - v(y_1, \dots, y_{n-1}, 0)| \leqslant \text{const} \, |y_n|^{\alpha}$$

and hence

$$\left| \frac{\partial}{\partial y_j} \int_{\mathcal{E}_{1/2}} H(y,z)v(z) \, dz \right| \leqslant \text{const} \int_{\mathcal{E}_{1/2}} |y - z|^{-n+\alpha} \, dz + \text{const} \, |y_n|^{\alpha}|y_n|^{-\alpha}.$$

Thus $v_{y_j}(y)$ is bounded in $\mathcal{E}_{1/4}$. To prove the properties of $\int_{\mathcal{E}_{1/2}} H(y,z) \, dz$ that we used, we write

$$\int_{\mathcal{E}_{1/2}} H(y,z) \, dz = \int_{\mathcal{E}_{1/2}} \sum_{i,k=1}^{n} \zeta(\bar{a}_{ik}\gamma)_{z_i z_k} \, dz + 2 \int_{\mathcal{E}_{1/2}} \sum_{i,k=1}^{n} \zeta_{z_i}(\bar{a}_{ik}\gamma)_{z_k} \, dz$$

$$+ \int_{\mathcal{E}_{1/2}} \sum_{i,k=1}^{n} \zeta_{z_i z_k}\bar{a}_{ik}\gamma \, dz - \int_{\mathcal{E}_{1/2}} \sum_{i=1}^{n} (\bar{a}_i \zeta\gamma)_{z_i} \, dz + \int_{\mathcal{E}_{1/2}} \bar{c}\zeta\gamma \, dz.$$

By Theorem 2 in Section 4.1 those integrals containing only $\gamma(y,z)$ are uniformly Hölder-continuously differentiable, and the integrals containing only first derivatives of $\gamma(y,z)$ can be transformed by integration by parts into integrals which contain only $\gamma(y,z)$ [boundary integrals vanish, since in particular $\gamma(y,z) = 0$ for $z_n = 0$]. Hence they too are uniformly Hölder-continuously differentiable. There remains by using the formula (3.87) in Section 3.8

$$\int_{\varepsilon_{\frac{1}{2}}} \sum_{i,k=1}^{n} \zeta(z)[\bar{a}_{ik}(z)\gamma(y,z)]_{z_i z_k}\, dz = \int_{\varepsilon_{\frac{1}{2}}} \sum_{i,k=1}^{n} \zeta(z)[(\bar{a}_{ik}(z) - \bar{a}_{ik}(y))\gamma(y,z)]_{z_i z_k}\, dz$$

$$= \int_{z_n=0} \sum_{i,k=1}^{n} \zeta(z)[(\bar{a}_{ik}(z) - \bar{a}_{ik}(y))\gamma(y,z)]_{z_i}\nu_k\, dS$$

$$+ \text{ a uniformly Hölder-continuously differentiable integral over a domain.}$$

The boundary integral contains no singularity in its integrand and hence is Hölder-continuously differentiable; thus $\int_{\varepsilon_{\frac{1}{2}}} H(y,z)\, dz$ is Hölder-continuously differentiable and, except for the boundary integral, all parts of $\int_{\varepsilon_{\frac{1}{2}}} H(y,z)\, dz$ have bounded first derivatives (for they are uniformly Hölder continuous).

It remains to estimate the first derivatives of the boundary integral, that is,

$$I = \frac{\partial}{\partial y_j} \int_{z_n=0} \zeta(z) \sum_{i,k=1}^{n} [(\bar{a}_{ik}(z) - \bar{a}_{ik}(y))\gamma(y,z)]_{z_i}\nu_k\, dS$$

$$= - \int_{z_n=0} \zeta(z) \sum_{i,k=1}^{n} [(\bar{a}_{ik}(z) - \bar{a}_{ik}(y))\gamma(y,z)]_{z,\nu_i}\, dS.$$

Let $\bar{y}$ be the projection $(y_1, \ldots, y_{n-1}, 0)$ of the point y onto the hyperplane $z_n = 0$. The expression I is an integral over an $(n-1)$-dimensional domain in the hyperplane $z_n = 0$, namely

$$I = - \int_{\substack{|z-\bar{y}| \leqslant 2d \\ z_n=0}} \zeta(z) \sum_{i,k=1}^{n} [(\bar{a}_{ik}(z) - \bar{a}_{ik}(y))\gamma(y,z)]_{z_i\nu_j}\, dz_1 \cdots dz_{n-1},$$

$$|I| \leqslant \text{const} \int_{\substack{|z-\bar{y}| \leqslant 2d \\ z_n=0}} |z - y|^{-n+1}\, dz_1 \cdots dz_{n-1}.$$

Since $|z - y| \geqslant |z - \bar{y}|$ and $|z - y| \geqslant |y - \bar{y}| = |y_n|$, we have

$$|I| \leqslant \text{const}\, |y_n|^{-\alpha} \int_{\substack{|z-\bar{y}| \leqslant 2d \\ z_n=0}} |z - \bar{y}|^{-n+1+\alpha}\, dz_1 \cdots dz_{n-1} \leqslant \text{const}\, |y_n|^{-\alpha}.$$

Thus the desired estimate

$$\left| \frac{\partial}{\partial y_j} \int_{\varepsilon_{\frac{1}{2}}} H(y,z)\, dz \right| \leqslant \text{const}\, |y_n|^{-\alpha}$$

has been proved. This proof was communicated to the author by E. Wienholtz.

3.8   Tools

Since the proofs of Theorems 1 and 3 were finished in Section 3.7, the symbols used there are now deprived of their previous meaning. Consider the functions $a_{ik}(x) \in C^j$,

$a_i(x) \varepsilon\ C^1$, $c(x) \varepsilon\ C^0$, $f(x) \varepsilon\ C^0$ in a convex normal domain $\mathfrak{D}$, $i, k = 1, \ldots, n, j \geqslant 2$, and let the partial differential equation

$$\sum_{i,k=1}^{n} a_{ik}(x)u_{x_i x_k} + \sum_{i=1}^{n} a_i(x)u_{x_i} + c(x)u = f(x) \tag{3.80}$$

be uniformly elliptic in $\mathfrak{D}$, that is,

$$\sum_{i,k=1}^{n} a_{ik}(x)\xi_i\xi_k \geqslant c_1 \sum_{i=1}^{n} (\xi_i)^2$$

for all real numbers $\xi_1, \ldots, \xi_n$ and all $x \varepsilon\ \mathfrak{D}$, where c_1 is a positive constant.

For this differential equation we define a singularity function $S(y,x)$, where $x \varepsilon\ \mathfrak{D}$, $y \varepsilon\ \mathfrak{D}$, $x \neq y$ by

$$S(y,x) = \begin{cases} \dfrac{|A_{ik}(y)|^{\frac{1}{2}}}{(n-2)\omega_n} \Big[\sum\limits_{i,k=1}^{n} A_{ik}(y)(y_i - x_i)(y_k - x_k)\Big]^{(2-n)/2}, & n > 2, \\[3mm] -\dfrac{|A_{ik}(y)|^{\frac{1}{2}}}{2\pi} \log \Big[\sum\limits_{i,k=1}^{n} A_{ik}(y)(y_i - x_i)(y_k - x_k)\Big]^{\frac{1}{2}}, & n = 2, \end{cases} \tag{3.81}$$

where the $A_{ik}(y)$ are the coefficients of the inverse matrix $\mathfrak{a}^{-1}$ of $\mathfrak{a} = (a_{ik}(y))$ and $|A_{ik}(y)|$ denotes the determinant of $\mathfrak{a}^{-1}$. These $A_{ik}(y)$ are j times continuously differentiable in $\mathfrak{D}$ as functions of y. The singularity function $S(y,x)$ is a generalization of the singularity function defined in Section I-3.2 for the equation $\Lambda_n u = 0$:

$$s(y,x) = \frac{1}{(n-2)\omega_n} |y - x|^{2-n}, \qquad n > 2; \qquad s(y,x) = -\frac{1}{2\pi} \log |y - x|,$$
$$n = 2. \tag{3.82}$$

Now we want to point out some properties of $S(y,x)$. By hypothesis there are two positive constants c_1 and c_2 such that for all $y \varepsilon\ \mathfrak{D}$ and all real numbers $\xi_1, \ldots, \xi_n$

$$c_1 \sum_{i=1}^{n} (\xi_i)^2 \leqslant \sum_{i,k=1}^{n} a_{ik}(y)\xi_i\xi_k \leqslant c_2 \sum_{i=1}^{n} (\xi_i)^2. \tag{3.83}$$

Hence

$$\frac{1}{c_2} |y - x|^2 \leqslant \sum_{i,k=1}^{n} A_{ik}(y)(y_i - x_i)(y_k - x_k) \leqslant \frac{1}{c_1} |y - x|^2 \tag{3.84}$$

for all $y \varepsilon\ \mathfrak{D}$ and $x \varepsilon\ \mathfrak{D}$. From (3.81) we see that for fixed $y \varepsilon\ \mathfrak{D}$, $S(y,x)$ is arbitrarily often differentiable in the variables $x_1, \ldots, x_n$ when $x \neq y$. Furthermore, for fixed x, $S(y,x)$ and every one of these derivatives is j times continuously differentiable with respect to the variables $y_1, \ldots, y_n$ in $y \varepsilon\ \mathfrak{D}$, $y \neq x$, because the $A_{ik}(y)$ belong to C^j in $\mathfrak{D}$. Finally, from (3.84) we obviously have the estimates

$$|S(y,x)| \leqslant \frac{\text{const}}{|y - x|^{n-2}}, \qquad |S_{x_i}(y,x)| \leqslant \frac{\text{const}}{|y - x|^{n-1}}, \qquad |S_{x_i x_k}(y,x)| \leqslant \frac{\text{const}}{|y - x|^n},$$
$$|S_{y_i}(y,x)| \leqslant \frac{\text{const}}{|y - x|^{n-1}}, \qquad |S_{x_i y_k}(y,x)| \leqslant \frac{\text{const}}{|y - x|^n}, \qquad i, k = 1, \ldots, n, \tag{3.85}$$

for all $x \varepsilon\ \mathfrak{D}$, $y \varepsilon\ \mathfrak{D}$, $x \neq y$. In the case $n = 2$, however, the first estimate has to be replaced by

$$|S(y,x)| \leqslant \text{const} \, \big|\log |y - x|\big| + \text{const.} \tag{3.85a}$$

Now let y be a fixed point in $\mathfrak{D}$. The symmetric matrix $\mathfrak{A} = (a_{ik}(y))$ is positive definite and by means of another positive-definite symmetric matrix $\mathfrak{B}$ it admits the representation $\mathfrak{A} = \mathfrak{B}\mathfrak{B}$. Then $\mathfrak{A}^{-1} = \mathfrak{B}^{-1}\mathfrak{B}^{-1}$, and

$$\sum_{i,k=1}^{n} A_{ik}(y)(y_i - x_i)(y_k - x_k) = (\mathfrak{B}^{-1}(y - x), \mathfrak{B}^{-1}(y - x)) = |\mathfrak{B}^{-1}(y - x)|^2.$$

If we denote the determinant of $\mathfrak{B}$ by $|\mathfrak{B}|$, then from (3.81) we have

$$S(y,x) = \begin{cases} \dfrac{1}{(n - 2)\omega_n|\mathfrak{B}|} \, |\mathfrak{B}^{-1}(y - x)|^{2-n}, & n > 2, \\[2ex] -\dfrac{1}{2\pi|\mathfrak{B}|} \, \log |\mathfrak{B}^{-1}(y - x)|, & n = 2. \end{cases}$$

By the transformation $z = \mathfrak{B}^{-1}x$, which carries the point y into z_0: $z_0 = \mathfrak{B}^{-1}y$, the relation

$$S(y,x) = \frac{1}{|\mathfrak{B}|} \, s(z_0,z) \tag{3.86}$$

is obtained, where $s(z_0,z)$ is the singularity function of the potential equation that was introduced in Section I-3.2. It was given again in (3.82). Since $\Delta_n s = 0$, an easy calculation gives

$$\sum_{i,k=1}^{n} a_{ik}(y)S_{x_i x_k}(y,x) = 0. \tag{3.87}$$

From this the important estimate

$$\Big| \sum_{i,k=1}^{n} a_{ik}(x)S_{x_i x_k}(y,x)\Big| \leqslant \frac{\text{const}}{|y - x|^{n-1}}, \qquad x \in \mathfrak{D}, \qquad y \in \mathfrak{D}, \qquad x \neq y \tag{3.88}$$

follows, which goes further than (3.85); since $\mathfrak{D}$ is convex, we may apply the mean-value theorem to obtain

$$\Big| \sum_{i,k=1}^{n} a_{ik}(x)S_{x_i x_k}(y,x)\Big| = \Big| \sum_{i,k=1}^{n} (a_{ik}(x) - a_{ik}(y))S_{x_i x_k}(y,x)\Big| \leqslant \text{const}\,|x - y| \frac{\text{const}}{|y - x|^n}.$$

Corresponding results hold for the higher derivatives; for instance,

$$\Big| \sum_{i,k=1}^{n} a_{ik}(x)S_{x_i x_k v_j}(y,x)\Big| = \Big|\Big(\sum_{i,k=1}^{n} a_{ik}(x)S_{x_i x_k}(y,x)\Big)_{v_j}\Big|$$

$$= \Big|\Big(\sum_{i,k=1}^{n} (a_{ik}(x) - a_{ik}(y))S_{x_i x_k}(y,x)\Big)_{v_j}\Big|$$

$$\leqslant \Big| \sum_{i,k=1}^{n} (a_{ik}(x) - a_{ik}(y))_{v_j}S_{x_i x_k}(y,x)\Big|$$

$$+ \Big| \sum_{i,k=1}^{n} (a_{ik}(x) - a_{ik}(y))S_{x_i x_k v_j}(y,x)\Big| \leqslant \frac{\text{const}}{|y - x|^n}. \tag{3.89}$$

Finally, for all functions $\Phi(x) \in C^2$ in $\mathfrak{D}$ which vanish identically in a neighborhood of $\dot{\mathfrak{D}}$, we have the relation

$$\Phi(y) = - \int_{\mathfrak{D}} S(y,x) \Big\{ \sum_{i,k=1}^{n} a_{ik}(x)\Phi_{x_i x_k} + \sum_{i=1}^{n} a_i(x)\Phi_{x_i} + c(x)\Phi\Big\} \, dx$$

$$+ \int_{\mathfrak{D}} \Big\{ \sum_{i,k=1}^{n} (a_{ik}(x)S(y,x))_{x_i x_k} - \sum_{i=1}^{n} (a_i(x)S(y,x))_{x_i}$$

$$+ c(x)S(y,x)\Big\}\Phi(x) \, dx \tag{3.90}$$

for $y \in \mathfrak{D}$. For the proof, let $y \in \mathfrak{D}$ be fixed. Because of (3.87) the last integral in (3.90) can be written in the form

$$\int_{\mathfrak{D}} \left\{ \sum_{i,k=1}^{n} [(a_{ik}(x) - a_{ik}(y))S(y,x)]_{x_i x_k} - \sum_{i=1}^{n} (a_i(x)S(y,x))_{x_i} + c(x)S(y,x) \right\} \Phi(x)\, dx.$$

In spite of the singularity of the integrand at $x = y$, by the Gauss theorem this expression may be rewritten as

$$\int_{\mathfrak{D}} S(y,x) \left\{ \sum_{i,k=1}^{n} (a_{ik}(x) - a_{ik}(y))\Phi_{x_i x_k} + \sum_{i=1}^{n} a_i(x)\Phi_{x_i} + c(x)\Phi \right\} dx \qquad (3.91)$$

because the singularity is sufficiently weak, by (3.85) and (3.88). For the precise details of the calculation compare with Section 3.2. If now we confront the transformed second integral (3.91) from (3.90) with the first integral from (3.90), we see that it remains only to show

$$\Phi(y) = - \int_{\mathfrak{D}} S(y,x) \sum_{i,k=1}^{n} a_{ik}(y)\Phi_{x_i x_k}(x)\, dx.$$

Using the transformation $x = \mathfrak{B}z$ we set

$$\Phi(x) = \Phi(\mathfrak{B}z) = \Psi(z).$$

From (3.86), then

$$- \int_{\mathfrak{D}} S(y,x) \sum_{i,k=1}^{n} a_{ik}(y)\Phi_{x_i x_k}(x)\, dx = - \frac{1}{|\mathfrak{B}|} \int_{\tilde{\mathfrak{D}}} s(z_0,z) \sum_{i=1}^{n} \Psi_{z_i z_i}(z) |\mathfrak{B}|\, dz \qquad (3.92)$$

results by the change of variable $x = \mathfrak{B}z$ and $\tilde{\mathfrak{D}} = \mathfrak{B}^{-1}(\mathfrak{D})$. But by (I-3.8), the right-hand side has the value $\Psi(z_0)$, and this is the same as $\Phi(y)$.

Now we come to the *pseudo-fundamental solution* $\gamma(y,x)$ which was needed in Section 3.7. In general, by a pseudo-fundamental solution belonging to the differential equation (3.80) in the domain $\mathfrak{D}$ we understand a function $\Gamma(y,x)$ which may be represented by $\Gamma(y,x) = S(y,x) + \Phi(y,x)$, where $S(y,x)$ is the singularity function which was just introduced and $\Phi(y,x)$ belongs to C^1 in $\bar{\mathfrak{D}}$, C^2 in $\mathfrak{D}$ relative to x and satisfies the differential equation with constant coefficients $\sum_{i,k=1}^{n} a_{ik}(y)\Phi_{x_i x_k}(y,x) = 0$, all for fixed $y \in \mathfrak{D}$. This definition has far-reaching analogies to that in Section I-3.2. It will perhaps be well to look once more at the properties of the function $\gamma(y,x) = s(y,x) + \Phi(y,x)$ which was defined there.

Let $\mathfrak{D}$ be the half-ball $|x - \dot{x}| \leq d$, $x_n \geq 0$. Here we give a special pseudo-fundamental solution $\gamma(y,x)$ of the differential equation (3.80) which satisfies $\gamma(y,x) = 0$ when x lies in the hyperplane $x_n = 0$. For this, fix $y \in \mathfrak{D}$. Again we have $\mathfrak{A} = (a_{ik}(y)) = \mathfrak{B}\mathfrak{B}$. Under the transformation $z = \mathfrak{B}^{-1}x$, y is transformed into $z_0 = \mathfrak{B}^{-1}y$. Now reflect z_0 in the image of the hyperplane $x_n = 0$ and let the reflected image be $\tilde{z}_0$. If we denote the inverse image of $\tilde{z}_0$ by $\tilde{y}$, that is, $\tilde{y} = \mathfrak{B}\tilde{z}_0$, then $\gamma(y,x)$ is given explicitly by $\gamma(y,x) = S(y,x) + \varphi(y,x)$ where

$$\varphi(y,x) = \begin{cases} - \dfrac{|A_{ik}(y)|^{1/2}}{(n-2)\omega_n} \left[\sum_{i,k=1}^{n} A_{ik}(y)(\tilde{y}_i - x_i)(\tilde{y}_k - x_k) \right]^{(2-n)/2}, & n > 2, \\[3mm] \dfrac{|A_{ik}(y)|^{1/2}}{2\pi} \log \left[\sum_{i,k=1}^{n} A_{ik}(y)(\tilde{y}_i - x_i)(\tilde{y}_k - x_k) \right]^{1/2}, & n = 2. \end{cases} \qquad (3.93)$$

For the proof we first show that for fixed $y \in \mathfrak{D}$, $\varphi(y,x)$ is arbitrarily often differentiable in the variables $x_1, \ldots, x_n$ for $x \in \mathfrak{D}$. For this it suffices to prove that none of the brackets in (3.93) vanishes when $x \in \mathfrak{D}$. Because of (3.84)

$$\frac{1}{c_1} |\bar{y} - x|^2 \leqslant \sum_{i,k=1}^{n} A_{ik}(y)(\bar{y}_i - x_i)(\bar{y}_k - x_k) \leqslant \frac{1}{c_2} |\bar{y} - x|^2,$$

and $\bar{y} \notin \mathfrak{D}$, for since z_0 and $\bar{z}_0$ are separated by the image of the hyperplane $x_n = 0$ their inverse images y and $\bar{y}$ are separated by the hyperplane $x_n = 0$.

Next we have to convince ourselves that $\sum\limits_{i,k=1}^{n} a_{ik}(y)\varphi_{x_i x_k}(y,x) = 0$, which can be done exactly as for (3.87). Similarly to (3.86) we finally obtain the relation

$$\varphi(y,x) = -\frac{1}{|\mathfrak{B}|} s(\bar{z}_0,z) \tag{3.94}$$

for $\varphi(y,x)$ if we set $x = \mathfrak{B}z, y = \mathfrak{B}z_0$. Hence

$$\gamma(y,x) = \frac{1}{|\mathfrak{B}|} \{s(z_0,z) - s(\bar{z}_0,z)\} \quad \text{where} \quad x = \mathfrak{B}z. \tag{3.95}$$

If x lies in the hyperplane $x_n = 0$, then $|z - z_0| = |z - \bar{z}_0|$ and $s(z_0,z) - s(\bar{z}_0,z)$ vanishes. Consequently $\gamma(y,x) = 0$ when x lies in the hyperplane $x_n = 0$.

It remains to show that $\gamma(y,x)$ has the properties stated in Section 3.7. The differentiability properties with respect to the variables $x_1, \ldots, x_n$ are trivial on account of our knowledge of $S(y,x)$ and $\varphi(y,x)$. Even relation (3.68) can be obtained easily by splitting the integrals by the decomposition

$$\gamma(y,x) = S(y,x) + \varphi(y,x)$$

and then transforming those which contain the regular part of $\varphi(y,x)$ by integration by parts and applying (3.90) to the remaining ones.

To obtain statements about the differentiability properties of $\gamma(y,x)$ with respect to the variables $y_1, \ldots, y_n$, we first verify the relation

$$\bar{y}_i = y_i - \frac{2y_n}{a_{nn}(y)} a_{ni}(y), \qquad i = 1, \ldots, n. \tag{3.96}$$

If this is used in (3.93), we see that, for fixed $x \in \mathfrak{D}$, $\gamma(y,x)$ and its derivatives in $x_1, \ldots,$ x_n are j times continuously differentiable with respect to the variables $y_1, \ldots, y_n$, and this is true for $y \in \mathfrak{D}, y \neq x$.

To be able finally to verify the estimates given in (3.66) and (3.67) we need the fact that the inequality

$$|y - x| \leqslant c_3 |\bar{y} - x| \quad \text{for} \quad \text{all } x \in \mathfrak{D}, \qquad y \in \mathfrak{D} \tag{3.97}$$

holds with a positive constant c_3. But since x and y are not separated by the hyperplane $x_n = 0$, their images $z = \mathfrak{B}^{-1}x, z_0 = \mathfrak{B}^{-1}y$ are not separated by the image of the hyperplane. Thus

$$|z - z_0| \leqslant |z - \bar{z}_0|,$$

and we have

$$|y - x| = |\mathfrak{B}(z_0 - z)| \leqslant c_4|z_0 - z| \leqslant c_4|\bar{z}_0 - z| = c_4|\mathfrak{B}^{-1}(\bar{y} - x)| \leqslant c_3|\bar{y} - x|.$$

4

WEYL'S LEMMA FOR EQUATIONS OF ELLIPTIC TYPE

4.1   Singular Integrals

In this section we do not assume that the domain $\mathfrak{D}$ is normal.

THEOREM 1.   Let $\mathfrak{D}$ be a bounded convex domain in R_n. Let $f(x)$ be integrable and bounded on $\mathfrak{D}$ and $H(y,x)$ be continuous in $x \in \mathfrak{D}$, $x \neq y$, y fixed in $\mathfrak{D}$. Suppose that for fixed α in $0 < \alpha < 1$,

$$|H(y,x)| \leqslant \text{const } |y - x|^{-n+1-\alpha}$$

for $y \in \mathfrak{D}$, $x \in \mathfrak{D}$, $y \neq x$. Furthermore let

$$|H(y^1,x) - H(y^2,x)| \leqslant \left\{ \frac{\text{const}}{|y^1 - x|^n} + \frac{\text{const}}{|y^2 - x|^n} \right\} |y^1 - y^2|^{1-\alpha} \tag{4.1}$$

for all $y^1 \in \mathfrak{D}$, $y^2 \in \mathfrak{D}$, $x \in \mathfrak{D}$ which satisfy $|y^1 - x| \geqslant 2|y^1 - y^2| > 0$. Then

$$g(y) = \int_{x\in\mathfrak{D}} f(x)H(y,x)\,dx$$

is uniformly Hölder continuous in $\mathfrak{D}$ and this is true for any fixed exponent γ from $0 < \gamma < 1 - \alpha$; that is, for γ satisfying $0 < \gamma < 1 - \alpha$ there is a constant $C(\gamma)$ such that

$$|g(y^1) - g(y^2)| \leqslant C(\gamma)|y^1 - y^2|^\gamma \quad \text{for} \quad \text{all } y^1, y^2 \in \mathfrak{D}.$$

Remark. Inequality (4.1) is satisfied if, for fixed $x \in \mathfrak{D}$, $H(y,x)$ is once continuously differentiable with respect to the y-variables on $\mathfrak{D}$, $y \neq x$, and if

$$|H_{y_i}(y,x)| \leqslant \text{const } |y - x|^{-n-\alpha} \tag{4.2}$$

for $y \in \mathfrak{D}$, $x \in \mathfrak{D}$, $y \neq x$, $i = 1, \ldots, n$.

Proof. Let y^1 and y^2 be in $\mathfrak{D}$, and $|y^1 - y^2| = h$, $h \neq 0$. Then

$$|g(y^1) - g(y^2)| \leqslant \int_{x \in \mathfrak{D}} |f(x)| \, |H(y^1,x) - H(y^2,x)| \, dx$$

$$\leqslant \text{const} \int_{\substack{x \in \mathfrak{D} \\ |x - y^1| \leqslant 2h}} |H(y^1,x) - H(y^2,x)| \, dx$$

$$+ \text{const} \int_{\substack{x \in \mathfrak{D} \\ |x - y^1| \geqslant 2h}} |H(y^1,x) - H(y^2,x)| \, dx$$

$$\leqslant \text{const} \int_{|x - y^1| \leqslant 2h} \{|y^1 - x|^{-n+1-\alpha} + |y^2 - x|^{-n+1-\alpha}\} \, dx$$

$$+ \text{const } h^{1-\alpha} \int_{d \geqslant |x - y^1| \geqslant 2h} \{|y^1 - x|^{-n} + |y^2 - x|^{-n}\} \, dx,$$

the last term following from (4.1) and where d is the diameter of $\mathfrak{D}$. Then it follows that

$$|g(y^1) - g(y^2)| \leqslant \text{const} \int_{|x - y^1| \leqslant 2h} |y^1 - x|^{-n+1-\alpha} \, dx + \text{const} \int_{|x - y^2| \leqslant 3h} |y^2 - x|^{-n+1-\alpha} \, dx$$

$$+ \text{const } h^{1-\alpha} \int_{d \geqslant |x - y^1| \geqslant 2h} |y^1 - x|^{-n} \, dx + \text{const } h^{1-\alpha} \int_{2d \geqslant |x - y^2| \geqslant h} |y^2 - x|^{-n} \, dx,$$

$$|g(y^1) - g(y^2)| \leqslant \frac{\text{const}}{1 - \alpha} h^{1-\alpha} + \text{const } h^{1-\alpha} \log \frac{d}{h} \leqslant C(\gamma) h^{\gamma}, \qquad 0 < \gamma < 1 - \alpha.$$

The validity of the remark is seen as follows: Since x does not lie in the segment from y^1 to y^2, for $|y^1 - x| \geqslant 2|y^1 - y^2| > 0$, by the mean-value theorem and from (4.2) we have

$$|H(y^1,x) - H(y^2,x)| \leqslant |y^1 - y^2| \sum_{i=1}^{n} |H_{y_i}(\tilde{y},x)| \leqslant |y^1 - y^2| \frac{\text{const}}{|\tilde{y} - x|^{n+\alpha}}. \tag{4.3}$$

Here in any case $|y^1 - \tilde{y}| < |y^1 - y^2|$ so that we have the estimate

$$|\tilde{y} - x| \geqslant |y^1 - x| - |y^1 - \tilde{y}| \geqslant \tfrac{1}{2}|y^1 - x| + \{\tfrac{1}{2}|y^1 - x| - |y^1 - y^2|\} \geqslant \tfrac{1}{2}|y^1 - x|$$

as long as $|y^1 - x| \geqslant 2|y^1 - y^2|$. If this is used in (4.3), then

$$|H(y^1,x) - H(y^2,x)| \leqslant |y^1 - y^2| \frac{\text{const}}{|y^1 - x|^{n+\alpha}} \leqslant |y^1 - y^2|^{1-\alpha} \frac{\text{const}}{|y^1 - x|^n}$$

follows, and this gives (4.1).

THEOREM 2.   Let $\mathfrak{D}$ be a bounded convex domain in R_n. Let $f(x)$ be integrable and bounded on $\mathfrak{D}$ and, for fixed $x \in \mathfrak{D}$, $H(y,x)$ be twice continuously differentiable with respect to the y-variables on $\mathfrak{D}$, $y \neq x$. Furthermore let $H(y,x)$ and $H_{y_i}(y,x)$ be continuous in $(y,x) \in \mathfrak{D} \times \mathfrak{D}$, $y \neq x$ $(i = 1, \ldots, n)$ and assume that for $y \in \mathfrak{D}$, $x \in \mathfrak{D}$, $y \neq x$, and fixed α, $0 < \alpha < 1$, we have

$$|H(y,x)| \leqslant \frac{\text{const}}{|y - x|^{n-2+\alpha}}, \qquad |H_{y_i}(y,x)| \leqslant \frac{\text{const}}{|y - x|^{n-1+\alpha}},$$

$$|H_{y_i y_k}(y,x)| \leqslant \frac{\text{const}}{|y - x|^{n+\alpha}}, \qquad i, k = 1, \ldots, n. \tag{4.4}$$

Then

$$g(y) = \int_{x \in \mathfrak{D}} f(x) H(y,x) \, dx$$

is once uniformly Hölder-continuously differentiable in $\mathfrak{D}$ and

$$g_{y_i}(y) = \int_{x \varepsilon \mathfrak{D}} f(x) H_{y_i}(y,x) \, dx, \qquad i = 1, \ldots, n. \tag{4.5}$$

Proof. Let i be fixed. By Theorem 1 and the corresponding remark,

$$\int_{x \varepsilon \mathfrak{D}} f(x) H_{y_i}(y,x) \, dx$$

is uniformly Hölder continuous in $\mathfrak{D}$. Hence

$$\int_a^{y_i} \int_{x \varepsilon \mathfrak{D}} f(x) H_{y_i}(y_1, \ldots, y_{i-1}, t, y_{i+1}, \ldots, y_n, x) \, dx \, dt \tag{4.6}$$

is once uniformly Hölder-continuously differentiable with respect to y_i as long as the points

$$y = (y_1, \ldots, y_{i-1}, y_i, y_{i+1}, \ldots, y_n)$$

and $\bar{a} = (y_1, \ldots, y_{i-1}, a, y_{i+1}, \ldots, y_n)$ lie in $\mathfrak{D}$. By interchanging the order of integration, we have from (4.6)

$$\int_{x \varepsilon \mathfrak{D}} f(x) \{H(y,x) - H(\bar{a},x)\} \, dx \equiv g(y) - g(\bar{a});$$

thus $g(y)$ is once uniformly Hölder-continuously differentiable with respect to y_i in $\mathfrak{D}$, and (4.5) holds.

THEOREM 3. Let $\mathfrak{D}$ be a bounded convex domain in R_n. Let $f(x)$ be uniformly Hölder continuous in $\mathfrak{D}$ and, for fixed $x \varepsilon \mathfrak{D}$, let $H(y,x)$ be twice continuously differentiable with respect to the y-variables on $\mathfrak{D}$, $y \neq x$. Furthermore let $H(y,x)$ and $H_{y_i}(y,x)$ be continuous in $(y,x) \varepsilon \mathfrak{D} \times \mathfrak{D}$, $y \neq x$, $(i = 1, \ldots, n)$, and assume that for $x \varepsilon \mathfrak{D}$, $y \varepsilon \mathfrak{D}$, $y \neq x$,

$$|H(y,x)| \leqslant \frac{\text{const}}{|y - x|^{n-1}}, \qquad |H_{y_i}(y,x)| \leqslant \frac{\text{const}}{|y - x|^n}, \qquad |H_{y_i y_k}(y,x)| \leqslant \frac{\text{const}}{|y - x|^{n+1}}$$
$$i, k = 1, \ldots, n. \tag{4.7}$$

Finally let

$$\int_{x \varepsilon \mathfrak{D}} H(y,x) \, dx$$

be once Hölder-continuously differentiable in $\mathfrak{D}$. Then $g(y) = \int_{x \varepsilon \mathfrak{D}} f(x) H(y,x) \, dx$

is once Hölder-continuously differentiable in $\mathfrak{D}$ and

$$g_{y_i}(y) = \int_{x \varepsilon \mathfrak{D}} [f(x) - f(y)] H_{y_i}(y,x) \, dx + f(y) \frac{\partial}{\partial y_i} \int_{x \varepsilon \mathfrak{D}} H(y,x) \, dx. \tag{4.8}$$

Proof. We have

$$g(y) = \int_{x \varepsilon \mathfrak{D}} [f(x) - f(y)] H(y,x) \, dx + f(y) \int_{x \varepsilon \mathfrak{D}} H(y,x) \, dx. \tag{4.9}$$

The function $f(x)$ may be approximated uniformly in $\mathfrak{D}$ by functions $f_m(x)$ which are twice continuously differentiable in $\mathfrak{D}$ and which are Hölder continuous in $\mathfrak{D}$ with the same coefficient C and the same exponent $1 - \alpha$ as $f(x)$. Thus the functions

$$g_m(y) = \int_{x\epsilon\mathfrak{D}} [f_m(x) - f_m(y)]H(y,x)\,dx + f_m(y)\int_{x\epsilon\mathfrak{D}} H(y,x)\,dx \qquad (4.10)$$

are defined. They are once (Hölder-) continuously differentiable in $\mathfrak{D}$ because, in particular,

$$\int_{x\epsilon\mathfrak{D}} [f_m(x) - f_m(y)]H(y,x)\,dx$$

is once Hölder-continuously differentiable in $\mathfrak{D}$ by Theorem 2, and we have

$$\frac{\partial g_m(y)}{\partial y_i} = \int_{x\epsilon\mathfrak{D}} [(f_m(x) - f_m(y))H(y,x)]_{y_i}\,dx + \frac{\partial}{\partial y_i} f_m(y)\int_{x\epsilon\mathfrak{D}} H(y,x)\,dx,$$

$$\frac{\partial g_m(y)}{\partial y_i} = \int_{x\epsilon\mathfrak{D}} [f_m(x) - f_m(y)]H_{y_i}(y,x)\,dx + f_m(y)\frac{\partial}{\partial y_i}\int_{x\epsilon\mathfrak{D}} H(y,x)\,dx.$$

A common estimate confirms that $\partial g_m(y)/\partial y_i$ converges uniformly in $\mathfrak{D}$ toward the right-hand-side of (4.8). Furthermore $\lim\limits_{m\to\infty} g_m(y) = g(y)$ and (4.8) results.

In spite of the uniform convergence and the Hölder continuity of $\partial g_m(y)/\partial y_i$ they need not be equicontinuous relative to m in the sense of Hölder. Hence the Hölder continuity of $g_{y_i}(y)$ has not been proved. But applying Theorem 1 to (4.8) we have

$$\begin{aligned}
|[f(x) &- f(y^1)]H_{y_i}(y^1,x) - [f(x) - f(y^2)]H_{y_i}(y^2,x)| \\
&= |[f(x) - f(y^1)][H_{y_i}(y^1,x) - H_{y_i}(y^2,x)] + [f(y^2) - f(y^1)]H_{y_i}(y^2,x)| \\
&\leqslant C|x - y^1|^{1-\alpha}|y^1 - y^2|\,\frac{\text{const}}{|y^1 - x|^{n+1}} + C|y^1 - y^2|^{1-\alpha}\,\frac{\text{const}}{|y^2 - x|^n}
\end{aligned}$$

if $|y^1 - x| \geqslant 2|y^1 - y^2| > 0$, from which the desired estimate follows by (4.1) which proves that $g_{y_i}(y)$ are Hölder continuous.

THEOREM 4.   (a) Let the numbers ρ and σ be such that $\rho < n, \sigma < n, \rho + \sigma > n$. Then there exists a constant c_1 depending on ρ and σ such that

$$\int_{z\epsilon R_n} \frac{dz}{|x - z|^\rho |z - y|^\sigma} \leqslant \frac{c_1}{|x - y|^{\rho+\sigma-n}} \qquad (4.11)$$

for all $x \in R_n,\, y \in R_n,\, x \neq y$.

    (b) Let the numbers ρ and σ be such that $0 < \rho < n,\, 0 < \sigma < n,\, \rho + \sigma < n$ and $\mathfrak{D}$ be a bounded domain in R_n. Then there exists a constant c_2 depending on ρ, σ, and $\mathfrak{D}$ such that

$$\int_{z\epsilon\mathfrak{D}} \frac{dz}{|x - z|^\rho |z - y|^\sigma} \leqslant c_2 \qquad (4.12)$$

for all $x \in R_n,\, y \in R_n$.

Proof. (a) It is clear that the integral in (4.11) exists. Its value does not change if we put $x = (0, 0, \ldots, 0)$ and $y = |x - y|(1, 0, \ldots, 0) = |x - y|\,e$. By the substitution $z = |x - y|\zeta$ we have

$$\int_{z\epsilon R_n} \frac{dz}{|z|^\rho |z - |x - y|e|^\sigma} = \int_{\zeta \epsilon R_n} \frac{|x - y|^n \, d\zeta}{|x - y|^{\rho+\sigma} |\zeta|^\rho |\zeta - e|^\sigma}$$

from which (4.11) follows.

(*b*) The integral in (4.12) is the same as

$$\int_{\substack{z\epsilon\mathfrak{D} \\ |x-z| \geqslant |z-y|}} \frac{dz}{|x - z|^\rho |z - y|^\sigma} + \int_{\substack{z\epsilon\mathfrak{D} \\ |x-z| \leqslant |z-y|}} \frac{dz}{|x - z|^\rho |z - y|^\sigma}.$$

Hence

$$\int_{z\epsilon\mathfrak{D}} \frac{dz}{|x - z|^\rho |z - y|^\sigma} \leqslant \int_{z\epsilon\mathfrak{D}} \frac{dz}{|z - y|^{\sigma+\rho}} + \int_{z\epsilon\mathfrak{D}} \frac{dz}{|x - z|^{\sigma+\rho}},$$

and the last two integrals define c_2.   For this use the theorem from Section 3.1.

4.2   Weyl's Lemma

THEOREM.   WEYL'S LEMMA.   In a domain $\mathfrak{D}$ of R_n let the differential operator

$$Kv(x) \equiv \sum_{i,k=1}^{n} \alpha_{ik}(x)v_{x_i x_k} + \sum_{i=1}^{n} \alpha_i(x)v_{x_i} + \gamma(x)v \tag{4.13}$$

be elliptic, that is, let $\sum_{i,k=1}^{n} \alpha_{ik}(x)\xi_i\xi_k > 0$ for all $\xi = (\xi_1, \ldots, \xi_n) \neq 0$ and all $x \in \mathfrak{D}$.

Assume that the coefficients $\alpha_{ik}(x) \in C^2$, $\alpha_i(x) \in C^1$, $\gamma(x) \in C^1$ in $\mathfrak{D}$ where the second derivatives of the $\alpha_{ik}(x)$ and the first derivatives of the $\alpha_i(x)$ shall be Hölder-continuous in $\mathfrak{D}$.   Furthermore let $\eta(x)$ be a function that is Hölder-continuous in $\mathfrak{D}$, that is, to any compact subdomain $\mathfrak{E}$ of $\mathfrak{D}$ there are constants H and α, $0 < \alpha < 1$, such that

$$|\eta(x^1) - \eta(x^2)| \leqslant H|x^1 - x^2|^\alpha \quad \text{for} \quad \text{all } x^1, x^2 \in \overline{\mathfrak{E}}.$$

If then $w(x)$ is a function, locally integrable in $\mathfrak{D}$ $\left(\text{that is, } \int_{\overline{\mathfrak{E}}} |w(x)| \, dx < \infty \text{ for}\right.$ every compact $\overline{\mathfrak{E}}$ in $\mathfrak{D}$) and if

$$\int_\mathfrak{D} w(x)K\Phi(x) \, dx = \int_\mathfrak{D} \eta(x)\Phi(x) \, dx \quad \text{for} \quad \text{all } \Phi(x) \in C_0^\infty(\mathfrak{D}), \tag{4.14}$$

then $w(x)$ coincides almost everywhere in $\mathfrak{D}$ with a function which is twice Hölder-continuously differentiable in $\mathfrak{D}$.[28]

---

[28] Here we stated a little more than was needed in Section 3.6.   This theorem and its proof were made available to the author by E. Wienholtz.   H. Weyl[20] only considered the case of the potential equation. After that a series of generalizations was treated.   We mention here F. E. Browder[21], G. Fichera[22], K. O. Friedrichs[23], L. Gårding[24], F. John[25], P. Lax[26], L. Nirenberg[27].

LEMMA.   Let the differential operator

$$Lu(x) \equiv \sum_{i,k=1}^{n} a_{ik}(x)u_{x_ix_k} + \sum_{i=1}^{n} a_i(x)u_{x_i} + c(x)u \tag{4.15}$$

be uniformly elliptic in S: $|x - x^0| < d$. Assume that the coefficients $a_{ik}(x) \in C^2$, $a_i(x) \in C^1$, $c(x) \in C^0$ in $\bar{S}$. Further let the functions $g_1(x)$, $\ldots$, $g_n(x)$ be Hölder-continuous in $\bar{S}$ with fixed coefficient and fixed exponent. If $v(x)$ is a function, integrable over S, and if

$$\int_S v(x)L\Phi(x)\,dx = \int_S \sum_{i=1}^{n} g_i(x)\Phi_{x_i}(x)\,dx \quad \text{for} \quad \text{all } \Phi(x) \in C_0^{\infty}(S), \tag{4.16}$$

then $v(x)$ coincides almost everywhere in S with a function which is once Hölder-continuously differentiable in S.

Proof of the Theorem.   It suffices to show that every point in $\mathfrak{D}$ possesses a neighborhood $\mathfrak{U}$ such that $w(x)$ coincides almost everywhere in $\mathfrak{U}$ with a function which is twice Hölder-continuously differentiable.   For if two such neighborhoods of two points have a nonempty intersection and if $w(x) = w_1(x)$ almost everywhere in one of them, $w(x) = w_2(x)$ almost everywhere in the other and $w_i(x) \in C^2$, then $w_1(x)$ and $w_2(x)$ coincide on that intersection since $w_1(x) = w(x) = w_2(x)$ almost everywhere and $w_1(x)$ and $w_2(x)$ are continuous.

After this remark, the theorem is a consequence of the lemma.   If we fix an arbitrary sphere $\bar{S}$ contained in $\mathfrak{D}$ and set

$$a_{ik}(x) = \alpha_{ik}(x), \qquad a_i(x) = \alpha_i(x), \qquad c(x) = \gamma(x), \qquad v(x) = w(x),$$

$$g_i(x) = -\delta_{i1}\int_{x^0_1}^{x_1} \eta(t, x_2, \ldots, x_n)\,dt \quad \text{with} \quad \delta_{ij} = \begin{cases} 1, & i = j, \\ 0, & i \neq j, \end{cases}$$

in $\bar{S}$, then the hypotheses of the lemma are satisfied.   In particular,

$$\int_S \eta(x)\Phi(x)\,dx = -\int_S g_{1x_1}(x)\Phi(x)\,dx = \int_S g_1(x)\Phi_{x_1}(x)\,dx, \qquad \Phi(x) \in C_0^{\infty}(S).$$

Hence by this lemma $w(x)$ coincides almost everywhere in S with a function $\bar{v}(x)$ which is once Hölder-continuously differentiable in S.

To obtain the higher differentiability properties of $\bar{v}(x)$, fix j and put $\tilde{D}\bar{v} = \bar{v}_{x_j}$. Then integration by parts gives

$$\int_S \tilde{D}\bar{v}K\Phi\,dx = -\int_S \left\{ \bar{v}\sum_{i,k=1}^{n} \tilde{D}(\alpha_{ik}\Phi_{x_ix_k}) + \bar{v}\sum_{i=1}^{n} \tilde{D}(\alpha_i\Phi_{x_i}) + \bar{v}\tilde{D}(\gamma\Phi) \right\}dx$$

for all $\Phi \in C_0^{\infty}(S)$.   Hence

$$\int_S \tilde{D}\bar{v}K\Phi\,dx = -\int_S \bar{v}K\,\tilde{D}\Phi\,dx - \int_S \left\{ \bar{v}\sum_{i,k=1}^{n} (\tilde{D}\alpha_{ik})\Phi_{x_ix_k} + \bar{v}\sum_{i=1}^{n} (\tilde{D}\alpha_i)\Phi_{x_i} + \bar{v}(\tilde{D}\gamma)\Phi \right\}dx.$$

Since $\bar{v}(x) = w(x)$ almost everywhere in S we may apply (4.14) to the first integral on the right-hand side; if in the second integral we rewrite the first part by integration by parts, we obtain

$$\int_S \tilde{D}\tilde{v}K\Phi \, dx = -\int_S \eta\tilde{D}\Phi \, dx$$
$$+ \int_S \Big\{ \sum_{i=1}^n \Big(\sum_{k=1}^n (\tilde{v}\tilde{D}\alpha_{ik})_{x_k} \Big)\Phi_{x_i} - \sum_{i=1}^n (\tilde{v}\tilde{D}\alpha_i)\Phi_{x_i} - (\tilde{v}\tilde{D}\gamma)\Phi \Big\} \, dx. \quad (4.17)$$

If we put

$$g_i(x) = -\delta_{ij}\eta(x) + \sum_{k=1}^n (\tilde{v}\tilde{D}\alpha_{ik})_{x_k} - \tilde{v}\tilde{D}\alpha_i$$
$$+ \delta_{ij} \int_{x^0_j}^{x_j} \tilde{v}\tilde{D}\gamma(x_1, \ldots, x_{j-1}, t, x_{j+1}, \ldots, x_n) \, dt \qquad (i = 1, \ldots, n),$$

then in a sphere $\bar{S}_1$, contained in S, the $g_i(x)$ are Hölder continuous with fixed coefficient and fixed exponent, and by (4.17) we have

$$\int_{S_1} \tilde{D}\tilde{v}K\Phi \, dx = \int_{S_1} \sum_{i=1}^n g_i\Phi_{x_i} \, dx \quad \text{for} \quad \text{all } \Phi \in C_0^\infty(S_1). \quad (4.18)$$

The expression $\tilde{D}\tilde{v}$ is integrable over S_1; hence by the lemma $\tilde{D}\tilde{v}$ coincides almost everywhere in S_1 with a function which is once Hölder-continuously differentiable in S_1; since $\tilde{D}\tilde{v} \in C^0$ it is identical with this one, consequently $\tilde{D}\tilde{v}$ is once, and $\tilde{v}$ therefore twice, Hölder-continuously differentiable in S_1, which was to be proved.

Proof of the Lemma. Let $S(y,x)$ be the singularity function defined in Section 3.8 for the operator (4.15) in S. If $\varphi(y) \in C_0^\infty(S)$, then

$$g(x) = \int_{y \in S} \varphi(y)S(y,x) \, dy \qquad (4.19)$$

belongs to C^2 in S.

We prove this only for the case $n > 2$. By Theorem 2 in Section 4.1 and because of the properties of $S(y,x)$ which were derived in Section 3.8, we have

$$g_{x_i}(x) = \int_{y \in S} \varphi(y)S_{x_i}(y,x) \, dy. \qquad (4.20)$$

If $|A_{ik}|$ again denotes the determinant of the matrix (A_{ik}), then

$$S_{x_i}(y,x) = \frac{|A_{ik}(y)|^{1/2}}{(n-2)\omega_n} \frac{\partial}{\partial x_j} \Big[\sum_{i,k=1}^n A_{ik}(y)(y_i - x_i)(y_k - x_k) \Big]^{-(n-2)/2}$$
$$= -\frac{|A_{ik}(y)|^{1/2}}{(n-2)\omega_n} \frac{\partial}{\partial y_j} \Big[\sum_{i,k=1}^n A_{ik}(y)(y_i - x_i)(y_k - x_k) \Big]^{-(n-2)/2}$$
$$- \frac{|A_{ik}(y)|^{1/2}}{2\omega_n} \Big[\sum_{i,k=1}^n A_{ik}(y)(y_i - x_i)(y_k - x_k) \Big]^{-n/2} \sum_{i,k=1}^n \frac{\partial A_{ik}(y)}{\partial y_j}(y_i - x_i)(y_k - x_k).$$

If we substitute this in (4.20), we obtain a sum of two integrals, one of which, by Theorem 2 in Section 4.1, is once continuously differentiable in the x-variables. By integration by parts the other can be transformed into an integral which, by Theorem 2, is continuously differentiable also; thus $g(x)$ is twice continuously differentiable in S.

The operator Lu from (4.15) and its formally adjoint operator

$$L^*u = \sum_{i,k=1}^n [a_{ik}(x)u]_{x_i x_k} - \sum_{i=1}^n [a_i(x)u]_{x_i} + c(x)u \qquad (4.21)$$

coincide in their principal parts, that is, in the coefficients of the $u_{x_i x_k}$. Hence $S(y,x)$ is also a singularity function for L^*u in $\bar{S}$, and by (3.90) and $L^{**} = L$ we have

$$\int_S \Phi(x)LS(y,x)\, dx = \Phi(y) + \int_S S(y,x)L^*\Phi(x)\, dx \tag{4.22}$$

for all $\Phi(x) \in C_0^\infty(S)$ and $y \in S$. From this we obtain the relation

$$\int_S \Phi(y)LS(y,x)\, dy = \Phi(x) + L\int_S \Phi(y)S(y,x)\, dy \tag{4.23}$$

for all $\Phi(y) \in C_0^\infty(S)$ and $x \in S$. The last term in (4.23) makes sense because of (4.19): By the theorem on the interchangeability of the order of integration, we have

$$\int_{x\epsilon S} \Psi(x) \int_{y\epsilon S} \Phi(y)LS(y,x)\, dy\, dx = \int_{y\epsilon S} \Phi(y) \int_{x\epsilon S} \Psi(x)LS(y,x)\, dx\, dy$$

for all functions $\Psi(x) \in C_0^\infty(S)$, and if (4.22) is applied to the last inner integral we obtain

$$\int_{x\epsilon S} \Psi(x) \int_{y\epsilon S} \Phi(y)LS(y,x)\, dy\, dx = \int_{y\epsilon S} \Phi(y) \left[\Psi(y) + \int_{x\epsilon S} S(y,x)L^*\Psi(x)\, dx \right] dy$$

$$= \int_{y\epsilon S} \Phi(y)\Psi(y)\, dy + \int_{x\epsilon S} [L^*\Psi(x)] \int_{y\epsilon S} \Phi(y)S(y,x)\, dy\, dx,$$

the latter again by interchanging the order of integration. Because of (4.19) and by integration by parts from this we obtain

$$\int_{x\epsilon S} \Psi(x) \int_{y\epsilon S} \Phi(y)LS(y,x)\, dy\, dx = \int_{x\epsilon S} \Psi(x)\Phi(x)\, dx + \int_{x\epsilon S} \Psi(x)L \int_{y\epsilon S} \Phi(y)S(y,x)\, dy\, dx,$$

and because of the arbitrariness of $\Psi(x) \in C_0^\infty(S)$ this gives relation (4.23) almost everywhere in S. But since both sides of (4.23) are continuous, (4.23) holds for all $x \in S$. To see the continuity of the left-hand side of (4.23) apply Theorem 1 from Section 4.1.

After these preparations we now come to the actual proof of the lemma. As in the previous proof it will suffice to show the desired property of $v(x)$ in the neighborhood of an arbitrary point of S. We fix such a point, let S_1 be a ball which has it for its center, $\bar{S}_1 \subset S$; furthermore we let S_2 be a ball concentric to S_1 and $\bar{S}_2 \subset S_1$. Assume the function $\rho(x)$ belongs to $C_0^\infty(S_1)$ and let $\rho(x) \equiv 1$ in S_2. By *Fubini's* theorem,[29]

$$w(y) = \int_{x\epsilon S_1} v(x)L\{\rho(x)S(y,x)\}\, dx - \int_{x\epsilon S_1} \sum_{i=1}^n g_i(x)\{\rho(x)S(y,x)\}_{x_i}\, dx \tag{4.24}$$

defines a function $w(y)$ almost everywhere in S_2 which is integrable over S_2.

This function $w(y)$ coincides with $v(y)$ almost everywhere in S_2 since for arbitrary $\Phi(y) \in C_0^\infty(S_2)$ we have

$$\int_{y\epsilon S_2} \Phi(y)w(y)\, dy = \int_{x\epsilon S_1} v(x) \int_{y\epsilon S_2} \Phi(y)L\{\rho(x)S(y,x)\}\, dy\, dx$$

$$- \int_{x\epsilon S_1} \sum_{i=1}^n g_i(x) \int_{y\epsilon S_2} \Phi(y)\{\rho(x)S(y,x)\}_{x_i}\, dy\, dx, \tag{4.25}$$

as follows by multiplication of (4.24) by $\Phi(y)$, integration over S_2, and interchanging the order of integration.

---

[29] After an idea of E. Heinz. In the known proofs of Weyl's lemma at this place we need not only $S(y,x)$ but also a fundamental solution for the differential operator L. Similar considerations were also made by G. Fichera[22].

The first term on the right-hand side of (4.25) may be decomposed into the sum

$$\int_{x\varepsilon\bar{S}_2} v(x) \int_{y\varepsilon S_2} \Phi(y)LS(y,x)\, dy\, dx + \int_{x\varepsilon S_1-\bar{S}_2} v(x)L\left\{\rho(x)\int_{y\varepsilon S_2}\Phi(y)S(y,x)\,dy\right\}dx. \quad (4.26)$$

Here we used the facts that $\rho(x) = 1$ in the first summand and that in the second summand every x possesses a neighborhood $\mathfrak{u}$ such that $S(y,x)$ is twice continuously differentiable in $\bar{S}_2 \times \bar{\mathfrak{u}}$; therefore differentiation and integration could be interchanged. We substitute relation (4.23) in the first summand in (4.26). Then (4.26) becomes

$$\int_{x\varepsilon\bar{S}_2} v(x)\Phi(x)\, dx + \int_{x\varepsilon S_1} v(x)L\left\{\rho(x)\int_{y\varepsilon S_2}\Phi(y)S(y,x)\,dv\right\}dx. \quad (4.27)$$

Here we noted that $\rho(x) = 1$ in S_2.

By Theorem 2 from Section 4.1, the second term on the right-hand side of (4.25) is the same as

$$-\int_{x\varepsilon S_1}\sum_{i=1}^{n} g_i(x)\frac{\partial}{\partial x_i}\left\{\rho(x)\int_{y\varepsilon S_2}\Phi(y)S(y,x)\,dy\right\}dx. \quad (4.28)$$

Now (4.27) and (4.28) together form the right-hand side of (4.25). Hence, with the abbreviation

$$\psi(x) = \rho(x)\int_{y\varepsilon S_2}\Phi(y)S(y,x)\,dy, \quad (4.29)$$

from (4.25) we have

$$\int_{y\varepsilon S_2}\Phi(y)[w(y) - v(y)]\, dy = \int_{x\varepsilon S_1} v(x)L\psi(x)\, dx - \int_{x\varepsilon S_1}\sum_{i=1}^{n} g_i(x)\psi_{x_.}(x)\, dx. \quad (4.30)$$

At the same time we found that $\psi(x)$ is twice continuously differentiable in S_1, which could also be seen directly from a comparison of (4.29) with (4.19), and $\psi(x)$ vanishes in a neighborhood of $\dot{S}_1$ because of $\rho(x) \varepsilon C_0^\infty(S_1)$. Hence $\psi(x)$, together with its first and second derivatives, may be approximated uniformly in S_1 by functions from $C_0^\infty(S_1)$, and by their derivatives, respectively. Thus, by (4.16), the right-hand side of (4.30) has the value zero. But now because of the arbitrariness of $\Phi(y) \varepsilon C_0^\infty(S_2)$, it follows from (4.30) that $w(y) = v(y)$ almost everywhere in S_2.

Thus, by (4.24), we have the integral equation

$$v(y) = \int_{x\varepsilon S_1} v(x)L\{\rho(x)S(y,x)\}\, dx - \int_{x\varepsilon S_1}\sum_{i=1}^{n} g_i(x)\{\rho(x)S(y,x)\}_{x_i}\, dx \quad (4.31)$$

almost everywhere in S_2, and, using (3.85) and (3.88), a simple estimate gives

$$|v(y)| \leqslant \int_{x\varepsilon S_1} |v(x)|\frac{c_1}{|y-x|^{n-1}}\, dx + c_2 \quad (4.32)$$

almost everywhere in S_2 with two constants c_1 and c_2. The last integral exists almost everywhere in S_2 by Fubini's theorem.

From (4.32) we want to conclude by iteration that $v(y)$ is bounded almost everywhere in a certain ball concentric to S_1. In order not to get into the "logarithmic case" with this iteration we weaken (4.32) to

$$|v(y)| \leqslant \int_{x\varepsilon S_1} |v(x)|\frac{c_1}{|y-x|^{n-\lambda}}\, dx + c_2 \quad (4.33)$$

almost everywhere in S_2 where $\lambda = \frac{1}{2}\sqrt{2}$. Ultimately the constants c_1 and c_2 depend only on the choice of the balls S_1 and S_2. Therefore, if S_3 is a ball concentric to S_2, $\bar{S}_3 \subset S_2$, then with new constants c_3 and c_4 we have

$$|v(y)| \leqslant \int_{x \varepsilon S_2} |v(x)| \frac{c_3}{|y-x|^{n-\lambda}}\, dx + c_4 \tag{4.34}$$

almost everywhere in S_3. Now we use (4.33) in (4.34) and continue the procedure as in Section 3.7. Thus we obtain a ball S_α concentric with S_1 and $v(y)$ is bounded almost everywhere in S_α; hence there is a function $h(x)$ integrable and bounded in S_α such that $v(x) = h(x)$ almost everywhere in S_α.

If $S_{\alpha'}$ is a ball concentric to S_α, $\bar{S}_{\alpha'} \subset S_\alpha$, then, in analogy to (4.31), but with a new function $\rho(x)$, we have

$$v(y) = \int_{x \varepsilon S_\alpha} h(x) L\{\rho(x)S(y,x)\}\, dx - \int_{x \varepsilon S_\alpha} \sum_{i=1}^{n} g_i(x)\{\rho(x)S(y,x)\}_{x_i}\, dx \tag{4.35}$$

almost everywhere in $S_{\alpha'}$. By Theorem 1 in Section 4.1 the right-hand side of (4.35) represents a function uniformly Hölder continuous in S_α; hence $v(y)$ coincides almost everywhere in $S_{\alpha'}$ with a function $k(y)$ uniformly Hölder continuous in $S_{\alpha'}$.

Finally let $S_{\alpha''}$ be a ball concentric to $S_{\alpha'}$ and $\bar{S}_{\alpha''} \subset S_{\alpha'}$. Then for suitable $\rho(x)$ we have

$$v(y) = \int_{x \varepsilon S_{\alpha'}} k(x) L\{\rho(x)S(y,x)\}\, dx - \int_{x \varepsilon S_{\alpha'}} \sum_{i=1}^{n} g_i(x)\{\rho(x)S(y,x)\}_{x_i}\, dx \tag{4.36}$$

almost everywhere in $S_{\alpha''}$. By Theorem 3 from Section 4.1, the right-hand side of (4.36) is once Hölder-continuously differentiable in $S_{\alpha'}$; hence from (4.36) it follows that $v(y)$ coincides almost everywhere in $S_{\alpha''}$ with a function once Hölder-continuously differentiable in $S_{\alpha''}$, which was to be shown.

It is not difficult to see that the right-hand side of (4.36) satisfies the hypotheses of Theorem 3 from Section 4.1; in particular, the integrals

$$\int_{x \varepsilon S_{\alpha'}} L\{\rho(x)S(y,x)\}\, dx \quad \text{and} \quad \int_{x \varepsilon S_{\alpha'}} \{\rho(x)S(y,x)\}_{x_i}\, dx, \qquad i = 1, \ldots, n,$$

are once Hölder-continuously differentiable in $S_{\alpha'}$. For these integrals may be written as sums of integrals which, except for the one summand

$$\int_{x \varepsilon S_{\alpha'}} \rho(x) LS(y,x)\, dx, \tag{4.37}$$

contain either $S(y,x)$ or $S_{x_i}(y,x)$, $i = 1, \ldots, n$ as a factor in the integrand. But since these integrands also contain $\rho(x)$ or derivatives of it as a factor they vanish in a neighborhood of $\dot{S}_{\alpha'}$; consequently by integration by parts these integrals may be put into the form

$$\int_{x \varepsilon S_{\alpha'}} \sigma(x) S(y,x)\, dx \quad \text{where} \quad \sigma(x) \varepsilon C^0 \text{ in } \bar{S}_{\alpha'},$$

which is once Hölder-continuously differentiable in $S_{\alpha'}$ by Theorem 2 from Section 4.1. The remaining part (4.37) may be written as

$$\rho(y) + \int_{x \varepsilon S_{\alpha'}} [L^*\rho(x)] S(y,x)\, dx$$

by means of (4.22), and this, too, is once Hölder-continuously differentiable in $S_{\alpha'}$.

REFERENCES

1. E. KAMKE, *Differentialgleichungen reeller Funktionen* (Leipzig: Akademische Verlagsgesellschaft Geest & Portig), 1945, p. 335.

2. H. LEWY, *Math. Ann.* **98,** 179 (1926).

3. K. FRIEDRICHS and H. LEWY, *Math. Ann.* **99,** 200 (1927).

4. R. COURANT and D. HILBERT, *Methoden der mathematischen Physik*, Part II (Berlin: Springer Verlag), 1937, p. 326 [or R. COURANT and D. HILBERT, *Methods of Mathematical Physics*, Vol. II (New York: Interscience Publishers, Inc.), 1962—Translator's note].

5. E. KAMKE, *Differentialgleichungen reeller Funktionen* (Leipzig: Akademische Verlagsgesellschaft Geest & Portig), 1945, p. 406.

6. J. CONLAN, *Arch. Ratl. Mech. Anal.* **3,** 355–380 (1959).

7. P. LAX, "Studies in Eigenvalue Problems," *Technical Report 14*, University of Kansas, Lawrence, Kansas, 1955.

8. C. LOEWNER, *J. Ratl. Mech.* **2,** 537–561 (1953).

9. W. HAACK and G. HELLWIG, *Math. Z.* **53,** 340–356 (1950).

10. E. HOLMGREN, *Ark. Mat. Astro. Fysik* **1,** 317–326 (1904); **6** (1911).

11. F. RELLICH, *Math. Ann.* **103,** 249–278 (1930); **112,** 490–492 (1936).

12. G. HELLWIG, *Math. Z.* **68,** 325–337 (1958).

13. J. HADAMARD, *Lectures on Cauchy's Problem in Linear Partial Differential Equations* (New York: Dover Publications), 1952.

14. G. HELLWIG, *Math. Z.* **66,** 371–388 (1957).

15. G. HELLWIG, *Math. Nachr.* **18,** 281–291 (1958).

16. W. MÄCHLER, *Comment. Math. Helv.* **5,** 256–304 (1933).

17. G. DOETSCH, *Tabellen zur Laplace-Transformation* (Berlin: Springer Verlag), 1947.

18. G. DOETSCH, *Handbuch der Laplace-Transformation*, three volumes (Basel and Stuttgart: Birkhäuser Verlag), 1950 and 1956.

19. R. V. CHURCHILL, *Modern Operational Mathematics in Engineering* (New York: McGraw-Hill Book Company, Inc.), 1944.

20. H. WEYL, *Duke Math. J.* **7,** 411–444 (1940).

21. F. E. BROWDER, *Comm. Pure Appl. Math.* **9,** 351–361 (1956).

22. G. FICHERA, *Convegno internationale sulle equarioni lineari alle derivate pariali*, Triste 1954 (Roma: Editione Cremonese), 1955, p. 174–227.

23. K. O. FRIEDRICHS, *Comm. Pure Appl. Math.* **6,** 299–326 (1953).

24. L. GÅRDING, *Kungl. Fysiograf. Sällsk. Lund Förhandl.* **20,** 250–253 (1950).

25. F. JOHN, *Comm. Pure Appl. Math.* **6,** 327–335 (1953).

26. P. LAX, *Comm. Pure Appl. Math.* **8,** 615–633 (1955).

27. L. NIRENBERG, *Comm. Pure Appl. Math.* **8,** 648–674 (1955).

Part V. Simple Tools from Functional Analysis Applied to Questions of Existence

We use only tools which can be found in any textbook on functional analysis. In notation and presentation we mainly follow F. Riesz and B. Sz.-Nagy[1]. Note, however, that we are interested only in real Banach and Hilbert spaces.

1

AUXILIARY TOOLS

1.1   Banach Space

A set $\mathfrak{B}$ of elements $u, v, w, \ldots$ is called a *(real) Banach space* if it satisfies the properties (a), (b), and (c).

(a) $\mathfrak{B}$ is a linear vector space over the field of real numbers; that is, the operations of addition and of multiplication by real numbers $a, b, c, \ldots$, are defined for its elements, and these operations satisfy the usual rules of vector algebra.

(b) To every element $u \in \mathfrak{B}$ there is assigned a real nonnegative number $\|u\|$, its norm, such that $\|u + v\| \leqslant \|u\| + \|v\|$, $\|au\| = |a|\,\|u\|$, and such that $\|u\| = 0$ if and only if u is the zero element; $\|u - v\|$ is called the distance between u and v.

(c) To every sequence of elements $u_1, u_2, \ldots \in \mathfrak{B}$ for which $\lim\limits_{n,m \to \infty} \|u_n - u_m\| = 0$ there is only one element $u \in \mathfrak{B}$ such that $\lim\limits_{n \to \infty} \|u_n - u\| = 0$. Instead of u we then also write $\lim\limits_{n \to \infty} u_n$.

In the following, by $\mathfrak{B}$ we always understand a Banach space. If by some prescription to every element $u \in \mathfrak{B}$ there is assigned an element $v \in \mathfrak{B}$, in a unique way, then we write $v = Au$ and call A an *operator* in $\mathfrak{B}$.

LEMMA 1.   Let A be an operator in $\mathfrak{B}$ and $0 \leqslant a < 1$ such that

$$\|Au - Av\| \leqslant a\|u - v\| \quad \text{for} \quad \text{all } u, v \in \mathfrak{B}; \tag{1.1}$$

then there exists a unique $z \in \mathfrak{B}$ such that $z = Az$.

Proof. Let the element $u_0 \in \mathfrak{B}$ be chosen arbitrarily. We form the sequence

$$u_1 = Au_0, \; u_2 = Au_1, \; u_3 = Au_2, \ldots, u_n = Au_{n-1}, \ldots. \tag{1.2}$$

Then by complete induction we have $\|u_n - u_{n+1}\| \leqslant a^n\|u_0 - u_1\|$. For $n = 1$ this follows from (1.1). Suppose it is proved for $n = k$; then

$$\|u_{k+1} - u_{k+2}\| = \|Au_k - Au_{k+1}\| \leqslant a\|u_k - u_{k+1}\| \leqslant aa^k\|u_0 - u_1\|, \tag{1.3}$$

which completes the induction. Furthermore, setting $m = n + p$, p a positive integer, we find

$$
\begin{aligned}
\|u_n - u_m\| &= \| u_n - u_{n+1} + u_{n+1} - u_{n+2} + \cdots + u_{n+p-1} - u_{n+p}\| \\
&\leqslant \|u_n - u_{n+1}\| + \|u_{n+1} - u_{n+2}\| + \cdots + \|u_{n+p-1} - u_{n+p}\| \quad (1.4) \\
&\leqslant (a^n + a^{n+1} + \cdots + a^{n+p-1})\|u_0 - u_1\| \leqslant \frac{a^n}{1-a}\|u_0 - u_1\|.
\end{aligned}
$$

The right-hand side tends to zero as $n \to \infty$. Thus $\lim\limits_{n,m \to \infty} \|u_n - u_m\| = 0$. By (c), there is a unique element $z \in \mathfrak{B}$, and for $\epsilon > 0$ an $N(\epsilon)$ such that $\|z - u_n\| < \epsilon/2$ for $n > N(\epsilon)$. For this z we then have

$$
\begin{aligned}
\|z - Az\| &= \|z - u_n + u_n - Az\| \leqslant \|z - u_n\| + \|u_n - Az\| \\
&= \|z - u_n\| + \|Au_{n-1} - Az\| \leqslant \|z - u_n\| + a\|u_{n-1} - z\| < \epsilon \quad (1.5)
\end{aligned}
$$

whenever $n > N(\epsilon) + 1$. On the left-hand side we have fixed elements so that $\|z - Az\| = 0$ follows and, by (b), this implies $z = Az$. Finally let z_1, z_2 be two distinct elements for which $z_i = Az_i$ and $\|z_1 - z_2\| > 0$. Then

$$
\|z_1 - z_2\| = \|Az_1 - Az_2\| \leqslant a\|z_1 - z_2\|, \quad \text{or} \quad 1 \leqslant a, \quad (1.6)
$$

which is a contradiction. Thus everything is proved.

We call to the reader's attention that Lemma 1 becomes false if, instead of (1.1), only $\|Au - Av\| < \|u - v\|$ is required.

Example. Let the elements of $\mathfrak{B}$ be the real numbers u with $-\infty < u < \infty$ and $\|u\| = |u|$. We set

$$
Au = u + \frac{\pi}{2} - \arctan u. \quad (1.7)
$$

Since $\arctan u < \pi/2$, there is no u for which $u = Au$. Furthermore we have

$$
\|Au - Av\| = |Au - Av| = |u - v - (\arctan u - \arctan v)|. \quad (1.8)
$$

By the mean-value theorem of the calculus of differentiation we obtain

$$
\arctan u - \arctan v = (u - v) \frac{1}{1 + (v + \theta(u - v))^2} \quad \text{where} \quad 0 < \theta < 1, \quad (1.9)
$$

and altogether

$$
|Au - Av| = |u - v| \left| 1 - \frac{1}{1 + (v + \theta(u - v))^2} \right| < |u - v|. \quad (1.10)
$$

1.2 Hilbert Space

A set $\mathcal{H}$ of elements $u, v, w, \ldots$ is called a *(real) Hilbert space* if $\mathcal{H}$ possesses properties (a) and (c), and the property (b_1).

(b_1) To every pair of elements u, $v \in \mathcal{H}$ there is assigned a real number in a unique way; this is called the scalar product and denoted by (u,v), which satisfies the following rules: $(au,v) = a(u,v)$ for any real number a; $(u + v, w) = (u,w) + (v,w)$; $(u,v) = (v,u)$; $(u,u) > 0$ for $u \neq$ the zero element; $(u,u) = 0$ for $u =$ the zero element.

In the following, by $\mathcal{H}$ we always understand a Hilbert space. If we put $\|u\| = (u,u)^{\frac{1}{2}}$, then ($b_1$) contains the well-known inequalities

$$|(u,v)| \leqslant \|u\| \, \|v\|, \qquad \|u + v\| \leqslant \|u\| + \|v\|, \tag{1.11}$$
$$\|u - v\| \leqslant \|u - w\| + \|w - v\|.$$

They are called *Schwarz's*, *Minkowski's*, and the *triangle inequality*, respectively. Hence by (b) we may interpret $\|u\|$ as norm and so, in particular, $\mathcal{H}$ is a Banach space $\mathcal{B}$. If $(u,v) = 0$ we say the elements u and v are orthogonal.

An operator A in $\mathcal{H}$ is defined as in Section 1.1. But in $\mathcal{H}$ we only consider *linear operators*: A is a prescription in $\mathcal{H}$ which assigns to every $u \in \mathcal{H}$ an element $w = Au \in \mathcal{H}$ in a unique way so that

$$A(au + bv) = aAu + bAv \tag{1.12}$$

for all u, $v \in \mathcal{H}$ and real numbers a, b. Hence the additional word "linear" will always be omitted in the case of operators in $\mathcal{H}$.

The operator A in $\mathcal{H}$ is called *completely continuous* if every infinite sequence

$$\{u_j\} \equiv u_1, u_2, \ldots \in \mathcal{H} \quad \text{where} \quad \|u_j\| \leqslant K$$

contains a subsequence $\{u_{j_i}\}$ for which $\lim\limits_{i \to \infty} \|Au_{j_i} - u\| = 0$ for a suitable $u \in \mathcal{H}$. With (c), we also say the sequence $\{Au_{j_i}\}$ converges to u and write $\lim\limits_{i \to \infty} Au_{j_i} - u$.

LEMMA 2. If the operator A in $\mathcal{H}$ can be approximated arbitrarily well by completely continuous operators A_ϵ in $\mathcal{H}$, then A is completely continuous. More precisely, if to every $\epsilon > 0$ there is a completely continuous operator A_ϵ in $\mathcal{H}$ such that for all $u \in \mathcal{H}$

$$\|Au - A_\epsilon u\| \leqslant \epsilon \|u\|, \tag{1.13}$$

then A is completely continuous.

Proof. We consider a null sequence $\{\epsilon_i\}$ and the corresponding sequence of completely continuous operators $\{A_{\epsilon_i}\}$. Furthermore we let $\{u_n\} \subset \mathcal{H}$ be an arbitrary sequence of elements with $\|u_n\| \leqslant K$. Because of the complete continuity of A_{ϵ_1} we may select a subsequence $u_{11}, u_{12}, u_{13}, \ldots$ from $\{u_n\}$ so that the sequence $\{A_{\epsilon_1} u_{1n}\}$ converges. For the same reason, from this subsequence we may select another subsequence $u_{21}, u_{22}, u_{23}, \ldots$ so that $\{A_{\epsilon_2} u_{2n}\}$ converges, and so on. Finally we obtain a sequence of sequences

$$u_{11}, u_{12}, u_{13}, u_{14}, \ldots$$
$$u_{21}, u_{22}, u_{23}, u_{24}, \ldots$$
$$u_{31}, u_{32}, u_{33}, u_{34}, \ldots$$
$$u_{41}, u_{42}, u_{43}, u_{44}, \ldots$$
$$\cdots$$

each of which is a subsequence of the preceding one.   The diagonal sequence

$$v_1 = u_{11}, \ v_2 = u_{22}, \ v_3 = u_{33}, \ \ldots$$

then has the property that $\{A_{\epsilon_k}v_n\}$ converges for every k.   If we prescribe an arbitrary $\epsilon > 0$, then by (1.13), for sufficiently large n and m, setting $w_n = Av_n$, $w_m = Av_m$, we have

$$\|w_n - w_m\| = \|Av_n - Av_m\| \leqslant \|Av_n - A_{\epsilon_k}v_n\| + \|A_{\epsilon_k}v_n - A_{\epsilon_k}v_m\|$$
$$+ \|Av_m - A_{\epsilon_k}v_m\| \leqslant 2\epsilon_k K + \|A_{\epsilon_k}v_n - A_{\epsilon_k}v_m\| < \epsilon; \quad (1.14)$$

for we may choose k so large that $2\epsilon_k K < \epsilon/2$, and for k fixed in this way and n and m sufficiently large we have $\|A_{\epsilon_k}v_n - A_{\epsilon_k}v_m\| < \epsilon/2$.

By (c) there is an element $v \ \varepsilon \ \mathcal{H}$ such that $\|w_n - v\| < \epsilon$ for $n > N(\epsilon)$.   In other words, we have proved that $\lim\limits_{n\to\infty} Av_n = v$ and consequently the complete continuity of A in $\mathcal{H}$.

Completely continuous operators are of great importance in analysis, since the *Fredholm alternative* is true for them, and the *eigenvalue problem* can be settled in a simple way.

The operator A^* is called the *adjoint* of A in $\mathcal{H}$ if

$$(Au,v) = (u,A^*v)$$

for all u, $v \ \varepsilon \ \mathcal{H}$.   If in particular $A = A^*$, then A is said to be *symmetric*.   The operator A^* is always linear and completely continuous if A is.

LEMMA 3.   FREDHOLM ALTERNATIVE.   Let A be completely continuous in $\mathcal{H}$. For the equations

$$u - Au = f, \qquad v - A^*v = h; \qquad u, v, f, h \ \varepsilon \ \mathcal{H} \qquad (1.15)$$

there exist the alternatives: Either (1) for arbitrary f and h, equations (1.15) have unique solutions u, $v \ \varepsilon \ \mathcal{H}$, or (2) each one of the two homogeneous equations

$$u - Au = 0, \qquad\qquad\qquad (1.16)$$
$$v - A^*v = 0, \qquad\qquad\qquad (1.16a)$$

has a nonzero solution.[1]

Addendum.   In the case (2) the numbers of linearly independent[2] solutions of (1.16) and (1.16a) are finite and equal, and equations (1.15) possess solutions if and only if f is orthogonal to all solutions v of (1.16a) and h orthogonal to all solutions u of (1.16): $(f,v) = 0$ and $(h,u) = 0$, respectively.

The eigenvalue problem for symmetric, completely continuous operators can be settled completely.   Every $\varphi \ \varepsilon \ \mathcal{H}$ distinct from the zero element which satisfies the equation $A\varphi = \mu\varphi$ with a suitable real number μ is called an *eigenelement* and μ an *eigenvalue* of A in $\mathcal{H}$.

---

[1] From the linearity of A and A^* [compare (1.12)] it follows that the zero element is a solution.
[2] We use the definition of vector algebra.

LEMMA 4. If A is a symmetric completely continuous operator in $\mathcal{H}$ which does not map the whole space $\mathcal{H}$ onto the zero element, then A possesses at least one and at most countably many eigenvalues $\mu_1, \mu_2, \mu_3, \ldots$ [3] distinct from zero, which we order by their absolute values: $|\mu_1| \geqslant |\mu_2| \geqslant |\mu_3| \geqslant \ldots$. Their number is either finite or countably infinite. In the latter case they form a null sequence. The corresponding eigenelements can be chosen so that they form an orthonormal system: $(\varphi_i, \varphi_j) = \delta_{ij}$, where $\delta_{ij} = 1$ if $i = j$ and $\delta_{ij} = 0$ if $i \neq j$. Every element of the form Av with arbitrary $v \in \mathcal{H}$ may be expanded in this orthonormal system:

$$Av = \sum_i (Av, \varphi_i)\varphi_i = \sum_i (v, A\varphi_i)\varphi_i = \sum_i \mu_i(v, \varphi_i)\varphi_i. \tag{1.17}$$

By this lemma the equation

$$u - \kappa Au = f \tag{1.18}$$

can easily be solved in explicit form.

First assume that κ is not the reciprocal of an eigenvalue. By (1.17) we have

$$u = f + \kappa Au = f + \kappa \sum_i \mu_i(u, \varphi_i)\varphi_i, \tag{1.19}$$

from which

$$(u, \varphi_k) = (f, \varphi_k) + \kappa\mu_k(u, \varphi_k), \tag{1.20}$$

$$(u, \varphi_k) = \frac{1}{1 - \kappa\mu_k}(f, \varphi_k) \tag{1.21}$$

result immediately, and finally, using (1.19), we obtain

$$u = f + \kappa \sum_i \frac{\mu_i}{1 - \kappa\mu_i}(f, \varphi_i)\varphi_i. \tag{1.22}$$

It is easy to show that this series converges and that it represents a solution of (1.18).

Now let κ be the reciprocal of an eigenvalue μ of multiplicity r and let

$$\varphi_s, \varphi_{s+1}, \ldots, \varphi_{s+r-1}$$

be the corresponding pairwise orthonormal eigenelements; then, by (1.20), f must necessarily be orthogonal to these eigenelements of the eigenvalue μ. Using Lemma 3 we easily confirm that this condition is also sufficient, and we verify that then

$$u = f + \kappa \sum_i c_i(f, \varphi_i)\varphi_i, \tag{1.23}$$

where
$$c_i = \begin{cases} \dfrac{\mu_i}{1 - \kappa\mu_i} & \text{for } i \neq s, s+1, \ldots, s+r-1 \\ \text{arbitrary} & \text{for } i = s, s+1, \ldots, s+r-1, \end{cases}$$

is a solution of (1.18).

---

[3] If to some μ_i there belong several linearly independent eigenelements, then this μ_i is to be written in the sequence with that multiplicity. The multiplicity of μ_i is finite when $\mu_i \neq 0$.

1.3   Bounded Linear Functionals in Hilbert Space

If to every element u of a Hilbert space $\mathcal{H}$ there is assigned a real number $L(u)$ in a unique way so that for all $u,\ v \in \mathcal{H}$ and any real c

$$L(u + v) = L(u) + L(v), \tag{1.24}$$
$$L(cu) = cL(u), \tag{1.25}$$
$$|L(u)| \leqslant \text{const } \|u\|, \tag{1.26}$$

then the correspondence $L(u)$ is called a *bounded linear functional* in $\mathcal{H}$. For fixed $w \in \mathcal{H}$, (w,u) is a bounded linear functional in $\mathcal{H}$. On the other hand, we have Lemma 5.

LEMMA 5.   If $L(u)$ is a bounded linear functional in $\mathcal{H}$, then there exists a $w \in \mathcal{H}$ for which

$$L(u) = (w,u) \quad \text{for} \quad \text{all } u \in \mathcal{H}. \tag{1.27}$$

Proof. Relation (1.26) gives the existence of $\alpha = \sup\limits_{\|u\| \neq 0} |L(u)|/\|u\|$. We have

$$|L(u)| \leqslant \alpha\|u\| \quad \text{for} \quad \text{all } u \in \mathcal{H}. \tag{1.28}$$

If $\alpha = 0$, then w equal to the zero element would have the desired properties, so suppose that $\alpha \neq 0$. Then $\alpha = \sup\limits_{\|v\| = 1} L(v)$, since to every u for which $\|u\| \neq 0$ there is a $v = \pm u/\|u\|$ satisfying $\|v\| = 1$ and

$$L(v) = L\left(\frac{\pm u}{\|u\|}\right) = \left| L\left(\frac{u}{\|u\|}\right) \right| = \frac{|L(u)|}{\|u\|},$$

the latter by (1.25). Now let v_m, $\|v_m\| = 1$ be a sequence of elements in $\mathcal{H}$ such that $\lim\limits_{m \to \infty} L(v_m) = \alpha$. This sequence converges since

$$\|v_m - v_l\|^2 = 2 - 2(v_m,v_l) = 2 + \tfrac{1}{2}\|v_m - v_l\|^2 - \tfrac{1}{2}\|v_m + v_l\|^2,$$
$$\|v_m - v_l\|^2 = 4 - \|v_m + v_l\|^2 \leqslant 4 - \frac{1}{\alpha^2}\left(L(v_m + v_l)\right)^2 \quad \text{by} \quad (1.28),$$
$$\|v_m - v_l\|^2 \leqslant 4 - \frac{1}{\alpha^2}\left\{L(v_m)^2 + L(v_l)^2 + 2L(v_m)L(v_l)\right\}.$$

The right-hand side tends to zero as $m,\ l \to \infty$. Thus there is a $v \in \mathcal{H}$, $\|v\| = 1$, such that $\lim\limits_{m \to \infty} v_m = v$, and $L(v) = \alpha$, the latter because of

$$|L(v) - \alpha| = \lim\limits_{m \to \infty} |L(v) - L(v_m)| = \lim\limits_{m \to \infty} |L(v - v_m)| \leqslant \lim\limits_{m \to \infty} \text{const } \|v - v_m\| = 0.$$

Now, as can easily be verified, for arbitrary $u \in \mathcal{H}$,

$$L(u) - (\alpha v,\ u) = \left(\alpha v,\ \frac{L(u)}{\alpha}\ v - u\right). \tag{1.29}$$

If we set $\alpha v = w$, it will be obvious that (1.27) follows from (1.29) as soon as we have shown that

$$\left(v, \frac{L(u)}{\alpha} v - u\right) = 0$$

for all $u \in \mathcal{H}$. To do this we set

$$\frac{L(u)}{\alpha} v - u = z.$$

Then $L(z) = 0$; thus everything will be proved if we show that $(v,z) = 0$ for all z with $L(z) = 0$. If z is the null element, this is trivial. For other z with $L(z) = 0$, we have

$$\alpha = L(v) = L(v + \lambda z) \leqslant \alpha\|v + \lambda z\|,$$
$$\alpha^2 \leqslant \alpha^2\{1 + 2\lambda(v,z) + \lambda^2(z,z)\},$$

and thus

$$2\lambda(v,z) + \lambda^2(z,z) \geqslant 0$$

for arbitrary real λ. If we select

$$\lambda = -\frac{(v,z)}{(z,z)},$$

then

$$-2\frac{(v,z)^2}{(z,z)} + \frac{(v,z)^2}{(z,z)} \geqslant 0$$

Hence, $(v,z) = 0$.

The conclusion of Lemma 5 may be extended. If to every two elements u, v of the Hilbert space $\mathcal{H}$ there is assigned a real number $B[u,v]$ in a unique way such that

$$B[u, v + w] = B[u,v] + B[u,w], \qquad B[u + n, w] = B[u,w] + B[n,w], \qquad (1.30)$$
$$B[u,cv] = cB[u,v] = B[cu,v], \qquad (1.31)$$
$$|B[u,v]| \leqslant \text{const } \|u\| \, \|v\|, \qquad (1.32)$$
$$B[u,u] \geqslant \gamma\|u\|^2 \quad \text{for} \quad \text{fixed } \gamma > 0, \qquad (1.33)$$

for all u, v, w in $\mathcal{H}$ and real numbers c, then $B[u,v]$ does not necessarily have all properties of the inner product (u,v). In particular, $B[u,v]$ need not be symmetric in u and v. Although in the proof of Lemma 5 we used the symmetry of the inner product, we have Lemma 6 without it.

LEMMA 6.   REPRESENTATION THEOREM OF LAX AND MILGRAM. If $L(u)$ is a bounded linear functional in $\mathcal{H}$ and $B[u,v]$ has properties (1.30) to (1.33), then there is an $h \in \mathcal{H}$ for which

$$L(u) = B[h,u] \quad \text{for} \quad \text{all } u \in \mathcal{H}.$$

Proof. Let S be the set of all $s \in \mathcal{H}$ for which there exists a $q \in \mathcal{H}$ such that

$$(s,u) = B[q,u]$$

for all $u \in \mathcal{H}$.

By Lemma 5 there is a $w \in \mathcal{H}$ for which $L(u) = (w,u)$ for all $u \in \mathcal{H}$. It only remains to show that w belongs to S. This is done by proving $S = \mathcal{H}$.

The set S is not empty, since it contains the zero element. It is a Hilbert space; since S is a subset of a Hilbert space, it suffices to show that, with s and t, as and $s + t$ also belong to S [Axiom (a)] and that S satisfies Axiom (c). Axiom (b_1) is satisfied automatically. Concerning Axiom (a), let

$$(s,u) = B[q,u] \quad \text{and} \quad (t,u) = B[r,u] \quad \text{for} \quad \text{all } u \in \mathcal{H}.$$

Then

$$(as,u) = a(s,u) = aB[q,u] = B[aq,u] \quad \text{by} \quad (1.31),$$
$$(s + t, u) = (s,u) + (t,u) = B[q,u] + B[r,u] = B[q + r, u] \quad \text{by} \quad (1.30).$$

Concerning Axiom (c), let $\{s_m\}$ be a sequence in S satisfying $\lim\limits_{m,l \to \infty} \|s_m - s_l\| = 0$. Thus $\lim\limits_{m \to \infty} \|s_m - s\| = 0$ for some $s \in \mathcal{H}$; we have to show $s \in S$. The sequence of the corresponding q_m converges; namely, for all $u \in \mathcal{H}$ we have

$$B[q_m - q_l, u] = (s_m - s_l, u),$$

and hence by (1.33) and (1.11) and $\gamma > 0$,

$$\gamma\|q_m - q_l\|^2 \leqslant B[q_m - q_l, q_m - q_l] = (s_m - s_l, q_m - q_l)$$
$$\leqslant \|s_m - s_l\| \cdot \|q_m - q_l\|,$$

from which $\lim\limits_{m,l \to \infty} \|q_m - q_l\| = 0$ results; thus there is a $q \in \mathcal{H}$ satisfying $\lim\limits_{m \to \infty} \|q_m - q\| = 0$ and

$$|(s,u) - B[q,u]| = |(s - s_m,u) - B[q - q_m,u]|$$
$$\leqslant \|s - s_m\| \|u\| + \text{const} \|q - q_m\| \|u\|.$$

The right-hand side, for every $u \in \mathcal{H}$, tends to zero as $m \to \infty$. Hence

$$(s,u) = B[q,u]$$

for all $u \in \mathcal{H}$, which implies $s \in S$; thus S satisfies Axiom (c).

If r is an arbitrary fixed element in $\mathcal{H}$, then (r,s) defines a bounded linear functional in the Hilbert space S, since in particular according to Schwarz's inequality (1.11) we have $|(r,s)| \leqslant \text{const} \|s\|$. By Lemma 5 there is a $t \in S$ such that $(r,s) = (t,s)$ for all $s \in S$. Thus

$$(r - t, s) = 0 \quad \text{for} \quad \text{all } s \in S. \tag{1.34}$$

On the other hand, $B[r - t, u]$ is a bounded linear functional in $\mathcal{H}$, as we see from (1.30), (1.31), and (1.32); thus Lemma 5 provides that there is a $v \in \mathcal{H}$ for which

$$B[r - t, u] = (v,u) \quad \text{for} \quad \text{all } u \in \mathcal{H}. \tag{1.35}$$

Of course v belongs to S. If in (1.35) we put $u = r - t$, then by (1.33) and (1.34) we have

$$\gamma\|r - t\|^2 \leqslant B[r - t, r - t] = (v, r - t) = 0.$$

Hence $r = t$ and therefore $r \in S$. But $r \in \mathcal{H}$ was arbitrary; thus $S = \mathcal{H}$, which concludes the proof.

Lemma 6 contains as a special case the lemma used in Section IV-3.5 in the third step. As we easily see, the space $\overset{0}{H}_1(\mathfrak{D})$ with the inner product $(u,v)_1^{\mathfrak{D}}$ introduced there is a Hilbert space, and, for fixed $f(x) \, \varepsilon \, C^0$ in $\mathfrak{D}$,

$$- \int_{\mathfrak{D}} f(x)\Phi(x)\, dx$$

is a bounded linear functional in $\overset{0}{H}_1(\mathfrak{D})$.   In particular, by Schwarz's inequality we have

$$\left| - \int_{\mathfrak{D}} f(x)\Phi(x)\, dx \right| \leqslant \sqrt{\int_{\mathfrak{D}} f^2(x)\, dx} \, \sqrt{\int_{\mathfrak{D}} \Phi^2(x)\, dx}$$

$$\leqslant \sqrt{\int_{\mathfrak{D}} f^2(x)\, dx} \, \sqrt{\int_{\mathfrak{D}} \Phi^2(x)\, dx + \int_{\mathfrak{D}} \sum_{i=1}^{n} \Phi_{x_i}^2 \, dx} = \text{const } \|\Phi\|_1^{\mathfrak{D}}$$

for all $\Phi \, \varepsilon \, \overset{0}{H}_1(\mathfrak{D})$.

2

SCHAUDER'S TECHNIQUE OF PROOF FOR EXISTENCE PROBLEMS IN ELLIPTIC DIFFERENTIAL EQUATIONS

This technique is exceedingly well suited to settle existence questions for boundary-value problems of elliptic differential equations under very weak hypotheses on the coefficients. Such techniques are necessary if we want to treat the existence problems for quasi-linear or still more general equations. The technique of Schauder is based on the use of exceedingly sharp *a priori* estimates for the solution.[4] In Section III-1.4 we met the simplest such estimate, with which, however, not very much can be done. We outline the method in a presentation which is not too general and without giving the complete hypotheses. However, the precise proof will not be any simpler than the one given in Section IV-3.

2.1 Posing the Problem

We consider a normal domain $\mathfrak{D}$ in R_n and for the points of R_n we use vector notation $x = (x_1, \ldots, x_n)$. The function $u(x)$ is said to be Hölder continuous in $\mathfrak{D}$ if for any two points $x^1, x^2 \in \mathfrak{D}$

$$|u(x^1) - u(x^2)| \leqslant H|x^1 - x^2|^\alpha \tag{2.1}$$

for fixed numbers H and α, where $0 < \alpha < 1$, which are independent of x^1, x^2. Then we write $u \in C^\alpha$ in $\mathfrak{D}$. This is equivalent to the fact that

$$h_\alpha[u] = \sup_{\substack{x^1, x^2 \in \mathfrak{D} \\ x^1 \neq x^2}} \frac{|u(x^1) - u(x^2)|}{|x^1 - x^2|^\alpha} \tag{2.2}$$

has a finite value. For $u(x) \in C^j$ in $\mathfrak{D}$, let $D^j u$ be *any* jth derivative of u. Then for $u(x) \in C^2$ we put

$$\|u\|_2 = \max_{x \in \mathfrak{D}} |u(x)| + \max_{x \in \mathfrak{D}} |D^1 u(x)| + \max_{x \in \mathfrak{D}} |D^2 u(x)|. \tag{2.3}$$

Finally we have to consider the set of all $u(x) \in C^{2+\alpha}$ in $\mathfrak{D}$, that is, $D^2 u \in C^\alpha$. For these u we set

$$\|u\|_{2+\alpha} = \|u\|_2 + h_\alpha[D^2 u]. \tag{2.4}$$

Then $\|u\|_{2+\alpha} \geqslant 0$ and $= 0$ if and only if $u \equiv 0$.

---

[4] The literature is so extensive that we only point out the summarizing book of C. Miranda, Part III, Reference[8], in which a simplified technique of proof for obtaining such estimates is given, and the paper by L. Nirenberg (see Part III, Reference[9]).

Furthermore for any real number a we have

$$\|u + v\|_{2+\alpha} \leqslant \|u\|_{2+\alpha} + \|v\|_{2+\alpha}, \qquad \|au\|_{2+\alpha} = |a| \, \|u\|_{2+\alpha}. \tag{2.5}$$

Hence by Section 1.1 the set of all functions $u \, \varepsilon \, C^{2+\alpha}$ in $\mathfrak{D}$ forms a Banach space $\mathfrak{B}$ if the norm of u is defined as $\|u\|_{2+\alpha}$. The validity of (c) for these functions is well-known but is not immediate.

In a domain $\mathfrak{D}$ with sufficiently smooth boundary $\dot{\mathfrak{D}}$ we consider the most general linear partial differential equation

$$Du = f \quad \text{where} \quad Du \equiv \sum_{i,k=1}^{n} a_{ik}(x)u_{x_ix_k} + \sum_{i=1}^{n} a_i(x)u_{x_i} + a(x)u \tag{2.6}$$

with sufficiently "well-behaved" coefficients. In particular we require that a_{ik}, a_i, a, $f \, \varepsilon \, C^{\alpha}$ in $\mathfrak{D}$ where

$$|a_{ik}|, \; |a_i|, \; |a| \leqslant K \quad \text{and} \quad a \leqslant 0 \quad \text{in} \quad \mathfrak{D}.$$

Furthermore let Du be uniformly elliptic in $\mathfrak{D}$:

$$\sum_{i,k=1}^{n} a_{ik}(x)\xi_i\xi_k \geqslant m \sum_{i=1}^{n} \xi_i^2 \quad \text{for} \quad \text{some fixed } m > 0 \tag{2.7}$$

for all $\xi = (\xi_1, \ldots, \xi_n)$ and all $x \, \varepsilon \, \mathfrak{D}$.

Proposition. $Du = f$ where $u = 0$ on $\dot{\mathfrak{D}}$ has exactly one solution $u \, \varepsilon \, C^{2+\alpha}$ in $\mathfrak{D}$.

2.2 Outline of Proof

The first and third requirements in the case $a \leqslant 0$ were already settled in Section III-1.1 so that only the problem of existence remains.

First Step. The boundary-value problem is embedded in a family of such problems:

$$D_t u = tDu + (1 - t) \, \Delta_n u = f \quad \text{in} \quad \mathfrak{D}, \qquad u = 0 \quad \text{on} \quad \dot{\mathfrak{D}}, \qquad 0 \leqslant t \leqslant 1. \tag{2.8}$$

For $t = 1$ and $D_1 \equiv D$ this is the problem which is to be solved. For $t = 0$ it is the "simple" problem $\Delta_n u = f$, $u = 0$ on $\dot{\mathfrak{D}}$, which we consider as settled.

We denote the set of real numbers $0 \leqslant t \leqslant 1$ by N. The subset T of N is defined as follows: $t \, \varepsilon \, T$ if for every $f(x) \, \varepsilon \, C^{\alpha}$ there exists a solution $u(x) \, \varepsilon \, \mathfrak{B}$ of the problem (2.8). If we can show $T \equiv N$, then everything will be proved, since then $t = 1 \, \varepsilon \, T$.

Since N is connected, this proof is given in three steps by showing that (1) T is not empty; (2) the set T is open relative to N [that is, if $t_0 \, \varepsilon \, T$, then all t-values satisfying $|t - t_0| < \epsilon(t_0)$ and $0 \leqslant t \leqslant 1$ also belong to T for sufficiently small $\epsilon > 0$ which may depend on t_0]; (3) T is closed relative to N (that is, if $\{t_j\} \, \varepsilon \, T$ is a sequence which converges to t, then $t \, \varepsilon \, T$). We agree to consider (1) to be already settled.

Second Step. If we put $\|u\|_0 = \max |u(x)|$, $x \, \varepsilon \, \mathfrak{D}$, then in Section III-1.4 for $Du = f$ and $u = 0$ on $\dot{\mathfrak{D}}$ we have the estimate $\|u\|_0 \leqslant M\|f\|_0$ where $M = M(K,m)$. From this the estimate for (2.8),

$$\|u\|_0 \leqslant \tilde{M}\|f\|_0 \quad \text{where} \quad \tilde{M} = \tilde{M}(\tilde{K},\tilde{m}), \qquad \tilde{K} = \max(K,1), \qquad \tilde{m} = \min(m,1), \quad (2.9)$$

follows immediately.   A much sharper *a priori* estimate of Schauder's for (2.8) is

$$\|u\|_{2+\alpha} \leqslant C\|f\|_\alpha \quad \text{where} \quad C = C(\tilde{K},\tilde{m}) \tag{2.10}$$

and $\|f\|_\alpha = h_\alpha[f]$.

Let $t_0 \in T$.   If $u \in \mathfrak{B}$, then for every fixed $t \in N$ the function g_t defined by

$$g_t(x) \equiv (t - t_0)\{\Delta_n u - Du\} + f \tag{2.11}$$

belongs to C^α.   The problem

$$D_{t_0} v = g_t \quad \text{in} \quad \mathfrak{D} \quad \text{where} \quad v = 0 \quad \text{on} \quad \dot{\mathfrak{D}} \tag{2.12}$$

has exactly one solution $v(x) \in \mathfrak{B}$, since $t_0 \in T$.   Thus by (2.11) and (2.12), and for fixed $t \in N$, to every $u \in \mathfrak{B}$ there is assigned a $v \in \mathfrak{B}: v = Au$.   Since this correspondence depends on t we obtain a family of operators A_t in $\mathfrak{B}$.

All the elements $A_t u$ have the property that they vanish on $\dot{\mathfrak{D}}$.   If for fixed $t \in N$ we could prove the existence of a $z \in \mathfrak{B}$ satisfying $z = A_t z$, then z would vanish on $\dot{\mathfrak{D}}$ and in $\mathfrak{D}$ would satisfy the equation

$$D_{t_0} z = (t - t_0)\{\Delta_n z - Dz\} + f \tag{2.13}$$

which, by (2.8), would mean $D_t z = f$.   Then $t \in T$ would be proved.   Hence, to show (2) it suffices to find an $\epsilon > 0$ such that for every t, $|t - t_0| < \epsilon$ and $0 \leqslant t \leqslant 1$, there exists a $z \in \mathfrak{B}$ satisfying $z = A_t z$.

This is done as follows.   We choose $u_1, u_2 \in \mathfrak{B}$ arbitrary, put $v_1 = Au_1$ and $v_2 = Au_2$ (omit the index t from A_t) and set

$$g_1 = (t - t_0)\{\Delta_n u_1 - Du_1\} + f, \qquad g_2 = (t - t_0)\{\Delta_n u_2 - Du_2\} + f.$$

Then by (2.11) and (2.12) we have

$$D_{t_0}(v_1 - v_2) = g_1 - g_2 \tag{2.14}$$
$$g_1 - g_2 = (t - t_0)\{\Delta_n(u_1 - u_2) - D(u_1 - u_2)\}. \tag{2.15}$$

From (2.10) the estimate

$$\|v_1 - v_2\|_{2+\alpha} \leqslant C\|g_1 - g_2\|_\alpha \tag{2.16}$$

follows immediately.   Because of (2.15), we have an estimate of the form

$$\|g_1 - g_2\|_\alpha \leqslant |t - t_0|C_1\|u_1 - u_2\|_{2+\alpha}, \tag{2.17}$$

so that altogether

$$\|v_1 - v_2\|_{2+\alpha} = \|Au_1 - Au_2\|_{2+\alpha} \leqslant |t - t_0|CC_1\|u_1 - u_2\|_{2+\alpha} \tag{2.18}$$

follows.   The numbers C, C_1 are independent of t, u_1, and u_2.   If we choose $|t - t_0| \leqslant (2CC_1)^{-1}$, then (2.18) becomes

$$\|Au_1 - Au_2\|_{2+\alpha} \leqslant \tfrac{1}{2}\|u_1 - u_2\|_{2+\alpha}, \tag{2.19}$$

and Lemma 1 from Section 1.1 gives the existence of $z \in \mathfrak{B}$.

Third Step.   Let $\{t_j\} \subset T$ be a sequence of numbers converging to t.   Then the corresponding $u_j(x) \in \mathfrak{B}$ are solutions of

$$D_{t_j}u_j = f \quad \text{where} \quad u_j = 0 \quad \text{on} \quad \dot{\mathfrak{D}}. \tag{2.20}$$

From (2.10) it follows that $\|u_j\|_{2+\alpha} \leqslant C\|f\|_\alpha$ so that the uniform boundedness of $u_j(x)$ and its first and second derivatives follows.   Then it is possible—we omit the proof—to select a subsequence $\{u_{j_i}\}$ so that u_j, together with its first and second derivatives, converges uniformly.   Now $\lim\limits_{i \to \infty} u_{j_i}(x)$ belongs to $\mathfrak{B}$.   Consequently, in that subsequence we may pass to the limit in (2.20) under the D-operator.   Hence $t \in T$.

3

THE REGULAR EIGENVALUE PROBLEM

3.1   Posing the Problem

In the normal domain $\mathfrak{D}$ of R_n we consider the problem[5]

$$Du \equiv \Delta_n u - q(x)u + \lambda k(x)u = 0 \quad \text{where} \quad u = 0 \quad \text{on} \quad \dot{\mathfrak{D}}. \tag{3.1}$$

For the present we assume that λ is a real parameter: $-\infty < \lambda < \infty$. Obviously $u \equiv 0$ is always a solution of (3.1). But we are looking for solutions $u(x) \not\equiv 0$ of (3.1) which are εC^0 in $\bar{\mathfrak{D}}$ and εC^2 in $\mathfrak{D}$. Such solutions, as we shall show, exist only for suitable values of λ. We call these values *eigenvalues*, and we call the corresponding solutions $u \not\equiv 0$ *eigenfunctions*. For instance, $\lambda = 2$ is an eigenvalue in Example 1 from Section III-1.1.

For a first orientation we assume in (3.1) that $q(x)$, $k(x) \varepsilon C^0$ in $\bar{\mathfrak{D}}$ and $k(x) > 0$ in $\bar{\mathfrak{D}}$. Then (3.1) is called a *regular eigenvalue problem*. If however (1) the hypotheses made about q and k hold only in $\mathfrak{D}$ with the possible exception of finitely many points, or if (2) $\mathfrak{D}$ is not a normal domain but extends to infinity, or if (3) the cases (1) and (2) occur simultaneously, then (3.1) is called a *singular eigenvalue problem*. In the singular problems, however—all eigenvalue problems of atomic physics are of this type—the condition $u = 0$ on $\dot{\mathfrak{D}}$ has to be weakened or must be omitted if the necessity arises. Here we consider the regular case; assume $q(x) \geqslant 0$. Then Theorem 2 from Section III-1.1 tells us that there are no eigenvalues for $\lambda \leqslant 0$, for, by Section III-1.1, we have

$$a(x) \equiv \lambda k(x) - q(x) \leqslant 0, \tag{3.2}$$

and $u \equiv 0$ is the only solution.

To keep down the amount of work we assume $q(x) \equiv 0$ and make the precise *hypotheses:* Let $\dot{\mathfrak{D}}$ satisfy the hypotheses of Theorem 1 in Section IV-3.3. Assume that $k(x) \varepsilon C^1$ in $\bar{\mathfrak{D}}$ where $k(x) > 0$.

3.2   Equivalent Formulation of the Problem

First the existence of the Green's function of first kind for $\Delta_n u$ will be proved. In Section I-3.4 we succeeded only in giving this proof for spheres.

---

[5] The seemingly more general differential equation

$$\sum_{i=1}^{n} (p(x)u_{x_i})_{x_i} - q(x)u + \lambda k(x)u = 0$$

in the case $p(x) \not= 0$ in $\bar{\mathfrak{D}}$ may be reduced to an equation of type (3.1) in v, by introducing a new unknown function $v(x) = \sqrt{p(x)}\, u(x)$. Here we may assume that $p > 0$, multiplying by -1 if necessary.

THEOREM 1. There exists a function $g(x,y)$ for an arbitrary but fixed $x \in \mathfrak{D}$ with the properties

$$(a) \quad g(x,y) = s(x,y) + \Phi(x,y), \tag{3.3}$$

where $\Phi \in C^0$ in $\bar{\mathfrak{D}}$, $\in C^2$ in $\mathfrak{D}$, Φ_{y_i} bounded and $\Delta_n\Phi = 0$ in $\mathfrak{D}$, all understood with respect to y. The function $s(x,y)$ is the singularity function from Section I-3.2.

$$(b) \quad g(x,y) = 0 \quad \text{for} \quad y \in \dot{\mathfrak{D}}. \tag{3.4}$$

Proof. For Φ we have the boundary-value problem in y:

$$\Delta_n\Phi = 0 \quad \text{in} \quad \mathfrak{D}, \qquad \Phi(x,y) = -s(x,y) \quad \text{for} \quad y \in \dot{\mathfrak{D}}. \tag{3.5}$$

We define $\Psi(x,y) = -s(x,y)$ for $|y - x| \geqslant \delta > 0$ and extend Ψ in $|y - x| \leqslant \delta$ in a way analogous to (IV-3.7) so that $\Psi(x,y) \in C^3$ in $\bar{\mathfrak{D}}$. Here δ is chosen so small that the sphere $|y - x| \leqslant \delta$ lies in $\mathfrak{D}$. Then we form $w(x,y) = \Phi - \Psi$. For w the problem

$$\Delta_n w = -\Delta_n\Psi \quad \text{in} \quad \mathfrak{D}, \qquad w = 0 \quad \text{on} \quad \dot{\mathfrak{D}} \tag{3.6}$$

arises. Now $-\Delta_n\Psi = f \in C^1$, and Theorem 3 from Section IV-3.3 gives the existence of w with the desired properties. Furthermore Problem 2 from Section I-3.4 shows that $g(x,y)$ is symmetric:

$$g(x,y) = g(y,x).$$

THEOREM 2. Let $f(x) \in C^1$ in $\mathfrak{D}$ and bounded in $\bar{\mathfrak{D}}$; then

$$u(x) = -\int_{\mathfrak{D}} g(x,y)f(y)\,dy \tag{3.7}$$

is a solution of the problem

$$\Delta_n u = f \quad \text{in} \quad \mathfrak{D}, \qquad u = 0 \quad \text{on} \quad \dot{\mathfrak{D}} \tag{3.8}$$

and $u(x) \in C^0$ in $\bar{\mathfrak{D}}$, $\in C^2$ in $\mathfrak{D}$.

Proof. First, from Section IV-3.2 and from Lemma 3 of the next section it follows immediately by the decomposition (3.3) that $u(x) \in C^0$ in $\mathfrak{D}$. Furthermore,

$$\int_{\mathfrak{D}} \int_{\mathfrak{D}} |g(x,y)|\,dx\,dy$$

exists as is seen from this lemma and from the estimate in Section III-1.4. If $\Phi(x)$ is a function $\in C^\infty$ in $\mathfrak{D}$ which vanishes in a neighborhood of $\dot{\mathfrak{D}}$, then by Fubini's theorem

$$\int_{\mathfrak{D}} \Delta_n\Phi(x) \left(\int_{\mathfrak{D}} g(x,y)f(y)\,dy \right) dx = \int_{\mathfrak{D}} f(y) \left(\int_{\mathfrak{D}} g(x,y)\Delta_n\Phi(x)\,dx \right) dy$$

$$= -\int_{\mathfrak{D}} f(y)\Phi(y)\,dy,$$

the latter for the reason that $g(x,y) = g(y,x)$ and therefore, because of (I-3.8),

$$\int_{\mathfrak{D}} g(x,y)\Delta_n\Phi(x)\,dx = \int_{\mathfrak{D}} g(y,x)\Delta_n\Phi(x)\,dx = -\Phi(y),$$

since in (I-3.8) we only have to set $a = y$ and $\gamma(y,x) = g(y,x)$. Altogether we have

$$\int_{\mathfrak{D}} u(x)\Delta_n\Phi(x)\,dx = \int_{\mathfrak{D}} f(x)\Phi(x)\,dx \quad \text{for} \quad \text{all these } \Phi(x) \tag{3.9}$$

for the $u(x)$ which is defined by (3.7) and continuous in $\mathfrak{D}$ and so, by Weyl's lemma as given in Section IV-4.2, we have $u(x) \in C^2$ in $\mathfrak{D}$. Now by integration by parts we can change the left-hand side in (3.9) so that from (3.9) it follows that

$$\int_{\mathfrak{D}} \Phi(x)\dot{\Delta}_n u(x)\,dx = \int_{\mathfrak{D}} \Phi(x)f(x)\,dx. \tag{3.10}$$

Boundary integrals do not occur in this integration by parts, since Φ vanishes in a neighborhood of $\dot{\mathfrak{D}}$. Since (3.10) holds for all $\Phi(x)$ of the kind mentioned it follows that $\Delta_n u = f$ in $\mathfrak{D}$.

It is easy to see that $u = 0$ on $\dot{\mathfrak{D}}$. First,

$$u(x) = - \int_{\mathfrak{D}} g(y,x)f(y)\,dy$$

since $g(x,y) = g(y,x)$. Thus $u = 0$ on $\dot{\mathfrak{D}}$ because $g(y,x) = 0$ for $x \in \dot{\mathfrak{D}}$. Finally $u(x) \in C^0$ in $\mathfrak{D}$ results from the fact that $u(x)$ tends to zero when $x \in \mathfrak{D}$ approaches a point on $\dot{\mathfrak{D}}$. This can be seen as follows: Let x^0 be a fixed point on $\dot{\mathfrak{D}}$. We prescribe an $\epsilon > 0$ and put a ball with radius $\eta > 0$ about x^0 as center; denote its intersection with $\mathfrak{D}$ by S_η. Then if M is a bound for $|f(x)|$ in $\mathfrak{D}$, we have

$$|u(x)| \leqslant M \int_{S_\eta} |g(x,y)|\,dy + M \int_{\mathfrak{D}-S_\eta} |g(x,y)|\,dy.$$

By Section III-1.4 we have the estimate $|g(x,y)| \leqslant s(x,y)$ for the dimension $n \geqslant 3$, so at least for $n \geqslant 3$ we have

$$|g(x,y)| \leqslant \text{const } |x - y|^{2-n} \quad \text{for} \quad x \neq y. \tag{3.11}$$

Hence

$$|u(x)| \leqslant M \text{ const} \int_{S_\eta} \frac{dy}{|x - y|^{n-2}} + M \int_{\mathfrak{D}-S_\eta} |g(x,y)|\,dy. \tag{3.12}$$

By a suitable choice of $\eta > 0$ the first term can be made smaller than ϵ for all $x \in S_\eta$, for by the estimate from Section IV-3.1 we have

$$\int_{S_\eta} \frac{dy}{|x - y|^{n-2}} \leqslant \int_{|x-y|\leqslant \eta} \frac{dy}{|x - y|^{n-2}} \leqslant \frac{\omega_n}{2}\,\eta^2$$

for $x \in S_\eta$. For η fixed in this way the second integral in (3.12) tends to zero as $x \to x^0$; since $x \neq y$ (in particular, $|x - y| \geqslant \eta/4$) by (3.11), the integrand is bounded for $|x - x^0| \leqslant \eta/2$. By the Lebesgue convergence theorem the limit of the integral as $x \to x^0$ is

$$M \int_{\mathfrak{D}-S_\eta} (\lim_{x\to x^0} |g(x,y)|)\,dy.$$

But this is zero since $g(x,y) = g(y,x)$ and $g(y,x^0) = 0$. The case $n = 2$ is left to the reader. Thus $\lim_{x\to x^0} |u(x)| < \epsilon$. But x^0 and ϵ were arbitrary, and so $u(x) \in C^0$ in $\mathfrak{D}$ is proved.

If in Theorem 2 we put $f(x) = -\lambda k(x)u(x)$, then it is to be expected that the problem (3.1) with $q(x) \equiv 0$ is equivalent to the problem which then results from (3.8) in the following form:

$$Au = \mu u \quad \text{where} \quad Au \equiv \int_{\mathfrak{D}} g(x,y)k(y)u(y)\, dy, \qquad \mu = \frac{1}{\lambda}. \tag{3.13}$$

Here A is understood as an operator in the Hilbert space $\mathfrak{IC}$, where $\mathfrak{IC}$ consists of all functions $u(x)$ which are square integrable in the sense of Lebesgue over $\mathfrak{D}$ with the weight factor $k(x)$, and the scalar product on $\mathfrak{IC}$ is $(u,v) = \int_{\mathfrak{D}} u(x)v(x)k(x)\, dx$. The properties (a), (b_1) from Section 1 are obvious, and (c) for this set of functions is just the conclusion of the well-known Riesz-Fischer theorem.

THEOREM 3.   The problems (3.1) with $q(x) \equiv 0$ and (3.13) are equivalent.

Proof. (1) If $u(x) \in C^0$ in $\bar{\mathfrak{D}}$, $\in C^2$ in $\mathfrak{D}$ is a solution $u(x) \not\equiv 0$ of (3.1) with $q(x) \equiv 0$, then $\lambda \neq 0$ because of (3.2), and we have

$$\frac{1}{\lambda}u(x) \in C^0 \quad \text{in} \quad \bar{\mathfrak{D}}, \quad \in C^2 \quad \text{in} \quad \mathfrak{D}, \quad \Delta_n\left(\frac{1}{\lambda}u\right) = f \quad \text{in} \quad \mathfrak{D}, \quad \frac{1}{\lambda}u = 0 \quad \text{on} \quad \dot{\mathfrak{D}} \tag{3.14}$$

if we put $-k(x)u(x) = f(x)$. On the other hand, $f(x) \in C^1$ in $\mathfrak{D}$, $\in C^0$ in $\bar{\mathfrak{D}}$, and hence by Theorem 2

$$Au \in C^0 \quad \text{in} \quad \bar{\mathfrak{D}}, \quad \in C^2 \quad \text{in} \quad \mathfrak{D}, \quad \Delta_n(Au) = f \quad \text{in} \quad \mathfrak{D}, \quad Au = 0 \quad \text{on} \quad \dot{\mathfrak{D}}. \tag{3.15}$$

By the uniqueness theorem (Theorem 2 in Section III-1.1) it follows from (3.14) that $Au = (1/\lambda)u$; thus $u(x)$ is a solution of (3.13). For the function $v(x) = Au - (1/\lambda)u$ would satisfy the problem $\Delta_n v = 0$ in $\mathfrak{D}$, $v = 0$ on $\dot{\mathfrak{D}}$, from which $v(x) \equiv 0$ follows.

(2) The converse is more difficult. We have to show: If $u(x) \in \mathfrak{IC}$ is a solution of the operator equation $Au = \mu u$, then $u(x) \in C^0$ in $\bar{\mathfrak{D}}$, $\in C^2$ in $\mathfrak{D}$, and $u(x)$ solves (3.1) with $q(x) \equiv 0$. To do so we prove (a) that $u(x)$ is bounded in $\bar{\mathfrak{D}}$ and (b) that $u(x) \in C^2$ in $\mathfrak{D}$.

(a) and (b) together show that $f(x) \equiv -k(x)u(x) \in C^1$ in $\mathfrak{D}$ and that $f(x)$ is bounded in $\bar{\mathfrak{D}}$. Application of Theorem 2 then gives

$$\begin{aligned} Au \in C^0 \quad \text{in} \quad \bar{\mathfrak{D}}, & \qquad \in C^2 \quad \text{in} \quad \mathfrak{D}, \\ \Delta_n Au = -k(x)u(x), & \qquad Au = 0 \quad \text{on} \quad \dot{\mathfrak{D}}. \end{aligned} \tag{3.16}$$

Therefore $u(x)$ has the desired properties, since $Au = (1/\lambda)u$. It only remains to prove (a) and (b).

(a) We have

$$|u(x)| = \left|\frac{1}{\mu}Au\right| \leqslant |\lambda| \int_{\mathfrak{D}} |g(x,y)||k(y)||u(y)|\, dy \quad \text{for} \quad x \in \bar{\mathfrak{D}}.$$

If for $n \geqslant 3$ we use the estimate $0 \leqslant g(x,y) \leqslant s(x,y)$ (Section III-1.4), then

$$|u(x)| \leqslant \text{const} \int_{\mathfrak{D}} |x - y|^{2-n}|u(y)|\, dy \quad \text{for} \quad x \in \bar{\mathfrak{D}}. \tag{3.17}$$

If $n = 3$, from Schwarz's inequality it follows that

$$|u(x)| \leqslant \text{const} \left(\int_{\mathfrak{D}} |x - y|^{-2}\, dy\right)^{\frac{1}{2}} \left[\int_{\mathfrak{D}} u^2(y)\, dy\right]^{\frac{1}{2}}.$$

The first integral exists by Section IV-3.1 and the second because $u(x) \in \mathcal{K}$. Thus $u(x)$ is bounded in $\bar{\mathfrak{D}}$ for $n = 3$. The case $n = 2$ is just as simple. In the case $n > 3$ we first have to iterate (3.17), analogous to the procedure in the proof in Section IV-3.7, in order to arrive at square integrable singularities. After that, again by means of Schwarz's inequality, we obtain the boundedness of $u(x)$ in $\bar{\mathfrak{D}}$.

(b) The function $f(x) = -\lambda k(x)u(x)$ is integrable in $\mathfrak{D}$, since $k(x) \in C^0$ in $\bar{\mathfrak{D}}$ and $u(x) \in \mathcal{K}$. Indeed we have

$$\left| \int_{\mathfrak{D}} f(x)\, dx \right| \leqslant |\lambda| \left[\int_{\mathfrak{D}} k^2(x)\, dx \right]^{\frac{1}{2}} \left[\int_{\mathfrak{D}} u^2(x)\, dx \right]^{\frac{1}{2}}.$$

For every function $\Phi(x) \in C^\infty$ in $\mathfrak{D}$ which vanishes in a neighborhood of $\dot{\mathfrak{D}}$ we have by Fubini's theorem, similar to the proof of Theorem 2,

$$\lambda \int_{\mathfrak{D}} \Delta_n\Phi(x) \left[\int_{\mathfrak{D}} g(x,y)k(y)u(y)\, dy \right] dx = \lambda \int_{\mathfrak{D}} k(y)u(y) \left[\int_{\mathfrak{D}} g(x,y)\Delta_n\Phi(x)\, dx \right] dy$$

$$= -\lambda \int_{\mathfrak{D}} k(y)u(y)\Phi(y)\, dy.$$

The inner integral on the left-hand side represents $(1/\lambda)u(x)$. Therefore we have

$$\int_{\mathfrak{D}} u(x)\Delta_n\Phi(x)\, dx = -\lambda \int_{\mathfrak{D}} k(x)u(x)\Phi(x)\, dx.$$

From this it follows immediately that

$$\int_{\mathfrak{D}} u(x)[\Delta_n\Phi(x) + \lambda k(x)\Phi(x)]\, dx = 0$$

for all such $\Phi(x)$. By Weyl's lemma in Section IV-4.2 it follows that $u(x) \in C^2$ in $\mathfrak{D}$. Thus everything is proved.

3.3   Complete Continuity of the Operator

In (3.13) we put

$$\bar{u}(x) = u(x)\, \sqrt{k(x)}. \tag{3.18}$$

Then we obtain the operator equation (3.13) in the form

$$\tilde{A}\bar{u} = \mu\bar{u} \quad \text{where} \quad \tilde{A}\bar{u} \equiv \int_{\mathfrak{D}} g(x,y)\, \sqrt{k(x)}\, \sqrt{k(y)}\, \bar{u}(y)\, dy, \tag{3.19}$$

and correspondingly $\tilde{A}$ has to be considered in the Hilbert space $\mathcal{K}$ consisting of all functions $\bar{u}(x)$ which are square integrable over $\mathfrak{D}$ in the sense of Lebesgue: $\int_{\mathfrak{D}} \bar{u}^2(x)\, dx < \infty$. The scalar product in $\mathcal{K}$ is given by $(\bar{u},\bar{v}) = \int_{\mathfrak{D}} \bar{u}(x)\bar{v}(x)\, dx$. The goal of our further investigations is the central theorem we now state.

THEOREM.   The operator $\tilde{A}$ in $\mathcal{K}$ is symmetric and completely continuous.

The proof will be given by way of the following lemmas.

LEMMA 1. (a) Let $\vartheta(x)$, $\bar{w}(x)$ be two functions from $\tilde{\mathcal{H}}$. Then the operator

$$\tilde{B}\bar{u} = (\vartheta,\bar{u})\bar{w} \equiv \bar{w}(x) \int_{\mathfrak{D}} \vartheta(x)\bar{u}(x) \, dx \tag{3.20}$$

defined for all $\bar{u}(x) \, \varepsilon \, \tilde{\mathcal{H}}$ is completely continuous in $\tilde{\mathcal{H}}$.

(b) Let $\vartheta_i(x)$, $\bar{w}_i(x)$, $i = 1, \ldots, n$ be functions from $\tilde{\mathcal{H}}$. Then the operator

$$\tilde{B}\bar{u} = \sum_{i=1}^{n} (\vartheta_i,\bar{u})\bar{w}_i \equiv \sum_{i=1}^{n} \bar{w}_i(x) \int_{\mathfrak{D}} \vartheta_i(x)\bar{u}(x) \, dx, \tag{3.21}$$

defined for all $\bar{u}(x) \, \varepsilon \, \tilde{\mathcal{H}}$ is completely continuous in $\tilde{\mathcal{H}}$.

Proof. (a) Let $\{\bar{u}_j(x)\} \subset \tilde{\mathcal{H}}$ be a sequence of functions satisfying $\|\bar{u}_j(x)\| \leqslant K$. By the Bolzano-Weierstrass theorem we can select a subsequence $\{\bar{u}_{j_i}(x)\}$ so that the sequence of numbers $\{(\vartheta,\bar{u}_{j_i})\}$ converges. If its limit is denoted by γ, then by (3.20) we have

$$\|\tilde{B}\bar{u}_{j_i} - \gamma\bar{w}\| = |(\vartheta,\bar{u}_{j_i}) - \gamma| \, \|\bar{w}\| \to 0 \tag{3.22}$$

for $j_i \to \infty$. By Section 1.2 this means that $\tilde{B}$ is completely continuous in $\tilde{\mathcal{H}}$. Case (b) is obtained analogously.

LEMMA 2. Let $h(x,y)$ be continuous in the $2n$ variables $x_1, x_2, \ldots, x_n, y_1, y_2, \ldots, y_n$ for all $x \, \varepsilon \, \mathfrak{D}$, $y \, \varepsilon \, \mathfrak{D}$. We shall write $h(x,y) \, \varepsilon \, C^0$ in $\mathfrak{D} \times \mathfrak{D}$. Furthermore let $h(x,y)$ be bounded in $\mathfrak{D} \times \mathfrak{D}$. Then the operator

$$\check{C}\bar{u} = \int_{\mathfrak{D}} h(x,y)\bar{u}(y) \, dy \tag{3.23}$$

defined for all $\bar{u}(x) \, \varepsilon \, \tilde{\mathcal{H}}$ is completely continuous in $\tilde{\mathcal{H}}$.

Proof. Let Q be a cube in R_{2n}, described by

$$Q: -\rho < x_i < \rho, \qquad -\rho < y_i < \rho, \qquad i = 1, \ldots, n,$$

with the property that $\mathfrak{D} \times \mathfrak{D} \subset Q$. This can always be achieved by suitable choice of ρ. Now we extend $h(x,y)$ in $\bar{Q}$ by setting $h(x,y) \equiv 0$ outside of $\mathfrak{D} \times \mathfrak{D}$. In $\bar{Q}$ we expand $h(x,y)$ into a $2n$-dimensional Fourier series and consider the partial sum $\sum_{i=1}^{l} \bar{w}_i(x)\vartheta_i(y)$, where $\bar{w}_i(x)$ and $\vartheta_i(y)$ are functions from $\tilde{\mathcal{H}}$ which are sums of trigonometric functions in $\bar{Q}$. Then

$$\int_{\mathfrak{D}} \int_{\mathfrak{D}} \left(h(x,y) - \sum_{i=1}^{l} \bar{w}_i(x)\vartheta_i(y)\right)^2 dx \, dy < \epsilon^2 \quad \text{for} \quad l > N(\epsilon). \tag{3.24}$$

We now define an operator $\tilde{C}_\epsilon$ in $\tilde{\mathcal{H}}$ by

$$\tilde{C}_\epsilon\bar{u} = \sum_{i=1}^{l} (\vartheta_i,\bar{u})\bar{w}_i \equiv \sum_{i=1}^{l} \bar{w}_i(x) \int_{\mathfrak{D}} \vartheta_i(x)\bar{u}(x) \, dx. \tag{3.25}$$

By Lemma 1, $\tilde{C}_\epsilon$ is a completely continuous operator in $\mathcal{H}$. Using (3.24) and Schwarz's inequality we have

$$\|\tilde{C}\bar{u} - \tilde{C}_\epsilon\bar{u}\|^2 \equiv \int_{\mathfrak{D}} \left(\int_{\mathfrak{D}} \{h(x,y) - \sum_{i=1}^{l} \bar{w}_i(x)\bar{v}_i(y)\}\bar{u}(y)\,dy \right)^2 dx$$

$$\leqslant \epsilon^2 \int_{\mathfrak{D}} \bar{u}^2(y)\,dy = \epsilon^2\|\bar{u}\|^2. \tag{3.26}$$

Lemma 2 from Section 1.2 proves the complete continuity of $\tilde{C}$ in $\mathcal{H}$.

LEMMA 3.   The Green's function $g(x,y)$ belonging to the domain $\mathfrak{D}$ is continuous in the $2n$ variables $x_1, x_2, \ldots, x_n, y_1, y_2, \ldots, y_n$ for all $x \in \mathfrak{D}$, $y \in \mathfrak{D}$ satisfying $x \neq y$.

We give the proof of this lemma at the end of Section 3.4 in order not to interrupt our considerations here. In any case, by Theorem 1 from Section 3.2, $g(x,y)$ is continuous in the variables y for $x \neq y$ and because of symmetry also in the variables x for $x = y$. But from this continuity in the $2n$ variables does not follow automatically.

Proof of the Theorem. The complete continuity of the operator $\tilde{A}$ in (3.19) cannot be inferred immediately from Lemma 2, since $g(x,y)$ is discontinuous for $x = y$. For this purpose we first replace the operator $\tilde{A}$ by "smoother" operators.

Let $w_\sigma(\xi)$ be a *smoothing function* in the variable ξ where $0 \leqslant \xi < \infty$. By this we mean the following: Let $w_\sigma(\xi) \in C^0$ in $0 \leqslant \xi < \infty$ and with $\sigma > 0$ be defined such that

$$w_\sigma(\xi) = 0 \quad \text{for} \quad 0 \leqslant \xi \leqslant \frac{1}{2\sigma}, \qquad 0 \leqslant w_\sigma(\xi) \leqslant 1 \quad \text{for} \quad \frac{1}{2\sigma} < \xi < \frac{1}{\sigma},$$

$$w_\sigma(\xi) = 1 \quad \text{for} \quad \frac{1}{\sigma} \leqslant \xi < \infty. \tag{3.27}$$

Using this smoothing function we define

$$g_\sigma(x,y) = g(x,y)w_\sigma(|x - y|). \tag{3.28}$$

Then $g_\sigma(x,y)$ does not possess any singularities. It is a continuous function in the $2n$ variables $x_1, \ldots, y_n$. In analogy to (3.19), setting $\epsilon = 1/\sigma^2$ we define the operators

$$\tilde{A}_\epsilon\bar{u} \equiv \int_{\mathfrak{D}} g_\sigma(x,y)\,\sqrt{k(x)k(y)}\,\bar{u}(y)\,dy. \tag{3.29}$$

By Lemmas 2 and 3 they are completely continuous in $\mathcal{H}$. By (3.19) and a skillful application of Schwarz's inequality we find

$$\|\tilde{A}\bar{u} - \tilde{A}_\epsilon\bar{u}\|^2 \equiv \int_{\mathfrak{D}} \left(\int_{\mathfrak{D}} \{g(x,y) - g_\sigma(x,y)\}\,\sqrt{k(x)k(y)}\,\bar{u}(y)\,dy \right)^2 dx$$

$$\leqslant \text{const} \int_{\mathfrak{D}} \left(\int_{\mathfrak{D}} |g(x,y) - g_\sigma(x,y)|\,|\bar{u}(y)|\,dy \right)^2 dx$$

$$= \text{const} \int_{\mathfrak{D}} \left(\int_{\mathfrak{D}} \sqrt{|g(x,y) - g_\sigma(x,y)|}\,\sqrt{|g(x,y) - g_\sigma(x,y)|}\,|\bar{u}(y)|\,dy \right)^2 dx$$

$$\leqslant \text{const} \int_{\mathfrak{D}} \left(\int_{\mathfrak{D}} |g(x,y) - g_\sigma(x,y)|\,dy \int_{\mathfrak{D}} |g(x,y) - g_\sigma(x,y)|\bar{u}^2(y)\,dy \right) dx. \tag{3.30}$$

Furthermore, carefully taking note of (3.27), we have

$$\int_{\mathfrak{D}} |g(x,y) - g_\sigma(x,y)| \, dy = \int_{|y-x| \leqslant 1/\sigma} |g(x,y) - g_\sigma(x,y)| \, dy \leqslant 2 \int_{|y-x| \leqslant 1/\sigma} |g(x,y)| \, dy. \quad (3.31)$$

If we use the estimate for $g(x,y)$ in Section III-1.4 and the theorem from Section IV-3.1, then for $n \geqslant 3$ it follows that

$$\int_{|y-x| \leqslant 1/\sigma} |g(x,y)| \, dy \leqslant \text{const} \, \frac{1}{\sigma^2}. \quad (3.32)$$

If we finally put $1/\sigma^2 = \epsilon$ and interchange the order of integration we obtain

$$\|\tilde{A}\bar{u} - \tilde{A}_\epsilon \bar{u}\|^2 \leqslant \text{const} \, \epsilon \int_{\mathfrak{D}} \left(\int_{\mathfrak{D}} |g(x,y) - g_\sigma(x,y)| \, dx \right) \bar{u}^2(y) \, dy$$

$$\leqslant \text{const} \, \epsilon^2 \int_{\mathfrak{D}} \bar{u}^2(y) \, dy \leqslant \text{const} \, \epsilon^2 \|\bar{u}\|^2, \quad (3.33)$$

so that Lemma 2 from Section 1.2 provides that the complete continuity of $\tilde{A}$ in $\mathfrak{IC}$ follows. The reader will easily settle the case $n = 2$. The symmetry of $\tilde{A}$ in $\mathfrak{IC}$ is obvious, since the kernel

$$g(x,y) \, \sqrt{k(x)k(y)}$$

of the operator is symmetric. Thus the theorem is proved.

3.4 The Expansion Theorem

This theorem represents a very important achievement in analysis. The first step was made by Poincaré; the important connections between completely continuous operators in Hilbert space and questions of expansion are due to D. Hilbert.

THEOREM.   The eigenvalue problem (3.1) with $q(x) \equiv 0$ possesses infinitely many positive eigenvalues λ_1, λ_2, λ_3, . . . which we imagine ordered by their size: $0 < \lambda_1 \leqslant \lambda_2 \leqslant \lambda_3 \leqslant \cdots$ and for which $\lim_{i \to \infty} \lambda_i = \infty$. For the corresponding eigenfunctions we have: $\varphi_i(x) \, \varepsilon \, C^0$ in $\bar{\mathfrak{D}}$, εC^2 in $\mathfrak{D}$, $\varphi_i(x) = 0$ for $x \, \varepsilon \, \dot{\mathfrak{D}}$, $\varphi_i(x) \neq 0$, and

$$\Delta_n \varphi_i(x) + \lambda_i k(x) \varphi_i(x) = 0, \qquad \int_{\mathfrak{D}} \varphi_i(x) \varphi_j(x) k(x) \, dx = \delta_{ij} \quad (3.34)$$

where $\delta_{ij} = 1$ for $i = j$ and $= 0$ for $i \neq j$.

Every arbitrary function $u(x) \, \varepsilon \, C^3$ in $\mathfrak{D}$, εC^0 in $\bar{\mathfrak{D}}$ for which $u = 0$ on $\dot{\mathfrak{D}}$ and $\Delta_n u$ is bounded in $\mathfrak{D}$ may be expanded in these eigenfunctions:

$$u(x) = \sum_{i=1}^{\infty} u_i \varphi_i(x) \quad \text{where} \quad u_i = \int_{\mathfrak{D}} u(x) \varphi_i(x) k(x) \, dx. \quad (3.35)$$

The convergence of the series is to be understood in mean square:

$$\int_{\mathfrak{D}} \left\{ u(x) - \sum_{i=1}^{l} u_i \varphi_i(x) \right\}^2 k(x) \, dx < \epsilon \quad \text{for} \quad l > N(\epsilon). \quad (3.36)$$

Addendum. For $q(x) \, \varepsilon \, C^1$ in $\overline{\mathfrak{D}}$ and $q(x) \geqslant 0$ in $\mathfrak{D}$ the theorem remains valid literally if in (3.34) the more general differential equation (3.1) is used. For $n = 2$ and $n = 3$ the convergence in (3.35) is uniform and absolute in $\overline{\mathfrak{D}}$.

Proof. For the operator $\tilde{A}$ (3.19) in $\tilde{\mathfrak{K}}$ we have the theorem from Section 3.3. Hence Lemma 4 from Section 1.2 is applicable. Therefore the eigenvalue problem $\tilde{A}\tilde{u} = \mu\tilde{u}$ has at least one but at most countably many eigenvalues $\mu_1, \mu_2, \ldots$ with corresponding eigenfunctions $\tilde{\varphi}_i(x) \, \varepsilon \, \tilde{\mathfrak{K}}$ which satisfy the relation

$$(\tilde{\varphi}_i, \tilde{\varphi}_j) = \delta_{ij}. \tag{3.37}$$

If we put $\tilde{\varphi}_i(x) = \varphi_i(x) \sqrt{k(x)}$, then these $\varphi_i(x)$ are eigenfunctions of the eigenvalue problem (3.13). By the equivalence Theorem 3 from Section 3.2, these $\varphi_i(x)$ are εC^0 in $\overline{\mathfrak{D}}$, εC^2 in $\mathfrak{D}$ and they are eigenfunctions of the eigenvalue problem (3.1) with $q(x) \equiv 0$, and these eigenfunctions belong to the eigenvalues $\lambda_i = 1/\mu_i$. From (3.2), furthermore, $\lambda_i > 0$, and (3.37) gives

$$\int_{\mathfrak{D}} \varphi_i(x)\varphi_j(x)k(x) \, dx = \delta_{ij}. \tag{3.38}$$

For the arbitrary function $u(x)$ mentioned in the theorem we form $\Delta_n u(x) = -v(x)$ [where $-v(x)$ is an abbreviation for the left-hand side]. Since $u = 0$ on $\dot{\mathfrak{D}}$ and $v \, \varepsilon \, C^1$ in $\mathfrak{D}$ and is bounded in $\mathfrak{D}$, Theorem 2 from Section 3.2 and the uniqueness theorem (Theorem 2 in Section III-1.1) provide that $u(x)$ may be represented by

$$u(x) = \int_{\mathfrak{D}} g(x,y)v(y) \, dy. \tag{3.39}$$

If we put $\bar{u}(x) = u(x) \sqrt{k(x)}$, $v(x) = \bar{v}(x) \sqrt{k(x)}$, then

$$\bar{u}(x) = \int_{\mathfrak{D}} g(x,y) \sqrt{k(x)} \sqrt{k(y)} \, \bar{v}(y) \, dy \equiv \tilde{A}\bar{v}. \tag{3.40}$$

Hence using Lemma 4 from Section 1.2 we find the expansion

$$\bar{u}(x) = \sum_{i=1}^{\infty} (\bar{u}, \tilde{\varphi}_i)\tilde{\varphi}_i. \tag{3.41}$$

It is impossible that only finitely many eigenvalues μ_i and finitely many eigenfunctions $\tilde{\varphi}_i(x)$ are available, because then such an expansion could not be correct for this arbitrary function $\bar{u}(x)$. Therefore we must have $\lim_{i \to \infty} \mu_i = 0$, and this means $\lim_{i \to \infty} \lambda_i = \infty$. If in (3.41) we go back to $u(x)$ and set $\tilde{\varphi}_i(x) = \varphi_i(x) \sqrt{k(x)}$, then the expansion (3.35) follows. Thus the expansion theorem is proved.

The reader may convince himself that the detour involving the operator $\tilde{A}$ (3.19) was not necessary since the operator A (3.13) is already symmetric in $\mathfrak{K}$. For finer questions and for the applications, however, it is often important to work with symmetric integral operators which possess a symmetric kernel. This is the case for (3.19), but not for (3.13).

Proof of Lemma 3.   From Theorem 1 of Section 3.2 it is clear that the statement about the continuity has to be proved only for $\Phi(x,y)$.   We have

$$|\Phi(x^1,y^1) - \Phi(x^2,y^2)| \leqslant |\Phi(x^1,y^1) - \Phi(x^1,y^2)| + |\Phi(x^1,y^2) - \Phi(x^2,y^2)|. \qquad (3.42)$$

By the continuity of Φ with respect to y, the first expression on the right-hand side can be made arbitrarily small, say $< \epsilon/2$, if y^1,y^2 are sufficiently close.   The second expression represents the value of the function $\tilde{\Phi}(y) = \Phi(x^1,y) - \Phi(x^2,y)$ at the point y^2.   By (3.3) and (3.4) the function $\tilde{\Phi}(y)$ has the boundary values

$$\tilde{\Phi}(y) = s(x^2,y) - s(x^1,y) \quad \text{for} \quad y \in \dot{\mathfrak{D}}.$$

If x^1,x^2 are sufficiently close, then $|\tilde{\Phi}(y)| < \epsilon/2$ can be achieved for all $y \in \dot{\mathfrak{D}}$.   But $\tilde{\Phi}(y)$ is harmonic in $\mathfrak{D}$, so that it follows from the maximum-minimum principle of Section I-3.7 that $|\tilde{\Phi}(y)| < \epsilon/2$ for all $y \in \bar{\mathfrak{D}}$.   Thus the continuity of $\Phi(x,y)$ in the $2n$ variables is proved. Our conclusion now follows immediately from $g(x,v) = s(x,v) + \Phi(x,v)$.

4

ELLIPTIC SYSTEMS OF DIFFERENTIAL EQUATIONS

4.1 Posing the Problem

We consider the most general linear elliptic system of two first-order partial differential equations in two unknown functions. After the considerations in Section II-2.6 and with suitable hypotheses on the coefficients, we may assume that the system is in the *integrable normal form* (II-2.65):

$$U^1_{x_1} - U^2_{x_2} + A^1_1(x)U^1 + A^1_2(x)U^2 + C^1(x) = 0,$$
$$U^1_{x_2} + U^2_{x_1} + A^2_1(x)U^1 + A^2_2(x)U^2 + C^2(x) = 0,$$

where $U^1(x)$ and $U^2(x)$ are the unknown functions. We abbreviate the left-hand sides by D^1 and D^2 and furthermore we use the abbreviated notation

$$D^i \equiv L^i + \sum_{j=1}^{2} A^i_j U^j + C^i = 0, \qquad i = 1, 2$$

$$L^1(x) \equiv U^1_{x_1} - U^2_{x_2}, \qquad L^2(x) \equiv U^1_{x_2} + U^2_{x_1}. \tag{4.1}$$

Our considerations take place in a normal domain $\mathfrak{D}$ of R_2, and we set $x = (x_1, x_2)$. Since the principal part in (4.1) is exactly of the form of the *Cauchy-Riemann differential equations*, it is always possible to transform $\mathfrak{D}$ onto the unit circle $|x| = 1$ by a conformal transformation of $\mathfrak{D}$. We always make this simplification and assume $A^i_j(x)$, $C^i(x) \; \varepsilon \; C^0$ in $\mathfrak{D}$, εC^1 in $\dot{\mathfrak{D}}$.[6] For (4.1) we pose a boundary-value problem with the condition

$$\alpha_1(\sigma)U^1 + \alpha_2(\sigma)U^2 = \varphi(\sigma) \quad \text{on} \quad \dot{\mathfrak{D}} \tag{4.2}$$

where α_1, α_2, and φ belong to C^1 as functions of the arc length σ with $(\alpha_1)^2 + (\alpha_2)^2 > 0$, $0 \leqslant \sigma \leqslant 2\pi$. The study of such boundary-value problems brings up some interesting phenomena, and so it will be worthwhile to discuss some simple questions.

4.2 The Green's Function of the Second Kind

As a tool for our further investigations we consider the so-called *second boundary-value problem*

$$\Delta_2 u = 0, \qquad u_\nu = \psi \quad \text{on} \quad \dot{\mathfrak{D}}, \qquad \psi \; \varepsilon \; C^0 \quad \text{on} \quad \dot{\mathfrak{D}}. \tag{4.3}$$

---

[6] A study of such systems on a function-theoretic basis, as a generalization of the Cauchy-Riemann differential equations, was made by L. Bers[2].

Here, as in Section I-1.4, u_ν denotes the directional derivative in the direction of the outer normal on $\dot{\mathfrak{D}}$. We look for solutions $u(x) \in C^1$ in $\bar{\mathfrak{D}}$, $\in C^2$ in $\mathfrak{D}$ of (4.3).

UNIQUENESS THEOREM. Any two solutions of (4.3) differ at most by an additive constant.

Proof. If $u^1(x)$ and $u^2(x)$ are two solutions of (4.3), then $w(x) = u^1(x) - u^2(x)$ satisfies the problem $\Delta_2 w = 0$ in $\mathfrak{D}$, $w_\nu = 0$ on $\dot{\mathfrak{D}}$. The first Green's formula (I-1.16) gives $0 = \int_{\mathfrak{D}} w \Delta_n w \, dx = - \int_{\mathfrak{D}} |\text{grad } w|^2 \, dx$. From this it follows that grad $w = 0$ in $\bar{\mathfrak{D}}$ or $w \equiv$ const in $\bar{\mathfrak{D}}$.

THEOREM 1. A necessary and sufficient condition for the solvability of (4.3) is $\int_{\dot{\mathfrak{D}}} \psi(\sigma) \, d\sigma = 0$.[7]

Proof. Necessity. If a solution $u(x) \in C^1$ in $\bar{\mathfrak{D}}$, $\in C^2$ in $\mathfrak{D}$ exists, then from (I-1.18) it follows that

$$0 = \int_{\mathfrak{D}} \Delta_2 u \, dx = \int_{\dot{\mathfrak{D}}} u_\nu \, d\sigma = \int_{\dot{\mathfrak{D}}} \psi \, d\sigma. \tag{4.4}$$

Sufficiency. Since $\int_{\dot{\mathfrak{D}}} \psi \, d\sigma = 0$, a function $\varphi(\sigma) \in C^1$ is defined by

$$\varphi(\sigma) = - \int_{\sigma_0}^{\sigma} \psi(\sigma) \, d\sigma.$$

Poisson's formula in Section I-3.5, provides that the problem

$$\Delta_2 v = 0, \qquad v = \varphi \quad \text{on} \quad \dot{\mathfrak{D}} \quad \text{where} \quad \varphi \in C^0 \quad \text{on} \quad \dot{\mathfrak{D}} \tag{4.5}$$

is solvable. If in Section I-3.5 we had assumed $\varphi \in C^1$, then Poisson's formula would give $v(x) \in C^1$ in $\bar{\mathfrak{D}}$, $\in C^2$ in $\mathfrak{D}$.[8] From this $v(x)$ we construct a $u(x)$ as a solution of the system

$$v_{x_1} - u_{x_2} = 0, \quad v_{x_2} + u_{x_1} = 0. \tag{4.6}$$

Then $\Delta_2 u = 0$. If $\dot{\mathfrak{D}}$ is described by[9] $x_1 = x_1(\sigma)$ and $x_2 = x_2(\sigma)$, then $\nu = (\dot{x}_2, -\dot{x}_1)$ where $\cdot \equiv d/d\sigma$. Then on $\dot{\mathfrak{D}}$ we have

$$-\psi(\sigma) = \dot{\varphi}(\sigma) = v_{x_1}\dot{x}_1 + v_{x_2}\dot{x}_2 = u_{x_2}\dot{x}_1 - u_{x_1}\dot{x}_2 = -u_\nu. \tag{4.7}$$

It should be pointed out that, together with u, $u + \text{const}$ is also a solution of (4.3).

Next we construct the *Green's function of the second kind* $\bar{g}(x,y)$. The simplest attempt to translate the considerations from Section I-3.4 will fail, since we would have to require

$$\bar{g}(x,y) = s(x,y) + \bar{\Phi}(x,y) \quad \text{where} \quad \Delta_2\bar{\Phi} = 0 \quad \text{in} \quad \mathfrak{D}, \qquad \bar{g}_\nu = 0 \quad \text{on} \quad \dot{\mathfrak{D}}, \tag{4.8}$$

---

[7] A more common, and perhaps more correct notation is $\int_0^{2\pi} \psi(\sigma) \, d\sigma = 0$. The notation in the theorem, which we shall retain, is suggested if we specialize the formulas from Section I-1.2 to R_2.

[8] We need this result only for the circle in R_2. Therefore it is more convenient to prove it by function-theoretic means, so we omit the proof here.

[9] For our modest purposes we even have $x_1 = \cos \sigma$, $x_2 = \sin \sigma$.

which is understood with respect to $y \in \dot{\mathfrak{D}}$ with $x \in \mathfrak{D}$ fixed. For $\tilde{\Phi}(x,y)$ we would have this second boundary-value problem:

$$\Delta_2 \tilde{\Phi} = 0 \quad \text{in} \quad \mathfrak{D}, \qquad \tilde{\Phi}_\nu(x,y) = -s_\nu(x,y) \quad \text{on} \quad \dot{\mathfrak{D}}. \tag{4.9}$$

If in the theorem from Section I-3.3 we put in particular $u \equiv 1, f \equiv 0, \Phi \equiv 0, a = x$, then

$$1 = - \int_{\dot{\mathfrak{D}}} s_\nu(x,y) \, d\sigma.$$

By (4.9) this implies

$$\int_{\dot{\mathfrak{D}}} \tilde{\Phi}_\nu(x,y) \, d\sigma = 1.$$

But (4.9) is not solvable by Theorem 1. Hence, either the requirement $\Delta_2 \tilde{\Phi} = 0$ or $\tilde{g}_\nu = 0$ must be relinquished; we give up the first.

THEOREM 2.   Let $k(x) \in C^1$ in $\mathfrak{D}$ and $k(x) > 0$ be an arbitrary norming function. Then for fixed $x \in \mathfrak{D}$ the Green's function of the second kind $\tilde{g}(x,y)$ exists with the properties

(a) $\tilde{g}(x,y) = s(x,y) + \tilde{\Phi}(x,y)$ where $\tilde{\Phi} \in C^1$ in $\mathfrak{D}$, $\in C^2$ in $\mathfrak{D}$,

(b) $\Delta_2 \tilde{\Phi} = \dfrac{k(y)}{\displaystyle\int_{\mathfrak{D}} k(x) \, dx}$,[10]

(c) $\tilde{g}_\nu(x,y) = 0$ for $y \in \dot{\mathfrak{D}}$,

(d) $\displaystyle\int_{\mathfrak{D}} \tilde{g}(x,y) k(y) \, dy = 0$;

$$\tag{4.10}$$

it is determined uniquely and satisfies the symmetry condition $\tilde{g}(x,y) = \tilde{g}(y,x)$.

Proof. We put $\tilde{\Phi} = \tilde{\Phi}_1 + \tilde{\Phi}_2$, where

$$\Delta_2 \tilde{\Phi}_1 = \frac{k(y)}{\displaystyle\int_{\mathfrak{D}} k(x) \, dx}.$$

By Section IV-3.2, such a $\tilde{\Phi}_1$ is given by

$$\tilde{\Phi}_1(y) = - \frac{1}{\displaystyle\int_{\mathfrak{D}} k(x) \, dx} \int_{\mathfrak{D}} s(y,x) k(x) \, dx. \tag{4.11}$$

---

[10] It should be pointed out expressly that the values of $\Delta_2 \tilde{\Phi}$ must not be chosen completely arbitrarily. The reason for these difficulties with the Green's function of the second kind is that the homogeneous problem $\Delta_2 u = 0, u_\nu = 0$ on $\dot{\mathfrak{D}}$ has the solution $u \equiv \gamma$ where γ is a constant different from zero. If we require $\Delta_2 \tilde{\Phi} = f(y), \tilde{g}_\nu = 0$ on $\dot{\mathfrak{D}}$, then for $\tilde{\Phi}$ this gives the requirements $\Delta_2 \tilde{\Phi} = f(y), \tilde{\Phi}_\nu = -s_\nu$ on $\dot{\mathfrak{D}}$. If we apply the second Green's formula from Section I-1.4 to the functions $u = \gamma, v = \tilde{\Phi}$, we obtain

$$-\gamma \int_{\dot{\mathfrak{D}}} s_\nu \, d\sigma = \gamma \int_{\mathfrak{D}} f \, dy.$$

Since $- \int_{\dot{\mathfrak{D}}} s_\nu \, d\sigma = 1$ we have to choose f so that $\int_{\mathfrak{D}} f \, dy = 1$. Our choice

$$f(y) = \frac{k(y)}{\displaystyle\int_{\mathfrak{D}} k(x) \, dx}$$

takes this circumstance into account.

Therefore we require $\Delta_2\tilde{\Phi}_2 = 0$. On $\dot{\mathfrak{D}}$ we must have $\tilde{\Phi}_{2\nu} = -s_\nu - \tilde{\Phi}_{1\nu}$. By Theorem 1 we need only prove

$$\int_{\dot{\mathfrak{D}}} (-s_\nu - \tilde{\Phi}_{1\nu})\, d\sigma = 0.$$

Now

$$\int_{\dot{\mathfrak{D}}} -s_\nu(x,y)\, d\sigma = 1$$

and

$$\int_{\dot{\mathfrak{D}}} -\tilde{\Phi}_{1\nu}\, d\sigma = \frac{1}{\int_{\mathfrak{D}} k(x)\, dx} \int_{\mathfrak{D}} k(x)\left(\int_{\dot{\mathfrak{D}}} s_\nu(y,x)\, d\sigma\right) dx = -1. \tag{4.12}$$

The change in the order of integration can be justified easily. Thus the existence of $\tilde{\Phi}_2$ is proved. Together with $\tilde{g}$, $\tilde{g} + \text{const}$ also satisfies the first three conditions in (4.10). The fourth requirement determines the constant uniquely. The symmetry is shown as in Problem 2 from Section I-3.4.

THEOREM 3. The problem (4.3) with

$$\int_{\dot{\mathfrak{D}}} \psi(\sigma)\, d\sigma = 0$$

is solved by

$$u(x) = \int_{\dot{\mathfrak{D}}} \tilde{g}(x,y)\psi(\sigma)\, d\sigma + \frac{1}{\int_{\mathfrak{D}} k(x)\, dx} \int_{\mathfrak{D}} u(x)k(x)\, dx. \tag{4.13}$$

Proof. The existence of a solution $u(x) \varepsilon C^1$ in $\bar{\mathfrak{D}}$, εC^2 in $\mathfrak{D}$ is secured in Theorem 1. If in formula (Section I-3.9) we set $a = x$, $\gamma(x,y) = \tilde{g}(x,y)$, and take the limit indicated, then (4.13) follows. The last term in (4.13) is a constant. If we look for solutions of (4.3) which satisfy the additional normalizing condition

$$\int_{\mathfrak{D}} u(x)k(x)\, dx = 0,$$

the constant is zero and the solution is given by

$$u(x) = \int_{\dot{\mathfrak{D}}} \tilde{g}(x,y)\psi(\sigma)\, d\sigma \equiv \int_0^{2\pi} \tilde{g}(x; \cos\sigma, \sin\sigma)\psi(\sigma)\, d\sigma. \tag{4.14}$$

4.3 Hilbert's Lemma

If $U^1(y)$, $U^2(y)$ and $V^1(y)$, $V^2(y) \varepsilon C^1$ in $\bar{\mathfrak{D}}$ are arbitrary functions, then we have the identity

$$\int_{\mathfrak{D}} \{V^1(U^1_{y_1} - U^2_{y_2}) - V^2(U^1_{y_2} + U^2_{y_1}) + U^1(V^1_{y_1} - V^2_{y_2}) - U^2(V^1_{y_2} + V^2_{y_1})\}\, dy$$

$$= \int_{\dot{\mathfrak{D}}} \{U^1(V^1\dot{y}_2 + V^2\dot{y}_1) + U^2(V^1\dot{y}_1 - V^2\dot{y}_2)\}\, d\sigma \tag{4.15}$$

by the Gauss theorem in R_2. The boundary $\dot{\mathfrak{D}}$ in (4.15) is described by $y_1 = y_1(\sigma) = \cos\sigma$, $y_2 = y_2(\sigma) = \sin\sigma$. We set

$$V^1(y) = -g_{\nu_1}(x,y), \quad V^2(y) = g_{\nu_2}(x,y) \quad \text{for} \quad \text{fixed } x \in \mathfrak{D}, \tag{4.16}$$

where $g(x,y)$ is the *Green's function of the first kind* which was constructed explicitly in Section I-3.5 for circles (and for spheres in R_n). If we introduce these quantities in (4.15), then the circle $|y - x| \leqslant \rho$ with center x has to be excluded. If, as in Section I-3.2, we take the limit as $\rho \to 0$, then using the abbreviations in (4.1) we have

$$U^1(x) = - \int_{\dot{\mathfrak{D}}} g_\nu(x,y)U^1(y)\, d\sigma + \int_{\mathfrak{D}} \{g_{\nu_1}(x,y)L^1(y) + g_{\nu_2}(x,y)L^2(y)\}\, dy \tag{4.17}$$

where $\nu = (\dot{y}_2, -\dot{y}_1)$ is the outer normal.

If we set

$$V^1(y) = \tilde{g}_{\nu_2}(x,y), \qquad V^2(y) = \tilde{g}_{\nu_1}(x,y) \quad \text{for} \quad \text{fixed } x \in \mathfrak{D}, \tag{4.18}$$

then, by analogous considerations and by taking (4.10) into account, we obtain

$$U^2(x) = \int_{\dot{\mathfrak{D}}} \tilde{g}_\sigma(x,y)U^1(y)\, d\sigma + \int_{\mathfrak{D}} \left\{ \tilde{g}_{\nu_1}(x,y)L^2(y) - \tilde{g}_{\nu_2}(x,y)L^1(y) + \frac{k(y)U^2(y)}{\displaystyle\int_{\mathfrak{D}} k(x)\, dx} \right\} dy, \tag{4.19}$$

where $\tilde{g}_\sigma = \tilde{g}_\sigma(x; y_1(\sigma), y_2(\sigma))$ is the derivative with respect to the arc length σ.

LEMMA.[11]   For the functions $P_1(x)$, $P_2(x) \in C^0$ in $\mathfrak{D}$ suppose

$$\begin{aligned} \int_{\mathfrak{D}} \{g_{\nu_1}(x,y)P_1(y) + g_{\nu_2}(x,y)P_2(y)\}\, dy &= 0, \\ \int_{\mathfrak{D}} \{\tilde{g}_{\nu_1}(x,y)P_2(y) - \tilde{g}_{\nu_2}(x,y)P_1(y)\}\, dy &= 0, \end{aligned} \tag{4.20}$$

are satisfied for all $x \in \mathfrak{D}$. Then $P_1(x) \equiv P_2(x) \equiv 0$ in $\mathfrak{D}$.

Proof. It will be given for $P_1(x)$, $P_2(x) \in C^2$ in $\mathfrak{D}$. We follow D. Hilbert. Let $u^1(x)$ be an arbitrary solution of

$$\Delta_2 u^1 = P_{1x_1}(x) + P_{2x_2}(x). \tag{4.21}$$

Then by integration we can immediately determine a solution $u^2(x)$ for which

$$u^1_{x_1} - u^2_{x_2} = P_1(x), \qquad u^1_{x_2} + u^2_{x_1} = P_2(x), \tag{4.22}$$

because the integrability condition is satisfied. We have $u^1(x)$, $u^2(x) \in C^1$ in $\mathfrak{D}$, $\in C^2$ in $\mathfrak{D}$. Thus for these u^1, u^2 the identities (4.17) and (4.19) hold. According to (4.22) we have to put

$$L^i(y) = P_i(y), \qquad i = 1, 2.$$

By hypothesis (4.20) and after integration by parts they have the form

$$u^1(x) = - \int_{\dot{\mathfrak{D}}} g_\nu(x,y)u^1(y)\, d\sigma, \tag{4.23}$$

$$u^2(x) = - \int_{\dot{\mathfrak{D}}} \tilde{g}(x,y)u^1_\sigma\, d\sigma + \frac{1}{\displaystyle\int_{\mathfrak{D}} k(x)\, dx} \int_{\mathfrak{D}} k(x)u^2(x)\, dx. \tag{4.24}$$

[11] In this form in W. A. Hurwitz[3]; D. Hilbert[4] assumed $P_1(x)$, $P_2(x) \in C^2$ in $\bar{\mathfrak{D}}$.

From formula (Section I-3.13) and from Poisson's formula in Section I-3.5 it follows that $\Delta_2 u^1 = 0$. From (4.13) and (4.24),

$$\Delta_2 u^2 = 0, \qquad u_\nu^2 = -u_\sigma^1 \quad \text{on} \quad \dot{\mathfrak{D}}. \tag{4.25}$$

follows. But every harmonic function[12]

$$v(x) = \int_{x^0}^{x} - u_{x_2}^1 \, dx_1 + u_{x_1}^1 \, dx_2 + \text{const}$$

which is conjugate to $u^1(x)$ is a solution of $\Delta_2 v = 0$ in $\mathfrak{D}$, where $v_\nu = -u_\sigma^1$ on $\dot{\mathfrak{D}}$. Hence, by the uniqueness theorem in Section 4.2, from (4.25) it follows that u^2 is a conjugate harmonic function to u^1 and thus u^1 and u^2 satisfy the Cauchy-Riemann differential equations. On the other hand, relations (4.22) hold, which implies $P_1(x) \equiv P_2(x) \equiv 0$ in $\dot{\mathfrak{D}}$.

4.4 Equivalent Formulations of the Problem

We begin with the obvious remark that for (4.15) it suffices to assume U^j, $V^j \in C^1$ in $\mathfrak{D}$, εC^0 in $\dot{\mathfrak{D}}$, and L^1, $L^2 \in C^0$ in $\dot{\mathfrak{D}}$. Here L^1 and L^2 in (4.1) have to be imagined written once in the U^j and the other time in the V^j. We consider the boundary-value problem

$$D^i \equiv L^i \mid \sum_{j=1}^{2} A_j^i U^j + C^i = 0 \quad \text{in} \quad \mathfrak{D}, \qquad U^1 = \Phi \quad \text{on} \quad \dot{\mathfrak{D}}, \qquad \Phi = \Phi(\sigma) \in C^1, \tag{4.26}$$

as well as the problem

$$U^i(x) = \int_{\mathfrak{D}} \sum_{j=1}^{2} K_j^i(x,y) U^j(y) \, dy + K^i(x), \qquad i = 1, 2, \tag{4.27}$$

with the abbreviations

$$\begin{aligned}
K_1^1(x,y) &= -g_{\nu_1}(x,y) A_1^1(y) - g_{\nu_2}(x,y) A_1^2(y), \\
K_2^1(x,y) &= -g_{\nu_1}(x,v) A_2^1(y) - g_{\nu_2}(x,y) A_2^2(y), \\
K_1^2(x,y) &= -\tilde{g}_{\nu_1}(x,y) A_1^2(y) + \tilde{g}_{\nu_2}(x,y) A_1^1(y), \\
K_2^2(x,y) &= -\tilde{g}_{\nu_1}(x,y) A_2^2(y) + \tilde{g}_{\nu_2}(x,y) A_2^1(y) + \dfrac{k(y)}{\displaystyle\int_{\mathfrak{D}} k(x)\,dx},
\end{aligned} \tag{4.28}$$

$$K^1(x) = -\int_{\mathfrak{D}} \{g_{\nu_1}(x,y) C^1(y) + g_{\nu_2}(x,y) C^2(y)\} \, dy - \int_{\dot{\mathfrak{D}}} g_\nu(x,y) \Phi(\sigma) \, d\sigma,$$

$$K^2(x) = -\int_{\mathfrak{D}} \{\tilde{g}_{\nu_1}(x,y) C^2(y) - \tilde{g}_{\nu_2}(x,y) C^1(y)\} \, dy + \int_{\dot{\mathfrak{D}}} \tilde{g}_\sigma(x,y) \Phi(\sigma) \, d\sigma.$$

THEOREM 1. The problems (4.26) and (4.27) are equivalent with regard to solutions $U^i(x) \in C^0$ in $\dot{\mathfrak{D}}$, εC^1 in $\mathfrak{D}$ where L^1, $L^2 \in C^0$ in $\dot{\mathfrak{D}}$.

---

[12] The function $v(x)$ is found from the Cauchy-Riemann differential equations $u_{x_1}^1 - v_{x_2} = 0$, $u_{x_2}^1 + v_{x_1} = 0$, in the form

$$dv = v_{x_1} \, dx_1 + v_{x_2} \, dx_2 = -u_{x_2}^1 \, dx_1 + u_{x_1}^1 \, dx_2 \text{ hence } v(x) = \int - u_{x_2}^1 \, dx_1 + u_{x_1}^1 \, dx_2 + \text{const.}$$

Here the integral is to be understood as a line integral.

Proof. (1) If we have such solutions of (4.26), then L^1, L^2 in (4.17) and (4.19) may be replaced by $- \sum\limits_{j=1}^{2} A_j^i U^j - C^i$, and (4.27) arises.

(2) If we have such solutions of (4.27), we substitute them in (4.17) and (4.19), and subtract the first equation in (4.27) from (4.17), and the second equation in (4.27) from (4.19). We obtain relations (4.20) with $P_i \equiv D^i$ so that the lemma gives $D^i \equiv 0$. It merely remains to show that $U^1 = \Phi$ on $\dot{\mathfrak{D}}$. By the considerations for (4.23), the last term in $K^1(x)$ has the boundary values Φ. All other terms in the first equation in (4.27) have the zero boundary values. Because of the symmetry of $g(x,y)$ in x and y, g will vanish if x approaches a point of $\dot{\mathfrak{D}}$. The same also holds for $g_{v_i}(x,y)$ if $x \in \dot{\mathfrak{D}}$, and hence also for K_1^1, K_2^1, and the first terms of K^1. But the argument is not complete, because the limit in (4.27) has to be taken under the integral sign. We have already carried out such justifications repeatedly.

THEOREM 2.   The normalized boundary-value problem (4.26) with the additional condition $\int_{\mathfrak{D}} U^2(x)k(x)\,dx = 0$ is equivalent in the sense of Theorem 1 to

$$U^i(x) = \int_{\mathfrak{D}} \sum_{j=1}^{2} \tilde{K}_j^i(x,y) U^j(y)\,dy + K^i(x), \tag{4.29}$$

where we put

$$\tilde{K}_1^1 = K_1^1, \qquad \tilde{K}_2^1 = K_2^1, \qquad \tilde{K}_1^2 = K_1^2, \qquad \tilde{K}_2^2 = -\tilde{g}_{v_1}(x,y)A_2^2(y) + \tilde{g}_{v_2}(x,y)A_2^1(y).$$

Proof. (1) Because of the additional normalizing condition in (4.26), the last term in K_2^2 drops out. (2) Every solution of (4.29) has the property $\int_{\mathfrak{D}} U^2(x)k(x)\,dx = 0$. For this purpose we consider the first term in the second equation of (4.29). We find

$$\begin{aligned}
\int_{\mathfrak{D}} U^2(x)k(x)\,dx &= - \int_{\mathfrak{D}} k(x)\left(\int_{\mathfrak{D}} \tilde{g}_{v_1}(x,y)A_1^2(y)U^1(y)\,dy \right) dx + \cdots \\
&= - \int_{\mathfrak{D}} A_1^2(y)U^1(y)\left(\int_{\mathfrak{D}} \tilde{g}_{v_1}(x,y)k(x)\,dx \right) dy + \cdots \\
&= - \int_{\mathfrak{D}} A_1^2(y)U^1(y)\left(\frac{\partial}{\partial y_1} \int_{\mathfrak{D}} \tilde{g}(y,x)k(x)\,dx \right) dy + \cdots\cdots
\end{aligned} \tag{4.30}$$

This term vanishes because of property (*d*) from (4.10). The changes in the order of integration, and so forth, can be justified easily. This procedure is repeated for the other terms so that the desired additional property for $U^2(x)$ follows.

THEOREM 3.   The normalized boundary-value problem (4.26) with $\int_{\mathfrak{D}} U^2(x)k(x)\,dx = \gamma$, where γ is given arbitrarily is equivalent in the sense of Theorem 1 to

$$U^i(x) = \int_{\mathfrak{D}} \sum_{j=1}^{2} \tilde{K}_j^i(x,y) U^j(y)\,dy + \tilde{K}^i(x). \tag{4.31}$$

Here we put

$$\tilde{K}^1(x) = - \int_{\mathfrak{D}} \{g_{\nu_1}(x,y)\tilde{C}^1(y) + g_{\nu_2}(x,y)\tilde{C}^2(y)\} \, dy - \int_{\dot{\mathfrak{D}}} g_\nu(x,y)\Phi(\sigma) \, d\sigma,$$

$$\tilde{K}^2(x) = - \int_{\mathfrak{D}} \{\tilde{g}_{\nu_1}\tilde{C}^2 - \tilde{g}_{\nu_2}\tilde{C}^1\} \, dy + \int_{\dot{\mathfrak{D}}} \tilde{g}_\sigma(x,y)\Phi(\sigma) \, d\sigma + \frac{\gamma}{\int_{\mathfrak{D}} k(x) \, dx},$$

and

$$\tilde{C}^1(x) = C^1(x) + \frac{\gamma A_2^1(x)}{\int_{\mathfrak{D}} k(x) \, dx}, \qquad \tilde{C}^2(x) = C^2(x) + \frac{\gamma A_2^2(x)}{\int_{\mathfrak{D}} k(x) \, dx}.$$

Proof. If we put $\tilde{U}^1(x) = U^1(x)$ and $\tilde{U}^2(x) = U^2(x) - \dfrac{\gamma}{\int_{\mathfrak{D}} k(x) \, dx}$, then for $\tilde{U}^i(x)$ the

normalized boundary-value problem [written in $\tilde{U}^i(x)$]

$$\tilde{D}^i \equiv \tilde{L}^i + \sum_{j=1}^{2} A_j^i \tilde{U}^j + \tilde{C}^i = 0, \qquad \tilde{U}^1 = \Phi \quad \text{on} \quad \dot{\mathfrak{D}}, \qquad \int_{\mathfrak{D}} \tilde{U}^2(x)k(x) \, dx = 0 \qquad (4.32)$$

follows, for which Theorem 2 holds. If we invert the transformation the conclusion follows.

Relations (4.27), (4.29), and (4.31) can now be rewritten easily as an operator equation in Hilbert space. For this purpose we first consider the first term in (4.27) as an example.

$$U^1(x) - \int_{\mathfrak{D}} K_1^1(x,y) U^1(y) \, dy \equiv AU^1 \qquad (4.33)$$

Here we consider A as operator in the Hilbert space $\mathfrak{IC}^1$: the set of all functions $U^1(x)$ for which $\int_{\mathfrak{D}} (U^1(x))^2 \, dx < \infty$, with the scalar product $(U^1,V^1) = \int_{\mathfrak{D}} U^1(x)V^1(x) \, dx$.

THEOREM 4.   The operator A is completely continuous in $\mathfrak{IC}^1$.

Proof. We proceed as in Section 3.3 and use the smoothing function (3.27). Accordingly we put—in order to avoid confusion we replace the parameter σ in (3.27) by α—

$$K_{1\alpha}^1(x,y) = K_1^1(x,y)\omega_\alpha(|x - y|). \qquad (4.34)$$

Because of Lemma 2 from Section 3.3 the operators A_ϵ for $1/\alpha = \epsilon$

$$A_\epsilon U^1 \equiv \int_{\mathfrak{D}} K_{1\alpha}^1(x,y) U^1(y) \, dy \qquad (4.35)$$

are completely continuous. The operator A is approximated arbitrarily well by these operators. After the considerations in Section 3.3 we need only prove the formula corresponding to (3.32):

$$\int_{|y - x| \leqslant 1/\alpha} |K_1^1(x,y)| \, dy \leqslant \text{const } \epsilon \quad \text{where} \quad \frac{1}{\alpha} = \epsilon. \qquad (4.36)$$

 Elliptic systems of differential equations

But this is true, for by Problem 2 from Section I-3.6, g (and also $\bar{g}$), and with it $K_j^i(x,y)$, satisfies an estimate $|K_j^i(x,y)| \leqslant \text{const}/|x - y|$. If we apply Section IV-3.1 to this, then (4.36) follows immediately. Lemma 2 from Section 1.2 then gives the complete continuity of A, and this statement remains valid if in (4.33), instead of K_1^1, we use the kernel K_j^i.

If we have a solution $U^1(x) \; \varepsilon \; \mathcal{K}^1$ of (4.33), then in a way similar to that in Section 3.2, by iteration as in Section IV-3.7, it follows that $U^1(x)$ must be bounded. If we use this in the right-hand side of (4.33), then $U^1(x)$ on the left-hand side is clearly continuous, even Hölder continuous, and finally $U^1(x) \; \varepsilon \; C^0$ in $\bar{\mathfrak{D}}$, $\varepsilon \; C^1$ in $\mathfrak{D}$, $L^i \; \varepsilon \; C^0$ in $\bar{\mathfrak{D}}$. For this we must use the theorems from Section IV-4.1.

If we use vector and matrix notation, we obtain the desired operator equation for relations (4.27), (4.29), and (4.31). We put

$$u(x) = (U^1(x),U^2(x)), \qquad k(x) = (K^1(x),K^2(x)), \qquad \tilde{k}(x) = (\tilde{K}^1(x),\tilde{K}^2(x)), \quad (4.37)$$

$$\mathcal{K}(x,y) = \begin{pmatrix} K_1^1(x,y) & K_2^1(x,y) \\ K_1^2(x,y) & K_2^2(x,y) \end{pmatrix}, \qquad \tilde{\mathcal{K}}(x,y) = \begin{pmatrix} \tilde{K}_1^1(x,y) & \tilde{K}_2^1(x,y) \\ \tilde{K}_1^2(x,y) & \tilde{K}_2^2(x,y) \end{pmatrix}, \qquad (4.38)$$

and, for (4.31) for example, we have the operator equation

$$u(x) = \int_{\mathfrak{D}} \tilde{\mathcal{K}}(x,y)u(y) \, dy + \tilde{k}(x) \equiv \tilde{a}u + \tilde{k}, \qquad (4.39)$$

which will be considered in the Hilbert space $\mathcal{K}$:

$$\int_{\mathfrak{D}} (u,u) \, dx \equiv \int_{\mathfrak{D}} \{(U^1(x))^2 + (U^2(x))^2\}dx < \infty$$

with the scalar product $<u,v> = \int_{\mathfrak{D}} (u,v) \, dx$. Then $\tilde{a}$ is completely continuous in $\mathcal{K}$ and it possesses the above properties, as can be shown by iteration.

4.5   The Homogeneous First Boundary-Value Problem[13]

We consider (4.26) with $C^i \equiv 0$ and $\Phi = 0$ on $\dot{\mathfrak{D}}$. This problem always has the solution $U^1 \equiv U^2 \equiv 0$. We show that there are always solutions for which $(U^1)^2 + (U^2)^2 \not\equiv 0$. But first some examples. We divide (4.26) into two classes for which

$$(1) \;\; A_{2x_1}^1 + A_{2x_2}^2 \equiv 0 \;\; \text{ in } \;\; \mathfrak{D}, \qquad (2) \;\; A_{2x_1}^1 + A_{2x_2}^2 \not\equiv 0 \;\; \text{ in } \;\; \mathfrak{D}.$$

Example 1. *First Class.* We put $U^1 \equiv 0$. For U^2 we have

$$-U_{x_2}^2 + A_2^1 U^2 = 0, \qquad U_{x_1}^2 + A_2^2 U^2 = 0, \qquad (4.40)$$

and the integrability is satisfied for the system. Hence it has solutions $U^2(x) \not\equiv 0$ and they can be given explicitly. For a fixed point $x^0 \; \varepsilon \; \mathfrak{D}$ and arbitrary $x \; \varepsilon \; \mathfrak{D}$ we have

$$U^2(x) = \text{const} \cdot \exp \left\{ - \int_{x^0}^{x} [A_2^2(y) \, dy_1 - A_2^1(y) \, dy_2] \right\}, \qquad (4.41)$$

where the integral is to be understood as a line integral.

[13] We follow G. Hellwig[5]. For other methods see I. N. Vekua[6]. The presentation of I. N. Vekua is more general since it does not necessarily assume $\mathfrak{D}$ to be simply connected.

Example 2.   *Second Class.*   The boundary-value problem

$$L^1 + 2x_1 U^2 = 0, \qquad L^2 + 2x_2 U^2 = 0 \quad \text{where} \quad U^1 = 0 \quad \text{on} \quad |x| = 1 \qquad (4.42)$$

has the solutions $U^1 = \text{const} \, ((x_1)^2 + (x_2)^2 - 1)$, $U^2 = - \text{const}$, the constants being the same.

THEOREM 1.   The boundary-value problem (4.26) with $C^i \equiv 0$, $\Phi = 0$ on $\dot{\mathfrak{D}}$ has a system of solutions[14]

$$U^1_H(x), \qquad U^2_H(x) \quad \text{for which} \quad (U^1_H)^2 + (U^2_H)^2 \neq 0 \quad \text{in} \quad \mathfrak{D},$$

and hence the infinitely many solutions $U^1(x) = cU^1_H(x)$, $U^2(x) = cU^2_H(x)$, where c is an arbitrary constant.

Proof.   Let the conclusion of the theorem be false.   Then there is at least one system

$$L^i + \sum_{j=1}^{2} A^i_j U^j = 0, \qquad U^1 = 0 \quad \text{on} \quad \dot{\mathfrak{D}}, \qquad (4.43)$$

which only admits the solution $U^1 \equiv U^2 \equiv 0$.   We must have $(A^1_2)^2 + (A^2_2)^2 \neq 0$ in $\mathfrak{D}$, because otherwise $U^1 \equiv 0$, $U^2 \equiv 1$ would be a solution.

First Step.   By assumption, (4.43) only has the solution $U^1 \equiv U^2 \equiv 0$.   Hence all solutions of (4.43) satisfy the normalizing condition $\int_{\mathfrak{D}} U^2(x)k(x) \, dx = 0$.   By Theorem 2 and the further considerations from Section 4.4, this requirement implies that (4.43) is equivalent to

$$u(x) = \int_{\mathfrak{D}} \tilde{\mathfrak{K}}(x,y)u(y) \, dy \equiv \tilde{\mathfrak{a}}u, \qquad (4.44)$$

which only admits the solution $u \equiv 0$.

Second Step.   Let $\gamma \neq 0$.   We consider (4.43) with the normalizing condition

$$\int_{\mathfrak{D}} U^2(x)k(x) \, dx = \gamma \quad \text{with} \quad \gamma \neq 0.$$

By assumption this problem does not have a solution.   By Theorem 3 from Section 4.4 it is equivalent to

$$u(x) = \tilde{\mathfrak{a}}u + \tilde{k}, \qquad \tilde{k} = (\tilde{K}^1(x), \tilde{K}^2(x)) \quad \text{and} \quad \tilde{k} \neq 0.[15] \qquad (4.45)$$

We consider this operator equation in the space $\mathfrak{K}$ of Section 4.4 and apply Lemma 3 from Section 1.2, which implies the solvability of (4.45), and this is a contradiction.

---

[14] We even have $(U^1_H)^2 + (U^2_H)^2 > 0$ in $\mathfrak{D}$.   But the proof is more difficult.
[15] For if $\tilde{k} = 0$, then (4.45) would be solvable with $u = 0$, and this is impossible.

THEOREM 2.   The boundary-value problem (4.26) with $C^i \equiv 0$, $\Phi = 0$ on $\dot{\mathfrak{D}}$ has precisely infinitely many solutions from Theorem 1, namely $U^1(x) = cU_H^1(x)$, $U^2(x) = cU_H^2(x)$ where c is an arbitrary constant, and $(U_H^1)^2 + (U_H^2)^2 \neq 0$ in $\mathfrak{D}$.

We now have several possibilities for a proof.   It follows conveniently from the similarity principle of L. Bers[7], but the theorem of T. Carleman[8] would also suffice, or we could use the arguments of I. N. Vekua.   In our presentation, however, Theorem 2 does not contain any statement about existence because of Theorem 1; we merely have to obtain a contradiction from the assumption that there exists a family of solutions with two arbitrary constants.   Therefore a simple proof should be possible.   But such a proof does not seem to be available to us now.   In the following we make use of Theorem 2.

4.6   The Inhomogeneous First Boundary-Value Problem

THEOREM 1.   The boundary-value problem (4.26) is always solvable.   If U_I^1, U_I^2 is any solution εC^0 in $\mathfrak{D}$, εC^1 in $\mathfrak{D}$, $L^i \varepsilon C^0$ in $\mathfrak{D}$, then

$$U^1(x) = cU_H^1(x) + U_I^1(x), \qquad U^2(x) = cU_H^2(x) + U_I^2(x)$$

represent all solutions.

The proof is given by means of Theorem 2.

THEOREM 2.   There always exists a suitable normalizing function $k(x) \varepsilon C^1$ and > 0 in $\mathfrak{D}$ such that the homogeneous normalized boundary-value problem (4.26) with $C^i \equiv 0$, $\Phi \equiv 0$ and $\int_{\mathfrak{D}} U^2(x)k(x)\,dx = 0$ has the unique solution $U^1 \equiv U^2 \equiv 0$.

Proof of Theorem 1.   By Section 4.4, the normalized boundary-value problem (4.26) with $\int_{\mathfrak{D}} U^2(x)k(x)\,dx = 0$ is equivalent to

$$u = \int_{\mathfrak{D}} \tilde{\mathcal{K}}(x,y)u(y)\,dy + k(x) \equiv \tilde{a}u + k. \tag{4.46}$$

Here u must be independent of the choice of the normalizing function $k(x)$.   This function only entered through the Green's function $\bar{g}(x,y)$; our original problem (4.26) is independent of $k(x)$.   According to Theorem 2 we choose k so that the homogeneous problem (4.26) with $C^i \equiv 0$, $\Phi = 0$ on $\dot{\mathfrak{D}}$ and $\int_{\mathfrak{D}} U^2(x)k(x)\,dx = 0$ only has the solution $U^1 \equiv U^2 \equiv 0$.   This means that $u = \tilde{a}u$ only admits the solution $u = 0$.   Lemma 3 from Section 1.2 gives the solvability of (4.46).   Thus a particular solution of (4.26) with the additional property $\int_{\mathfrak{D}} U^2(x)k(x)\,dx = 0$ has been found.   If according to Theorem 2 from Section 4.5 we add all solutions of the homogeneous boundary-value problem, the conclusion follows.

Proof of Theorem 2.   It will be given by means of Lemmas 1 and 2.

LEMMA 1.   There is no boundary-value problem (4.26) with $C^i \equiv 0$, $\Phi = 0$ on $\dot{\mathfrak{D}}$ for which $U_H^2(x) \equiv 0$.

Proof.   Suppose there is such a problem.   Then, by Theorem 2 in Section 4.5,

$$U_{x_1}^1 + A_1^1 U^1 = 0, \qquad U_{x_2}^1 + A_1^2 U^1 = 0, \qquad U^1 = 0 \quad \text{on} \quad \dot{\mathfrak{D}} \tag{4.47}$$

would have to hold.   Now let $|x_2^0| < 1$ so that the line $x_2 = x_2^0$ intersects $\dot{\mathfrak{D}}$ in exactly two points (we have agreed that $\dot{\mathfrak{D}}$ should be the unit circle).   In the points of this line, as far as they belong to $\mathfrak{D}$, we consider the first equation in (4.47); it then becomes an ordinary differential equation which has the solution

$$U^1(x_1, x_2^0) = c \exp \left\{ - \int^{x_1} A_1^1(\xi, x_2^0) \, d\xi \right\}.$$

Since $U^1 = 0$ on $\dot{\mathfrak{D}}$, $U^1(x_1, x_2^0) \equiv 0$, and thus $U^1 \equiv 0$ in $\mathfrak{D}$ if these considerations are repeated for every such line in $\mathfrak{D}$.   Thus a system has been found which only has the solution $U_H^i \equiv 0$, and this is a contradiction to Theorem 1 from Section 4.5.   So the lemma is proved.

Hence the case $U_H^2 \equiv 0$ cannot occur.   Let $U_H^2 \gtrless 0$ in $\mathfrak{D}$.   Then

$$0 = \int_{\mathfrak{D}} U^2(x)k(x) \, dx = c \int_{\mathfrak{D}} U_H^2(x)k(x) \, dx \tag{4.48}$$

implies $c = 0$ and Theorem 2 is proved for every such $k(x)$.   If $U_H^2 \leqslant 0$ in $\mathfrak{D}$, we multiply (4.48) by -1.   There remains the case where U_H^2 can assume positive as well as negative values.   Then there is an $x^0 \varepsilon \mathfrak{D}$, where $U_H^2(x^0) = 2M > 0$.   Furthermore, there is a circle Γ in $\mathfrak{D}$: $|x - x^0| \leqslant \delta$ such that $U_H^2(x) > M$ in Γ.   Next we put $\max |U_H^2(x)| = N$ for $x \varepsilon \mathfrak{D}$.   If by $\mathfrak{D} - \Gamma$ we denote all points which belong to $\mathfrak{D}$ but not to Γ, then by Theorem 2 from Section 4.5 we have

$$c \int_{\Gamma} U_H^2(x)k(x) \, dx = -c \int_{\mathfrak{D}-\bar{\Gamma}} U_H^2(x)k(x) \, dx, \qquad |c|M \int_{\Gamma} k(x) \, dx \leqslant |c|N \int_{\mathfrak{D}-\bar{\Gamma}} k(x) \, dx.$$

$$\tag{4.49}$$

The following lemma shows that a $k(x)$ can be found so that

$$\int_{\Gamma} k(x) \, dx > 1, \qquad \frac{N}{M} \int_{\mathfrak{D}-\bar{\Gamma}} k(x) \, dx < 1. \tag{4.50}$$

Then (4.49) contains a contradiction, except for $|c| = 0$.   Thus Theorem 2 is proved.

LEMMA 2.   About the point $x^0 \varepsilon \mathfrak{D}$, which was fixed, put a circle Γ_1: $|x - x^0| \leqslant \eta$ so that $\Gamma_1 \supset \mathfrak{D}$.   Then

$$k(x) = \gamma \exp \left\{ \frac{1}{|x - x^0|^2 + \beta^2} \right\}, \tag{4.51}$$

where

$$\gamma = \frac{M}{(2\eta)^4 N e^{1/\delta^2}}$$

is a normalizing function for $\mathfrak{D}$. If β^2 in $0 < \beta^2 < \delta^2$ is chosen suitably, then (4.50) is satisfied.

Proof. Writing $|x - x^0| = r$, we have

$$\frac{N}{M}\int_{\mathbf{r}-\Gamma} k\,dx < \frac{N}{M}\int_{\Gamma_1-\Gamma} k\,dx = \gamma\frac{N}{M}\int_0^{2\pi}\int_\delta^\eta e^{1/(r^2+\beta^2)} r\,dr\,d\varphi = \frac{\pi\gamma N}{M}\int_{1/(\eta^2+\beta^2)}^{1/(\delta^2+\beta^2)} \frac{e^z}{z^2}\,dz$$

$$< \frac{\pi\gamma N}{M}(\eta^2+\beta^2)^2\int_{1/(\eta^2+\beta^2)}^{1/(\delta^2+\beta^2)} e^z\,dz < \frac{\pi\gamma N}{M}(2\eta^2)^2 e^{1/(\delta^2+\beta^2)} < \frac{\pi\gamma N}{M}(2\eta^2)^2 e^{1/\delta^2} = \frac{\pi}{4} < 1,$$

$$\tag{4.52}$$

$$\int_{\mathbf{r}} k\,dx = \gamma\int_0^{2\pi}\int_0^\delta e^{1/(r^2+\beta^2)} r\,dr\,d\varphi = \pi\gamma\int_{1/(\delta^2+\beta^2)}^{1/\beta^2} \frac{e^z}{z^2}\,dz$$

$$> \pi\gamma\beta^4\int_{1/(\delta^2+\beta^2)}^{1/\beta^2} e^z\,dz > \pi\gamma\beta^4(e^{1/\beta^2} - e^{1/\delta^2}),\tag{4.53}$$

where γ and δ are independent of β. We have

$$\lim_{\beta\to 0} \beta^4(e^{1/\beta^2} - e^{1/\delta^2}) = \infty.$$

Hence a β in $0 < \beta^2 < \delta^2$ can be found so that the left-hand side in (4.53) becomes larger than 1.

4.7   The General Boundary-Value Problem with Characteristic Zero

We begin with the study of the general problem (4.1) and (4.2) under the hypotheses made there. Since in the boundary condition (4.2) we assumed $(\alpha_1(\sigma))^2 + (\alpha_2(\sigma))^2 > 0$ on $\dot{\mathfrak{D}}$, we may without loss of generality suppose that

$$(\alpha_1(\sigma))^2 + (\alpha_2(\sigma))^2 = 1 \quad \text{on} \quad \dot{\mathfrak{D}}.\tag{4.54}$$

On $\dot{\mathfrak{D}}$, where $0 \leqslant \sigma < 2\pi$, let us consider $(U^1(\sigma), U^2(\sigma))$ and $(\alpha_1(\sigma), \alpha_2(\sigma))$ as two vectors u and α. During one transit of $\dot{\mathfrak{D}}$ in the positive sense ($\mathfrak{D}$ is always to our left) the vector $\alpha(\sigma)$ may carry out several complete revolutions in the positive or negative sense. However, because of the continuity of $\alpha_i(\sigma)$ on $\dot{\mathfrak{D}}$ they always have to be complete revolutions. The *winding number*, provided with the appropriate sign, of the vector field $\alpha(\sigma)$ on $\dot{\mathfrak{D}}$ is called the *characteristic* κ of α. The number κ is always an integer. The tangent of the angle of rotation is given by α_2/α_1. Hence we have

$$\kappa = \frac{1}{2\pi}\int_{\dot{\mathfrak{D}}} \frac{d}{d\sigma} \arctan \frac{\alpha_2(\sigma)}{\alpha_1(\sigma)}\,d\sigma \equiv \frac{1}{2\pi}\int_0^{2\pi} \frac{d}{d\sigma} \arctan \frac{\alpha_2(\sigma)}{\alpha_1(\sigma)}\,d\sigma.\tag{4.55}$$

LEMMA. The characteristic κ is zero if and only if there is a vector field $(\gamma_1(x), \gamma_2(x))$ in $\mathfrak{D}$ such that $\gamma_1 = \alpha_1$, $\gamma_2 = \alpha_2$ on $\dot{\mathfrak{D}}$ and $(\gamma_1)^2 + (\gamma_2)^2 > 0$ in $\mathfrak{D}$. Here we assume $\gamma_i(x) \in C^1$ in $\mathfrak{D}$, $\in C^2$ in $\mathfrak{D}$. Then $\gamma_1(x)$ and $\gamma_2(x)$ are called field functions.

Proof. (1) Suppose there exist field functions $\gamma_1(x)$ and $\gamma_2(x)$. Then, using the outer normal $\nu = (\dot{x}_2, -\dot{x}_1)$, by the Gauss integral theorem we have

$$\int_{\dot{\mathfrak{D}}} \frac{d}{d\sigma} \arctan \frac{\alpha_2}{\alpha_1}\, d\sigma = \int_{\dot{\mathfrak{D}}} \frac{d}{d\sigma} \arctan \frac{\gamma_2}{\gamma_1}\, d\sigma = \int_{\dot{\mathfrak{D}}} \left\{ \left(\arctan \frac{\gamma_2}{\gamma_1} \right)_{x_1} \dot{x}_1 \right.$$

$$\left. + \left(\arctan \frac{\gamma_2}{\gamma_1} \right)_{x_2} \dot{x}_2 \right\} d\sigma = \int_{\dot{\mathfrak{D}}} \left\{ - \left(\arctan \frac{\gamma_2}{\gamma_1} \right)_{x_1} \nu_2 + \left(\arctan \frac{\gamma_2}{\gamma_1} \right)_{x_2} \nu_1 \right\} d\sigma$$

$$= \int_{\mathfrak{D}} \left\{ - \left(\arctan \frac{\gamma_2}{\gamma_1} \right)_{x_1 x_2} + \left(\arctan \frac{\gamma_2}{\gamma_1} \right)_{x_2 x_1} \right\} dx = 0. \quad (4.56)$$

(2) Let $\kappa = 0$. First, by (4.54), we may put $\alpha_1 = \cos \delta(\sigma)$, $\alpha_2 = \sin \delta(\sigma)$. Then from (4.55) we get

$$\frac{1}{2\pi} \int_{\dot{\mathfrak{D}}} \frac{d}{d\sigma} \delta(\sigma)\, d\sigma = 0$$

so that $\delta(\sigma)$ is continuous on $\dot{\mathfrak{D}}$, and by our assumptions on α_1, α_2 we have even $\delta(\sigma) \in C^1$. Now $\delta(\sigma)$ is the angle formed by α and the x_1-axis. We consider the boundary-value problem

$$\Delta_2 \theta = 0 \quad \text{in} \quad \mathfrak{D}, \qquad \theta = \delta(\sigma) \quad \text{on} \quad \dot{\mathfrak{D}}; \quad (4.57)$$

it is well known that this has the unique solution $\theta(x) \in C^1$ in $\bar{\mathfrak{D}}$, $\in C^2$ in $\mathfrak{D}$.[16] If we put

$$\gamma_1(x) = \cos \theta(x), \qquad \gamma_2(x) = \sin \theta(x), \quad (4.58)$$

then field functions with the desired properties will have been constructed.

THEOREM. The boundary-value problem (4.1), (4.2) with characteristic $\kappa = 0$ is always solvable. The homogeneous problem ($C^i \equiv 0$, $\varphi = 0$ on $\dot{\mathfrak{D}}$) has

$$U^1(x) = c U_H^1(x), \qquad U^2(x) = c U_H^2(x)$$

for the totality of its solutions, with an arbitrary constant c. If $U_I^1(x)$, $U_I^2(x)$ is a solution of the inhomogeneous problem, then the totality of solutions of (4.1), (4.2) is given by $U^i(x) = c U_H^i(x) + U_I^i(x)$, $i = 1, 2$.

Proof. The problem will be reduced to Theorem 1 from Section 4.6. By the lemma, there exist field functions $\gamma_1(x)$ and $\gamma_2(x)$. Using these functions, we put

$$\hat{U}^1 = \gamma_1 U^1 + \gamma_2 U^2, \qquad \hat{U}^2 = -\gamma_2 U^1 + \gamma_1 U^2. \quad (4.59)$$

The inverse transformation is given by

$$U^1 = \frac{1}{E} (\gamma_1 \hat{U}^1 - \gamma_2 \hat{U}^2), \qquad U^2 = \frac{1}{E} (\gamma_2 \hat{U}^1 + \gamma_1 \hat{U}^2), \quad (4.60)$$

where $E(x) = (\gamma_1)^2 + (\gamma_2)^2 > 0$ in $\bar{\mathfrak{D}}$. We show that then $\hat{U}^i(x)$ must satisfy a problem of the form (4.26). First we have $\hat{U}^1 = \varphi(\sigma)$ on $\dot{\mathfrak{D}}$, and furthermore by (4.1),

$$D^1 \equiv \frac{1}{E} (\gamma_1 \hat{U}_{x_1}^1 - \gamma_2 \hat{U}_{x_1}^2 - \gamma_2 \hat{U}_{x_2}^1 - \gamma_1 \hat{U}_{x_2}^2) + (\cdots) \hat{U}^1 + (\cdots) \hat{U}^2 + C^1 = 0,$$

$$\quad (4.61)$$

$$D^2 \equiv \frac{1}{E} (\gamma_1 \hat{U}_{x_2}^1 - \gamma_2 \hat{U}_{x_2}^2 + \gamma_2 \hat{U}_{x_1}^1 + \gamma_1 \hat{U}_{x_1}^2) + (\cdots) \hat{U}^1 + (\cdots) \hat{U}^2 + C^2 = 0.$$

---

[16] By Poisson's theorem (Section I-3.5) in a sharpened version: If furthermore $\varphi(x) \in C^1$ on $|x| = R$, then $u(x) \in C^1$ in $|x| \leq R$.

By taking linear combinations of the D^i:

$$\hat{D}^1 = \gamma_1 D^1 + \gamma_2 D^2, \qquad \hat{D}^2 = -\gamma_2 D^1 + \gamma_1 D^2 \tag{4.62}$$

we obtain

$$\hat{D}^i \equiv \hat{L}^i + \sum_{j=1}^{2} \hat{A}^i_j \hat{U}^j + \hat{C}^i = 0 \quad \text{where} \quad \hat{U}^1 = \varphi(\sigma) \quad \text{on} \quad \dot{\mathfrak{D}}. \tag{4.63}$$

Here the coefficients $\hat{A}^i_j(x)$ and $\hat{C}^i(x)$ turn out to be εC^0 in $\mathfrak{D}$, εC^1 in $\mathfrak{D}$, where we used the properties of the field functions which were given in the lemma. We apply Theorem 1 from Section 4.6 to (4.63) and then transform back using (4.60). This concludes the proof.

4.8   The General Boundary-Value Problem with Arbitrary Integer Characteristic

We begin with an auxiliary formula and consider arbitrary functions $U^1(x)$, $U^2(x)$; $V^1(x)$, $V^2(x)$, where U^i, $V^i \,\varepsilon\, C^0$ in $\mathfrak{D}$, εC^1 in $\mathfrak{D}$, and $L^i \,\varepsilon\, C^0$ in $\mathfrak{D}$. Here L^i is to be formed with the U^j as well as the V^j. The Gauss theorem in R_2 then immediately gives

$$\int_{\dot{\mathfrak{D}}} \{(V^2 U^2 - V^1 U^1)\dot{x}_1 + (V^2 U^1 + V^1 U^2)\dot{x}_2\} \, d\sigma$$

$$= \int_{\mathfrak{D}} \{V^2(U^1_{x_1} - U^2_{x_2}) + V^1(U^1_{x_2} + U^2_{x_1}) + U^2(V^1_{x_1} - V^2_{x_2}) + U^1(V^1_{x_2} + V^2_{x_1})\} \, dx. \tag{4.64}$$

Let $U^i(x)$ be solutions of

$$D^i \equiv L^i + \sum_{j=1}^{2} A^i_j U^j + C^i = 0, \qquad i = 1, 2, \tag{4.65}$$

and $V^i(x)$ solutions of

$$\begin{aligned} \bar{D}^1 &\equiv \bar{L}^1 - A^2_2 V^1 - A^1_2 V^2 - F^1 = 0, \\ \bar{D}^2 &\equiv \bar{L}^2 - A^2_1 V^1 - A^1_1 V^2 - F^2 = 0. \end{aligned} \tag{4.66}$$

Here the $\bar{L}^i$ denote the L^i written for the V^j. Then (4.64) gives the integral relation

$$\int_{\dot{\mathfrak{D}}} \{V^1(-U^1\dot{x}_1 + U^2\dot{x}_2) + V^2(U^1\dot{x}_2 + U^2\dot{x}_1)\} \, d\sigma = \int_{\mathfrak{D}} \{U^1 F^2 + U^2 F^1 - V^1 C^2 - V^2 C^1\} \, dx. \tag{4.67}$$

The homogeneous system (4.66) (with $F^i \equiv 0$) is called the adjoint system to the homogeneous system (4.65) (that is, $C^i \equiv 0$). Note that the adjoint system of the adjoint system is again the homogeneous system (4.65).

LEMMA 1.   If two vector fields $\alpha(\sigma) = (\alpha_1(\sigma),\alpha_2(\sigma))$ and $\overset{0}{\alpha}(\sigma) = (\overset{0}{\alpha}_1(\sigma),\overset{0}{\alpha}_2(\sigma))$ with the same characteristic κ are given on $\dot{\mathfrak{D}}$, then

$$\begin{aligned} \alpha_1(\sigma) &= -\beta_2(\sigma)\overset{0}{\alpha}_2(\sigma) + \beta_1(\sigma)\overset{0}{\alpha}_1(\sigma), \\ \alpha_2(\sigma) &= \beta_1(\sigma)\overset{0}{\alpha}_2(\sigma) + \beta_2(\sigma)\overset{0}{\alpha}_1(\sigma), \end{aligned} \tag{4.68}$$

where $\beta(\sigma) = (\beta_1(\sigma),\beta_2(\sigma))$ is a vector field of characteristic zero. All vector components are assumed to be in C^1 on $\dot{\mathfrak{D}}$, and $|\alpha|, |\overset{0}{\alpha}| > 0$.

Proof. We have

Therefore
$$\frac{\alpha_2}{\alpha_1} = \left(\frac{\overset{0}{\alpha_2}}{\overset{0}{\alpha_1}} + \frac{\beta_2}{\beta_1}\right)\Big/\left(-\frac{\beta_2}{\beta_1}\frac{\overset{0}{\alpha_2}}{\overset{0}{\alpha_1}} + 1\right). \tag{4.69}$$

If we put $\tan \psi_1 = \overset{0}{\alpha_2}/\overset{0}{\alpha_1}$, $\tan \psi_2 = \beta_2/\beta_1$, and note

$$\tan (\psi_1 + \psi_2) = \frac{\tan \psi_1 + \tan \psi_2}{1 - \tan \psi_1 \tan \psi_2},$$

then the representation

$$\text{arc tan} \frac{\alpha_2}{\alpha_1} = \text{arc tan} \frac{\overset{0}{\alpha_2}}{\overset{0}{\alpha_1}} + \text{arc tan} \frac{\beta_2}{\beta_1} + j\pi, \qquad j = 0,\ \pm 1,\ \pm 2,\ \ldots \tag{4.70}$$

results. But noting (4.55), it follows immediately from (4.70) that the vector field β has characteristic zero. The possibility of the representation (4.68) now follows from the remark that in·(4.68) we can solve uniquely for the β_i.

LEMMA 2. The general boundary-value problem (4.1) and (4.2), where $\alpha = (\alpha_1,\alpha_2)$ is any vector field on $\dot{\mathfrak{D}}$ with characteristic κ, can be reduced to the special boundary-value problem (4.1), where $\overset{0}{\alpha_1}(\sigma)U^1 + \overset{0}{\alpha_2}(\sigma)U^2 = \varphi(\sigma)$ on $\dot{\mathfrak{D}}$. Here $\overset{0}{\alpha} = (\overset{0}{\alpha_1},\overset{0}{\alpha_2})$ is a special vector field of characteristic κ.

Proof. By Lemma 1 we obtain the representation

$$\varphi(\sigma) = \alpha_1 U^1 + \alpha_2 U^2 = \overset{0}{\alpha_1}(\beta_1 U^1 + \beta_2 U^2) + \overset{0}{\alpha_2}(-\beta_2 U^1 + \beta_1 U^2) \tag{4.71}$$

where β is a vector field of characteristic zero. By the lemma from Section 4.7, to the β_i there exist field functions $\gamma_i(x)$. If instead of the $U^i(x)$ we introduce new unknown functions $\hat{U}^i(x)$ by

$$\hat{U}^1 = \gamma_1 U^1 + \gamma_2 U^2, \qquad \hat{U}^2 = -\gamma_2 U^1 + \gamma_1 U^2, \tag{4.72}$$

the boundary conditions go over into

$$\overset{0}{\alpha_1}\hat{U}^1 + \overset{0}{\alpha_2}\hat{U}^2 = \varphi(\sigma) \quad \text{on} \quad \dot{\mathfrak{D}}. \tag{4.73}$$

If for the transformation (4.72) we use the theorem from Section 4.7, then for the $\hat{U}^i(x)$ a boundary-value problem of the same kind results, but with the special boundary conditions (4.73). Thus everything is proved.

Finally the integral theorem (4.67) also gives an important connection between boundary-value problems of characteristic κ and boundary-value problems of characteristic $-1 - \kappa$. This we now establish. Let the $U^i(x)$ be solutions of the boundary-value problem (4.1) with boundary condition (4.2)

$$\alpha_1(\sigma)U^1 + \alpha_2(\sigma)U^2 = \varphi(\sigma) \quad \text{on} \quad \dot{\mathfrak{D}}. \tag{4.74}$$

Let the vector field α have characteristic κ. We suppose we have already solved this boundary-value problem if that is at all possible. Then we would also know the quantity $\rho(\sigma)$ for which

$$-\alpha_2(\sigma)U^1(\sigma) + \alpha_1(\sigma)U^2(\sigma) = \rho(\sigma) \quad \text{on} \quad \dot{\mathfrak{D}}. \tag{4.75}$$

Then (4.74) and (4.75) give the representation

$$U^1 = \frac{\alpha_1 \varphi - \alpha_2 \rho}{\alpha_1^2 + \alpha_2^2}, \qquad U^2 = \frac{\alpha_1 \rho + \alpha_2 \varphi}{\alpha_1^2 + \alpha_2^2} \tag{4.76}$$

on $\dot{\mathfrak{D}}$ for the solutions $U^i(x)$. If for $V^1(x)$, $V^2(x)$ for the time being we take any solutions of (4.66), we may apply the integral theorem (4.67). Since $\dot{\mathfrak{D}}$ is described by $x_1(\sigma) = \cos \sigma$, $x_2(\sigma) = \sin \sigma$, using (4.76) we find the representation

$$\int_{\dot{\mathfrak{D}}} \frac{1}{(\alpha_1)^2 + (\alpha_2)^2} \{\rho(\epsilon_1 V^1 + \epsilon_2 V^2) + \varphi(-\epsilon_2 V^1 + \epsilon_1 V^2)\}\, d\sigma$$

$$= \int_{\mathfrak{D}} \{U^1 F^2 + U^2 F^1 - V^1 C^2 - V^2 C^1\}\, dx \tag{4.77}$$

for (4.67). Here we put

$$\begin{aligned}
\epsilon_1(\sigma) &= \alpha_1(\sigma) \cos \sigma - \alpha_2(\sigma) \sin \sigma, \\
\epsilon_2(\sigma) &= -\alpha_1(\sigma) \sin \sigma - \alpha_2(\sigma) \cos \sigma.
\end{aligned} \tag{4.78}$$

The vector field $\epsilon(\sigma) = (\epsilon_1(\sigma), \epsilon_2(\sigma))$ has characteristic $-1 - \kappa$ as is seen from the calculation

$$\frac{\epsilon_2}{\epsilon_1} = -\frac{\tan \sigma + \alpha_2/\alpha_1}{1 - (\alpha_2/\alpha_1) \tan \sigma} = \tan\{-(\sigma + \psi)\} \quad \text{where} \quad \tan \psi = \frac{\alpha_2}{\alpha_1}. \tag{4.79}$$

Then we have

$$\arctan \frac{\epsilon_2}{\epsilon_1} = -\sigma - \arctan \frac{\alpha_2}{\alpha_1} + j\pi, \qquad j = 0, \pm 1, \pm 2, \ldots, \tag{4.80}$$

from which the characteristic can be read off immediately by using (4.55). If now for the solutions $V^i(x)$ of (4.66) we pose the boundary-value problem

$$\epsilon_1(\sigma) V^1 + \epsilon_2(\sigma) V^2 = \eta(\sigma) \quad \text{on} \quad \dot{\mathfrak{D}}, \tag{4.81}$$

the desired connection is established.

The central theorem for the general boundary-value problem (4.1) and (4.2) is now stated.

THEOREM. (1) *The Homogeneous Problem* (4.1) *and* (4.2) *with* $\varphi = 0$ *on* $\dot{\mathfrak{D}}$, $C^i = 0$. If the vector field $\alpha(\sigma) = (\alpha_1(\sigma), \alpha_2(\sigma))$ on $\dot{\mathfrak{D}}$ has characteristic $\kappa < 0$, then the homogeneous problem has $U^1 \equiv U^2 \equiv 0$ as its only solution. If $\kappa \geqslant 0$, then the general solution of the homogeneous problem contains exactly $2\kappa + 1$ arbitrary constants.

(2) *The Inhomogeneous Problem.* If $\kappa < 0$, then for the solvability of the inhomogeneous problem it is necessary and sufficient that the quantities $\varphi(\sigma)$, $C^1(x)$, $C^2(x)$ satisfy exactly $-(2\kappa + 1)$ integral conditions which are given explicitly. If $\kappa \geqslant 0$, then the inhomogeneous problem is always solvable. The general solution contains exactly $2\kappa + 1$ constants.

We use the statements for $\kappa > 0$ in the theorem and prove only the statements for $\kappa < 0$. The case $\kappa = 0$ is settled already.

Proof. (1) If $\kappa < 0$, then $-1 - \kappa \geqslant 0$. We use the integral theorem (4.77). For U^i we use a solution of the homogeneous problem (4.1) with characteristic $\kappa < 0$.[17] Let $V^i(x)$ be solutions of the boundary problem (4.66) with the boundary condition (4.81) where $\eta(\sigma) = 0$. This is an inhomogeneous boundary-value problem with nonnegative characteristic, hence solvable for any choice of F^1, F^2. The integral theorem (4.77) then gives

$$\int_{\mathfrak{D}} (U^1 F^2 + U^2 F^1)\, dx = 0 \tag{4.82}$$

for any choice of $F^1(x)$, $F^2(x)$. Hence $U^1 \equiv U^2 \equiv 0$.

(2) Let the functions $V^i(x)$ be solutions of the homogeneous boundary-value problem (4.66) with $F^i \equiv 0$ and the boundary condition (4.81), where $\eta = 0$ on $\dot{\mathfrak{D}}$. The characteristic is $-1 - \kappa \geqslant 0$. Hence by the theorem this problem has the general solution

$$V^i(x) = \sum_{j=1}^{-2\kappa-1} c_j V_H^{i,j}(x), \qquad i = 1,\, 2. \tag{4.83}$$

Let $U^i(x)$ be a solution of the inhomogeneous problem (4.1) and (4.2). If we apply the integral theorem (4.77) to these $U^i(x)$ and to $V_H^{i,j}(x)$, we obtain

$$\int_{\dot{\mathfrak{D}}} \frac{\varphi}{(\alpha_1)^2 + (\alpha_2)^2} \{-\epsilon_2 V_H^{1,j} + \epsilon_1 V_H^{2,j}\}\, d\sigma + \int_{\mathfrak{D}} (V_H^{1,j} C^2 + V_H^{2,j} C^1)\, dx = 0. \tag{4.84}$$

These are $-(2\kappa + 1)$ integral conditions for $\varphi(\sigma)$, $C^1(x)$, $C^2(x)$; their validity is necessary for the solvability of the inhomogeneous boundary-value problem with $\kappa < 0$, and it is also sufficient—which we shall not show here.[18]

This theorem has many possible applications. It is the starting point for a systematic theory of isometric deformations of surfaces[11], and for an appropriate treatment of the shell problems in mechanics.[19]

REFERENCES

1. F. Riesz and B. Sz.-Nagy, *Functional Analysis* (New York: Frederick Ungar Publishing Company), 1955.

2. L. Bers, *Theory of Pseudo-Analytic Functions* (lecture notes), New York University, 1953.

3. W. A. Hurwitz, Dissertation, Göttingen, 1910.

4. D. Hilbert, *Grundzüge einer allgemeinen Theorie der linearen Integralgleichungen* (Leipzig: B. G. Teubner), 1924.

5. G. Hellwig, *Math. Z.* **55**, 276–283 (1952); **56**, 388–408 (1952).

6. I. N. Vekua, *Systems of First-Order Differential Equations of Elliptic Type and Boundary Value Problems* (Berlin: VEB Deutscher Verlag der Wissenschaften) 1956 [originally published in *Mat. Sbornik* **31**, No. 73, 217–314 (1952)—Translator's note].

[17] The solvability is clear, since, in any case, $U^i \equiv 0$ is a solution.

[18] We followed W. Haack[9]. The omitted proofs can be found there. For the general case of a multiply connected $\mathfrak{D}$ and for a different presentation of the theory see I. N. Vekua again, as well as J. Nitsche [10]. With W. Haack the κ used here has to be replaced by $-\kappa$.

[19] See the paper by I. N. Vekua[6], and also E. Behlendorff[12].

7. L. Bers, "Contributions to the Theory of Partial Differential Equations," *Ann. Math. Studies*, No. 33, p. 77 (Princeton: Princeton University Press), 1954.

8. T. Carleman, *Compt. rend. séances acad. sci.* (*Paris*) **197**, 471–474 (1933).

9. W. Haack, *Math. Nachr.* **7**, 1–30 (1952); **8**, 123–132 (1953).

10. J. Nitsche, *Math. Nachr.* **14**, 75–127 (1955).

11. W. Haack, *Elementare Differentialgeometrie* (Basel: Birkhäuser Verlag), 1954.

12. E. Behlendorff, *Z. Angew. Math. Mech.* **36**, 399–413 (1956).

SOLUTIONS

Section I-3.2. If we put $r = |a - x|$, then for $v(r)$ we find the equation

$$v'' + \frac{2}{r} v' + \kappa v = 0, \qquad ' = \frac{d}{dr},$$

which can also be brought into the form $(rv)'' + \kappa(rv) = 0$. All solutions of this equation are given by:

$$v = c_1 \frac{\sin \sqrt{\kappa}\, r}{r} + c_2 \frac{\cos \sqrt{\kappa}\, r}{r}.$$

Since the first part has no singularity at $r = 0$, we put $c_1 = 0$. As $\kappa \to 0$, the second part should transform to the known singularity function $(1/4\pi)|a - x|^{-1}$ of the potential equation. Consequently we have to put $c_2 = 1/4\pi$, and so we finally obtain

$$s(a,x) = \frac{\cos (\sqrt{\kappa}\, |a - x|)}{4\pi |a - x|}.$$

Section I-3.4 (Problem 2). Let x^1, x^2 with $x^1 \neq x^2$ be two fixed points from $\mathfrak{D}$. In the second Green's formula (I-1.17) we put $u(y) = g(x^1,y)$, $v(y) = g(x^2,y)$. Because of the singularities of g we must remove the spheres $\bar{S}_1$: $|y - x^1| \leq \rho$ and $\bar{S}_2$: $|y - x^2| \leq \rho$ from $\mathfrak{D}$. We choose ρ so small that $\bar{S}_1$, $\bar{S}_2 \subset \mathfrak{D}$ and furthermore $x^1 \notin \bar{S}_2$ and $x^2 \notin \bar{S}_1$.

If for abbreviation we put $g_1(y) = g(x^1,y)$, $g_2(y) = g(x^2,y)$, then because of $\Delta_n g_1 = \Delta_n g_2 = 0$ we obtain

$$0 = \int_{\mathfrak{D} - \bar{S}_1 - \bar{S}_2} \{g_1 \Delta_n g_2 - g_2 \Delta_n g_1\}\, dy = \int_{\dot{\mathfrak{D}}} \{g_1 g_{2\nu} - g_2 g_{1\nu}\}\, dS$$

$$+ \int_{|y - x^1| = \rho} \{g_1 g_{2\nu} - g_2 g_{1\nu}\}\, dS + \int_{|y - x^2| = \rho} \{g_1 g_{2\nu} - g_2 g_{1\nu}\}\, dS.$$

Here ν denotes the outer normal of the domains $\mathfrak{D}$ or $|y - x^1| < \rho$ or $|y - x^2| < \rho$, respectively. Since $g_1 = g_2 = 0$ on $\dot{\mathfrak{D}}$, the integral over $\dot{\mathfrak{D}}$ vanishes. By the considerations in Section I-3.3 we find

$$\lim_{\rho \to 0} \int_{|y - x^1| = \rho} \{g_1 g_{2\nu} - g_2 g_{1\nu}\}\, dS = g_2(x^1),$$

$$\lim_{\rho \to 0} \int_{|y - x^2| = \rho} \{g_1 g_{2\nu} - g_2 g_{1\nu}\}\, dS = -g_1(x^2).$$

If in the Green's formula we take the limit $\rho \to 0$, then

$$0 = g_2(x^1) - g_1(x^2) \quad \text{or} \quad g(x^2,x^1) = g(x^1,x^2)$$

follows, so that everything is proved.

Section I-3.6 (Problem 2).   After the considerations in Section I-3.6, setting $w = F(z)$, $\omega = F(z^1)$ we have

$$g(x,y) = g(z^1,z) = -\frac{1}{2\pi} \log \left| \frac{\omega - w}{1 - \omega\bar{w}} \right| = -\frac{1}{2\pi} \operatorname{Re} \log \frac{\omega - w}{1 - \omega\bar{w}}$$

$$= -\frac{1}{2\pi} \operatorname{Re} \{ \log (\omega - w) - \log (1 - \omega\bar{w}) \},$$

and since $(\operatorname{Re} \{ \cdot \cdot \cdot \})_{x_1} = \operatorname{Re}(\{ \cdot \cdot \cdot \}_{x_1}))$, where $' \equiv d/dz^1$, we have

$$g_{x_1}(x,y) = -\frac{1}{2\pi} \operatorname{Re} \left\{ \frac{\omega'}{\omega - w} + \frac{\omega'\bar{w}}{1 - \omega\bar{w}} \right\}$$

$$= -\frac{1}{2\pi} \operatorname{Re} \left\{ \frac{\omega'}{\omega - w} \left(1 + \bar{w} \frac{\omega - w}{1 - \omega\bar{w}} \right) \right\},$$

$$|g_{x_1}(x,y)| \leqslant \frac{1}{2\pi} \frac{1}{|z^1 - z|} \cdot \left| \frac{\omega'(z^1 - z)}{\omega - w} \right| \cdot \left| 1 + \bar{w} \frac{\omega - w}{1 - \omega\bar{w}} \right|.$$

Now $w = F(z)$ is a mapping of $\mathfrak{D}$ onto the unit disk. Thus $|w| \leqslant 1$, $|\omega| \leqslant 1$. Furthermore $(\omega - w)/(1 - \omega\bar{w})$ is a mapping of the unit disk onto itself so that

$$|1 + \bar{w}[(\omega - w)/(1 - \omega\bar{w})]| \leqslant 2$$

follows.   Next we show that

$$\left| \frac{\omega - w}{\omega'(z^1 - z)} \right| = \left| \frac{F(z^1) - F(z)}{F'(z^1)(z^1 - z)} \right| \geqslant \beta > 0 \quad \text{for} \quad \text{all } z^1, z \quad \text{from} \quad \mathfrak{D}.$$

For this purpose we consider the function

$$H(z^1,z) = \begin{cases} \dfrac{F(z^1) - F(z)}{F'(z^1)(z^1 - z)} & \text{for} \quad z^1 \neq z, \\ 1 & \text{for} \quad z^1 = z. \end{cases}$$

This $H(z^1,z)$ is continuous in the two complex variables z^1, z and different from zero for all z^1, z in $\mathfrak{D}$.   For $F(z)$ is a conformal mapping and therefore $F'(z) \neq 0$ in $\mathfrak{D}$; furthermore $F(z^1) - F(z) \neq 0$, since distinct points $z^1 \neq z$ have distinct image points $F(z^1) \neq F(z)$. But then

$$|g_{x_1}(x,y)| \leqslant \frac{1}{2\pi\beta} \frac{2}{|z^1 - z|} = \frac{1}{\pi\beta} \frac{1}{|x - y|}.$$

Analogous results hold for $|g_{x_2}(x,y)|$.

Section I-3.7 (Problem 1).   There exists a point x^0 such that $u(x^0) \neq 0$.   Put a sphere of radius r about x^0.   Then we have

$$\int_{R_n} u^2(x)\, dx \geqslant \int_{|x - x^0| \leqslant r} u^2(x)\, dx = \int_0^r \left(\int_{|x - x^0| = \rho} u^2(x)\, dS \right) d\rho.$$

Schwarz's inequality gives

$$\left(\int_{|x - x^0| = \rho} 1 \cdot u(x)\, dS \right)^2 \leqslant \int_{|x - x^0| = \rho} dS \cdot \int_{|x - x^0| = \rho} u^2(x)\, dS = \rho^{n-1}\omega_n \int_{|x - x^0| = \rho} u^2(x)\, dS$$

so that the estimate

$$\int_{R_n} u^2(x)\, dx \geqslant \frac{1}{\omega_n} \int_0^r \rho^{1-n} \left(\int_{|x - x^0| = \rho} u(x)\, dS \right)^2 d\rho$$

follows. Now to the last integral apply the mean-value property for the harmonic function $u(x)$. We find

$$u^2(x^0) = \left(\frac{1}{\omega_n \rho^{n-1}}\right)^2 \left(\int_{|x-x^0|=\rho} u(x)\, dS\right)^2$$

and thus

$$\int_{R_n} u^2(x)\, dx \geq \omega_n u^2(x^0) \int_0^r \rho^{n-1}\, d\rho = \frac{\omega_n}{n} u^2(x^0) r^n.$$

This inequality is true for every r. The conclusion follows as $r \to \infty$.

Section I-3.7 (Problem 2). Obvious for the case grad $u \equiv$ const. Let grad $u \not\equiv$ const, and let the maximum of $|\text{grad } u(x)|$ be attained at the point $x^0 \in \mathfrak{D}$. We choose the coordinate system so that at the point x^0 the vector in direction of the x_1-axis and of length $|\text{grad } u(x^0)|$ coincides with the vector grad $u(x^0)$. Then $u_{x_1}(x^0) = |\text{grad } u(x^0)|$. Now $u(x)$ was assumed to be harmonic in $\mathfrak{D}$. From our hypotheses also

$$\Delta_n u_{x_1}(x) = 0$$

follows, so that also $u_{x_1}(x)$ is harmonic. According to the proof of Theorem 1 in Section I-3.7, the maximum and minimum of u_{x_1} do not lie in $\mathfrak{D}$. Hence in a neighborhood of x^0 we can find points x such that

$$u_{x_1}(x) > u_{x_1}(x^0) = |\text{grad } u(x^0)|.$$

In particular then

$$(u_{x_1}(x))^2 + (u_{x_2}(x))^2 + \cdots + (u_{x_n}(x))^2 = |\text{grad } u(x)|^2 \geq (u_{x_1}(x))^2 > |\text{grad } u(x^0)|^2,$$

which is a contradiction. Thus everything is proved.

Section III-1.1. Let the problem (1.9) have two solutions $u^1(x)$ and $u^2(x)$. Then the difference $u(x) = u^1 - u^2$ satisfies the problem $Du = 0$ in $\mathfrak{D}$, $u = 0$ on $\dot{\mathfrak{D}}$. We put $u(x) = w(x)v(x)$. Then we have

$$u_{x_i} = w_{x_i}v + wv_{x_i}, \qquad u_{x_i x_k} = w_{x_i x_k}v + w_{x_i}v_{x_k} + w_{x_k}v_{x_i} + wv_{x_i x_k},$$

$$Du = v\, Dw + w\left\{ \sum_{i,k=1}^n a_{ik}v_{x_i x_k} + \sum_{i,k=1}^n \left(\frac{2a_{ik}}{w} w_{x_k} + a_i\right) v_{x_i} \right\} = 0.$$

Now $Dw = 0$; hence $\{\cdot\cdot\cdot\} = 0$ follows. This is a differential equation for $v(x)$ in which the coefficient of $v(x)$ is zero. Furthermore $v = 0$ on $\dot{\mathfrak{D}}$. By Theorem 2, then, $v(x) \equiv 0$ follows, and hence $u(x) \equiv 0$.

Section III-1.4. First we assume $n \geq 3$ and suppose $x \in \mathfrak{D}$. Then by Section I-3.4

$$g(x,y) = s(x,y) + \Phi(x,y), \qquad \Delta_n \Phi = 0$$

with respect to the y-variables, and $\Phi = -s$ for $y \in \dot{\mathfrak{D}}$. We have $s(x,y) > 0$ for $y \in \dot{\mathfrak{D}}$. Hence $\Delta_n \Phi = 0$, $\Phi < 0$ for $y \in \dot{\mathfrak{D}}$. Now Φ is harmonic and therefore has its maximum on $\dot{\mathfrak{D}}$. Thus $\Phi(x,y) < 0$ in $\mathfrak{D}$. With that, $g(x,y) < s(x,y)$ for $x \in \mathfrak{D}$, $y \in \dot{\mathfrak{D}}$, $x \neq y$ is shown. Furthermore, $g(x,y)$ is zero on $\dot{\mathfrak{D}}$ and positive on the boundary of a sufficiently small ball about the point $y = x$. By the maximum principle it follows that $0 \leq g(x,y)$, first for the remaining domain but then also for $x \in \mathfrak{D}$, $y \in \dot{\mathfrak{D}}$.

The case $n = 2$ is more difficult, since $s(x,y) = -(1/2\pi) \log |x - y|$ does not necessarily have a fixed sign in $\mathfrak{D}$. Let d be the largest distance between any two points x^1, $x^2 \in \mathfrak{D}$.

Then the function

$$\Gamma(x,y) = \Phi(x,y) - \frac{1}{2\pi} \log d$$

has the boundary values $(1/2\pi) \log |x - y| - (1/2\pi) \log d = (1/2\pi) \log (|x - y|/d)$. We have $|x - y|/d < 1$ and therefore $\Gamma(x,y) < 0$ for $x \in \mathfrak{D}, y \in \dot{\mathfrak{D}}$. Furthermore Γ is harmonic in y so that by the maximum principle $\Gamma(x,y) < 0$ for $x \in \mathfrak{D}$ and $y \in \bar{\mathfrak{D}}$ follows. We find

$$g(x,y) = s(x,y) + \Phi(x,y) + \frac{1}{2\pi} \log d - \frac{1}{2\pi} \log d < s(x,y) - \frac{1}{2\pi} \log d.$$

Analogously we find $0 \leqslant g(x,y)$, and finally we obtain the conclusion using

$$c = -(1/2\pi) \log d.$$

Section IV-1.8. The boundary-value problem to be solved for the general telegraph equation, in a slightly different notation, reads

$$\alpha^2 U_{xx} - U_{tt} - 2\beta U_t = 0, \qquad U(x,0) = U_0(x), \qquad U_t(x,0) = U_1(x),$$

where $-\infty < x < \infty$, $-\infty < t < \infty$. We are striving to determine a solution at the world point (space-time)(x_0,t_0). If we set $U(x,t) = e^{-\beta t}u(x,t)$, the new problem

$$\alpha^2 u_{xx} - u_{tt} + \beta^2 u = 0, \qquad u(x,0) = U_0(x), \qquad u_t(x,0) = U_1(x) + \beta U_0(x)$$

arises. If we introduce the new independent variable $x_1 = x - \alpha t$, $x_2 = x + \alpha t$ in this, then the new initial-value problem

$$u_{x_1 x_2} + au = 0 \quad \text{where} \quad a = \frac{\beta^2}{4\alpha^2}$$

arises with the new initial curve $k: x_2 - x_1 = 0$ on which the prescriptions

$$u(x_1,x_1) = U_0(x_1), \; u_{x_2}(x_1,x_1) - u_{x_1}(x_1,x_1) = \frac{1}{\alpha} (U_1(x_1) + \beta U_0(x_1))$$

are to be made. For this initial-value problem we deduce Riemann's solution formula. Finally the transformations carried out are inverted again.

According to the general theory we look for a solution $u(x)$ at the arbitrary point $y = (y_1,y_2)$. For this we need the Riemann function $W(x,y)$. It has to satisfy the requirements

$$W_{x_1 x_2} + aW = 0, \qquad W_{x_1} = 0 \quad \text{on} \quad c_2, \qquad W_{x_2} = 0 \quad \text{on} \quad c_1, \quad \text{and} \quad W(y,y) = -1.$$

Equivalent requirements are

$$W_{x_1 x_2} + aW = 0, \qquad W = -1 \quad \text{for} \quad x_1 = y_1 \quad \text{as well as for} \quad x_2 = y_2.$$

We set $W(x,y) = \Phi(\xi)$, where $\xi = (x_1 - y_1)(x_2 - y_2)$. The partial differential equation for $W(x,y)$ then becomes

$$\xi \Phi''(\xi) + \Phi'(\xi) + a\Phi = 0, \qquad ' = \frac{d}{d\xi}.$$

If further we put $\eta = 2\sqrt{a\xi}$ and $\Phi(\xi) = \Psi(\eta)$, then the Bessel differential equation

$$\ddot{\Psi}(\eta) + \frac{1}{\eta} \dot{\Psi}(\eta) + \Psi(\eta) = 0, \qquad \cdot = \frac{d}{d\eta}$$

arises with the general solution $\Psi(\eta) = C_1 J_0(\eta) + C_2 N_0(\eta)$. The requirement $W = -1$ for $x_1 = y_1$ and $x_2 = y_2$ here gives $\Psi(0) = -1$. If we note that $J_0(0) = 1$, then we find $C_1 = -1$, $C_2 = 0$. Thus the Riemann function

$$W(x,y) = -J_0(2\sqrt{a(x_1 - y_1)(x_2 - y_2)})$$

has been found explicitly. From the general theory we know that it is determined uniquely so that it does not depend on the special construction. Now we can use the general solution formula (1.106). Here k is to be described by $x_2 = x_1$. We find the point y^1 (Figure 32) by proceeding from y along c_2 up to the intersection with k. In our case $y^1 = (y_2, y_2)$, and correspondingly $y^2 = (y_1, y_1)$. Now (1.106) becomes

$$u(y) = \tfrac{1}{2}\{U_0(y_1) + U_0(y_2)\} + \int_{y_1}^{y_2} \{P(x,y) - Q(x,y)\}\, dx_1$$

where $P(x,y) - Q(x,y) = \tfrac{1}{2} W(x,y)(u_{x_2}(x) - u_{x_1}(x)) - \tfrac{1}{2} u(x)(W_{x_2}(x,y) - W_{x_1}(x,y))$, everything being valid for $x_2 = x_1$. If we take the initial conditions into account, then

$$P(x,y) - Q(x,y) = \frac{1}{2\alpha} W(x,y)(U_1(x_1) + \beta U_0(x_1)) - \tfrac{1}{2} U_0(x_1)(W_{x_2}(x,y) - W_{x_1}(x,y))$$

follows, again for $x_2 = x_1$. Using $\dot{J}_0(\eta) = -J_1(\eta)$, calculation of the last term gives

$$W_{x_2}(x,y) - W_{x_1}(x,y) = a(y_2 - y_1)\frac{J_1(2\sqrt{a(x_1 - y_1)(x_1 - y_2)})}{\sqrt{a(x_1 - y_1)(x_1 - y_2)}}.$$

Our solution formula (with τ as integration variable) now gives

$$u(y) = \tfrac{1}{2}\{U_0(y_1) + U_0(y_2)\} + \frac{1}{2\alpha}\int_{y_1}^{y_2} J_0(2\sqrt{a(\tau - y_1)(\tau - y_2)})(U_1(\tau) + \beta U_0(\tau))\, d\tau$$

$$+ \frac{a}{2}(y_2 - y_1)\int_{y_1}^{y_2}\frac{J_1(2\sqrt{a(\tau - y_1)(\tau - y_2)})}{\sqrt{a(\tau - y_1)(\tau - y_2)}} U_0(\tau)\, d\tau.$$

Now we invert the transformations which were carried out at the beginning. Let the point $y = (y_1, y_2)$ correspond to the world point (x_0, t_0); that is, we put

$$y_1 = x_0 - \alpha t_0, \qquad y_2 = x_0 + \alpha t_0.$$

Noting the meaning of a, we then obtain

$$\sqrt{a(\tau - y_1)(\tau - y_2)} = \frac{\beta}{2\alpha}\sqrt{(x_0 - \tau)^2 - \alpha^2 t_0^2},$$

and thus finally

$$U(x_0, t_0) = \frac{e^{-\beta t_0}}{2}\{U_0(x_0 - \alpha t_0) + U_0(x_0 + \alpha t_0)\}$$

$$+ \frac{e^{-\beta t_0}}{2\alpha}\int_{x_0 - \alpha t_0}^{x_0 + \alpha t_0} J_0\left(\frac{\beta}{\alpha}\sqrt{(x_0 - \tau)^2 - \alpha^2 t_0^2}\right)(U_1(\tau) + \beta U_0(\tau))\, d\tau$$

$$+ \frac{\beta t_0\, e^{-\beta t_0}}{2}\int_{x_0 - \alpha t_0}^{x_0 + \alpha t_0}\frac{J_1\left(\frac{\beta}{\alpha}\sqrt{(x_0 - \tau)^2 - \alpha^2 t_0^2}\right)}{\sqrt{(x_0 - \tau)^2 - \alpha^2 t_0^2}} U_0(\tau)\, d\tau.$$

This is the desired solution of the given initial-value problem.

BIBLIOGRAPHY

The field of differential equations of elliptic type has been surveyed almost completely up to the year 1954 in the reports listed here.

L. Lichtenstein, "Neuere Entwicklungen in der Potentialtheorie," *Enzyklopädie der mathematischen Wissenschaften*, Vol. 2, 3.1 (Leipzig: B. G. Teubner), 1909.

L. Lichtenstein, "Neuere Entwicklungen der Theorie der partiellen Differentialgleichungen zweiter Ordnung vom elliptischen Typus," *Enzyklopädie der mathematischen Wissenschaften*, Vol. 2, 3.2 (Leipzig: B. G. Teubner), 1924.

C. Miranda, *Equazioni alle derivate parziali di tipo ellittico* (Berlin: Springer Verlag), 1955.

In other fields the situation is not so favorable. A great many references to recent literature can be found in the paper of

P. C. Rosenbloom, "Linear partial differential equations," *Surveys in Applied Mathematics*, Vol. 5 (New York: John Wiley & Sons), 1958.

A survey is also provided by the following conference reports:

"Contributions to the theory of partial differential equations," *Ann. Math. Studies No. 33* (Princeton: Princeton University Press), 1954.

Convegno Internazionale sulle equazioni lineari alle derivate parziali (Trieste 1954) (Rome: Edizioni Cremonese), 1955.

Proceedings of the Conference on Differential Equations (University of Maryland Book Store), 1956.

Transactions of the Symposium on Partial Differential Equations, (Berkeley 1955) (New York: Interscience Publishers, Inc.), 1956; reprinted from *Comm. Pure Appl. Math.* Vol. IX (1956).

A survey of differential equations of mixed type and their applications can be found in

L. Bers, "Mathematical Aspects of Subsonic and Transonic Gas Dynamics," *Surveys in Applied Mathematics*, Vol. 3 (New York: John Wiley & Sons), 1958.

For the whole field of partial differential equations and, in particular, for the connections with calculus of variations, see also

R. Courant and D. Hilbert, *Methods of Mathematical Physics*, Vol. II (New York: Interscience Publishers, Inc.), 1962 and Vol. II of the German edition (Berlin, Springer Verlag), 1937.

A different approach to certain boundary-value problems for elliptic differential equations is given in

S. Bergman and M. Schiffer, *Kernel Functions and Elliptic Differential Equations in Mathematical Physics* (New York: Academic Press, Inc.), 1953.

For numerical questions, which have not been treated here, we refer among others to

L. Collatz, *The Numerical Treatment of Differential Equations*, 3rd ed. (Berlin: Springer Verlag), 1960.

Mathematische Leitfäden (Fortsetzung)

Kategorien und Funktoren
Von Dr. rer. nat. B. PAREIGIS, o. Prof. an der Universität München
192 Seiten mit 49 Aufgaben und zahlreichen Beispielen. Kart. DM 42,—

Nichteuklidische Elementargeometrie der Ebene
Von Prof. Dr. Dr. h. c. O. PERRON
134 Seiten mit 70 Bildern. Ln. DM 34,—

Topologie
Eine Einführung
Von Dr. rer. nat. Dr. h. c. H. SCHUBERT, o. Prof. an der Universität Düsseldorf
4. Auflage. 328 Seiten mit 23 Bildern, 121 Aufgaben und zahlreichen Beispielen. Kart. DM 4

Lineare Operatoren in Hilberträumen
Von Dr. rer. nat. J. WEIDMANN, Prof. an der Universität Frankfurt/M.
368 Seiten mit 221 Aufgaben und 93 Beispielen. Kart. DM 58,—

Preisänderungen vorbehalten

B. G. Teubner Stuttgart